T0136075

Climate Change in Wildlands

Pioneering Approaches to Science and Management

Edited by

Andrew J. Hansen, William B. Monahan,
S. Thomas Olliff, and David M. Theobald

Washington | Covelo | London

ISLAND PRESS is a trademark of The Center for Resource Economics.

No copyright claim is made in the works of William B. Monahan, S. Thomas Olliff, Isabel W. Ashton, Ben Bobowski, Karl Buermeyer, Robert Al-Chokhachy, John E. Gross, Nathaniel Hitt, Virginia Kelly, Matthew A. Kulp, Kristin Legg, Shuang Li, Forrest Melton, Cristina Milesi, Jeffrey T. Morisette, Ramakrishna Nemani, Ashley Quackenbush, Daniel Reinhart, Ann Rodman, David Thoma, and Jun Xiong, employees of the federal government.

Island Press would like to thank Katie Dolan for generously supporting the publication of this book.

Library of Congress Control Number: 2015952374

Printed on recycled, acid-free paper ⊕

Manufactured in the United States of America

10 9 8 7 6 5 4 3 2 1

Keywords: Island Press, climate change, wildlands, climate adaptation planning, Climate-Smart Conservation, resource management, federal lands, climate science, land use, Rocky Mountains, Appalachian Mountains, vulnerability assessment, exposure, sensitivity, potential impact, adaptive capacity, vegetation response, coldwater fish, whitebark pine, sugar maple, landscape conservation cooperative, climate science center, national park.

CONTENTS

Foreword   xi
Woody Turner

Acknowledgments   xiii

Chapter 1. Why Study Climate Change in Wildlands?   1
            Andrew J. Hansen

PART 1: APPROACHES FOR CLIMATE ADAPTATION
PLANNING

Chapter 2. Effectively Linking Climate Science and Management   17
            John E. Gross and S. Thomas Olliff

Chapter 3. Challenges and Approaches for Integrating Climate   33
            Science into Federal Land Management
            S. Thomas Olliff and Andrew J. Hansen

PART 2: CLIMATE AND LAND USE CHANGE

Chapter 4. Analyses of Historical and Projected Climates to Support 55
            Climate Adaptation in the Northern Rocky Mountains
            John E. Gross, Michael Tercek, Kevin Guay, Marian Talbert,
            Tony Chang, Ann Rodman, David Thoma, Patrick Jantz,
            and Jeffrey T. Morisette

Chapter 5.   Historical and Projected Climates as a Basis for Climate   78
             Change Exposure and Adaptation Potential across the
             Appalachian Landscape Conservation Cooperative
             Kevin Guay, Patrick Jantz, John E. Gross,
             Brendan M. Rogers, and Scott J. Goetz

Chapter 6.   Assessing Vulnerability to Land Use and Climate   95
             Change at Landscape Scales Using Landforms and
             Physiographic Diversity as Coarse-Filter Targets
             David M. Theobald, William B. Monahan, Dylan Harrison-
             Atlas, Andrew J. Hansen, Patrick Jantz, John E. Gross,
             and S. Thomas Olliff

PART 3: ECOLOGICAL CONSEQUENCES
AND VULNERABILITIES

Chapter 7.   Potential Impacts of Climate and Land Use Change   119
             on Ecosystem Processes in the Great Northern and
             Appalachian Landscape Conservation Cooperatives
             Forrest Melton, Jun Xiong, Weile Wang, Cristina Milesi,
             Shuang Li, Ashley Quackenbush, David M. Theobald,
             Scott J. Goetz, Patrick Jantz, and Ramakrishna Nemani

Chapter 8.   Potential Impacts of Climate Change on Vegetation   151
             for National Parks in the Eastern United States
             Patrick Jantz, William B. Monahan, Andrew J. Hansen,
             Brendan M. Rogers, Scott Zolkos, Tina Cormier,
             and Scott J. Goetz

Chapter 9.   Potential Impacts of Climate Change on Tree Species   174
             and Biome Types in the Northern Rocky Mountains
             Andrew J. Hansen and Linda B. Phillips

Chapter 10. Past, Present, and Future Impacts of Climate on the   190
             Vegetation Communities of the Greater Yellowstone
             Ecosystem across Elevation Gradients
             Nathan B. Piekielek, Andrew J. Hansen, and Tony Chang

Chapter 11. Vulnerability of Tree Species to Climate Change in the   212
Appalachian Landscape Conservation Cooperative
*Brendan M. Rogers, Patrick Jantz, Scott J. Goetz,
and David M. Theobald*

Chapter 12. Likely Responses of Native and Invasive Salmonid   234
Fishes to Climate Change in the Rocky Mountains
and Appalachian Mountains
*Bradley B. Shepard, Robert Al-Chokhachy, Todd Koel,
Matthew A. Kulp, and Nathaniel Hitt*

PART 4: MANAGING UNDER CLIMATE CHANGE

Chapter 13. Approaches, Challenges, and Opportunities   259
for Achieving Climate-Smart Adaptation
*S. Thomas Olliff, William B. Monahan, Virginia Kelly,
and David M. Theobald*

Chapter 14. Perspectives on Responding to Climate Change   279
in Rocky Mountain National Park
Ben Bobowski, Isabel W. Ashton, and William B. Monahan

Chapter 15. Case Study: Whitebark Pine in the Greater   304
Yellowstone Ecosystem
Karl Buermeyer, Daniel Reinhart, and Kristin Legg

Chapter 16. Insights from the Greater Yellowstone Ecosystem   327
on Assessing Success in Sustaining Wildlands
Andrew J. Hansen and Linda B. Phillips

Chapter 17. Synthesis of Climate Adaptation Planning   354
in Wildland Ecosystems
*Andrew J. Hansen, David M. Theobald, S. Thomas Olliff,
and William B. Monahan*

About the Contributors   368

Land managers have a responsibility to preserve the ecological integrity of the landscapes they manage. Climate change is a global phenomenon affecting these landscapes in complex ways. Adaptation to climate change to sustain ecological integrity requires understanding how this global phenomenon will play out at the landscape scale. Thus, managers and members of the public who care about natural systems share a need to translate the effects of global climate change to the scale of a park, forest, refuge, or other management unit. Complicating this translation are coincident and interacting changes in land use. The challenge of maintaining the ecological integrity of natural landscapes under the dual onslaught of climate change and land use change is a seemingly hopeless task for those managing locally without direct access to the regional, much less global, picture. Bringing the bigger picture to land managers responding to a rapidly changing world is possible, as the following pages make clear. Required is an integration of the latest observation technologies, including satellites imaging globally at landscape resolutions (i.e., pixel sizes ranging from tens to hundreds of square meters) and an interoperable framework for climate and ecological models to relate climate and land use changes to ecological responses.

This book represents the capstone of the Landscape Climate Change Vulnerability Project (LCCVP). NASA funded the LCCVP in 2011 to explore the potential of using a range of observations from satellites to in situ sensors with a wide array of models, all operating on a powerful computer platform, to forecast the vulnerability of ecosystems and individual species to the inseparable twin threats of climate change and land use change. The project focus has been landscapes managed by federal land managers. More specifically, the project provided analyses, decision support tools, and dialogue on climate adaptation for managers working in two landscape conservation cooperatives, or LCCs: the Great Northern LCC and the Appalachian LCC. Both LCCs center on mountain ranges, typically seen as areas of rapid biological movement under changing climate. However, the contrasting responses of life in these two North American mountain chains to recent climate reflect the complex and unique ways in which similar but different ecosystems react locally to a global phenomenon. These differing responses point to the necessity of downscaling the impacts of climate change and land use change as closely as possible to the local level, because nationally—even regionally—modeled responses are insufficient to address management needs.

This book is for those attempting to maintain natural ecosystems and the species within them through a period of interacting climate change and land use change. It is also for anyone interested in understanding the effects of these changes on wildlands. Despite the challenging goals, there is much hope in these pages. We are developing the tools to allow humans to predict and visualize the impacts of change, natural and anthropogenic, on the world around us—the world on which we all depend for countless environmental goods and services. As a species, we perhaps stand alone in the degree to which our actions affect the entire planet. A global footprint carries with it great responsibility for how one steps out in the world. Yet stepping responsibly requires better understanding of how natural systems function and respond to change. The authors assemble cutting-edge research and twenty-first-century technologies to offer stewards and users of public lands and surrounding private lands a detailed view of the future for North American mountain ecosystems. In so doing, they show the way for those interested in conserving natural systems on a changing planet.

Woody Turner
Program Scientist for Biological Diversity and Program Manager for Ecological Forecasting in the NASA Headquarters Science Mission Directorate
Washington, D.C.

July 2015

ACKNOWLEDGMENTS

Many people and organizations contributed to bringing this book to fruition. Funding was provided by the NASA Applied Sciences Program (10-BIOCLIM10-0034, 05-DEC05-S2-0010), the NASA Earth and Space Science Fellowship (15-EARTH15R-0003), the NASA Land Cover Land Use Change Program (05-LC/LUC05-2-0089), the North Central Climate Sciences Center (G13AC00394, G14AP00181, G15AP00074), and NSF EPSCoR Track-I EPS-1101342 (INSTEP 3). Program managers administrating this support include Woody Turner, Jay Skiles, Garik Gutman, and Jeff Morisette. Grant administration at Montana State University was enabled by Julie Geyer, Joan McDonald, Barbara Bungee, and Sondra Torma. In-kind support for salaries of federal employees on the research team was provided by the DOI Great Northern Landscape Conservation Cooperative and the DOI National Park Service Inventory and Monitoring Program, Climate Change Response Programs, and the Intermountain Region.

The work benefited from discussions with many scientists, including Richard Waring, Steve Running, Cathy Whitlock, Jeff Morisette, Dennis Ojima, Robert Keane, Dave Roberts, Joseph Barsugli, Gary Tabor, Ben Poulter, Brian Miller, Shannon McNeeley, Nick Fisichelli, Gregor Schuurman, Jeff Connor, Tammy Cook, Ann Rodman, Roy Renkin, P. J. White, Doug Smith, Molly Cross, Arjun Adhikari, Kathryn Ireland, and Regan Nelson. Linda Phillips coordinated communication among the book's authors, production of graphics, and formatting of chapters. Barbara Dean and Erin Johnson, editors from Island Press, guided us through the process from an initial query letter through production of the book.

Several federal resource specialists closely collaborated with the research team (see table 3-1 in chapter 3). Most central in linking the science with agency managers were Virginia Kelly (Greater Yellowstone Coordinating Committee), Ben Bobowski (Rocky Mountain National Park), Dave Hallac and Ann Rodman (Yellowstone National Park), Kristin Legg (NPS I&M Greater Yellowstone Network), Matt Marshall (NPS I&M Eastern Rivers and Mountains Network), Jim Renfro (Great Smoky Mountains National Park), Jim Schaberl (Shenandoah National Park), Karl Buermeyer and Dan Reinhart (Greater Yellowstone Coordinating Committee Whitebark Pine Subcommittee), Richard Evans and Leslie Morelock (Delaware Water Gap National Recreational Area), Robert Emmott (NPS I&M Appalachian Highlands Network), Jim Comiskey (NPS I&M Mid-Atlantic Network), Sue Consolo Murphy and Kelly McClosky (Grand Teton National Park), and Mike Britten (NPS I&M Rocky Mountain Network).

The authors of the chapters invested heavily in telling the stories of climate change in the Rockies and the Appalachian Mountains and were responsive to the many rounds of revisions the book underwent. We sincerely thank all of these individuals and organizations.

Chapter 1

Why Study Climate Change in Wildlands?

Andrew J. Hansen

Most nations around the world set aside some lands from where people live and work for the benefit of nature. Wildland ecosystems are those lands occupied chiefly by native plants and animals, not intensively used as urban or residential areas, and not intensively managed for the production of domesticated plants or animals (Kalisz and Wood 1995). Public parks, forests, grasslands, seashores, and other wildland ecosystems are central to the global strategy for the conservation of nature. These areas are also vital to the well-being of people. They provide essential ecosystem services, such as provisioning of food and water, supporting pollination and nutrient cycling, regulating floods and other disturbances, and providing aesthetic and recreational services (Wilkie et al. 2006; Friedman 2014).

While humans have benefited substantially from these services, impacts associated with our activities, particularly habitat loss from land conversion, climate change, and exotic species introductions, are driving major losses in biodiversity and subsequent disruption of ecosystem services (Millennium Ecosystem Assessment 2005). A major challenge facing humankind is how to sustain ecosystem services in the face of population growth and climate change. This challenge is particularly large in wildlands because they have become magnets for human development on their peripheries (Theobald and Romme 2007; Wittemyer et al. 2008; Radeloff et al. 2010). The challenge is also great because many wildlands are set in mountains or deserts that are undergoing particularly high rates of climate change (Hansen et al. 2014).

This book aims to link science and management to better understand human-caused change in wildland ecosystems and to better inform management to sustain wildland ecosystems and ecosystem services. Within the United States, the federal agencies that manage most of our wildlands have been charged with consideration of climate change since only 2009. Consequently, we focus on the challenge of understanding and managing wildland ecosystems under climate change but do so in the context of land use that is changing simultaneously.

Climate Change in Mountain Wildlands

The US Rocky Mountains in Montana, Wyoming, and Colorado are known for soaring summits, expansive public lands, iconic wildlife, blue-ribbon trout streams, and long, cold, snowy winters. In many ways, the wildland ecosystems of the region are framed by the harshness of the climate (fig. 1-1). The higher elevations are too cold and snowy for many plant species to tolerate; consequently, rates of ecological productivity are low, and the highest diversity of plant and wildlife species are at lower elevations with more equitable climate (Hansen et al. 2000). At first blush, one might think that climate warming would benefit ecosystems that are so limited by winter conditions. The interactions among climate, ecosystems, and plant and animal species are complex, however, especially in the context of human land use. There are many direct and indirect interactions that can lead to threshold changes and surprises. Understanding these interactions is essential to managing these wildlands to sustain native species and ecosystem services for people.

Forests at the highest elevations are dominated by pines, particularly whitebark pine (*Pinus albicaulis*) and lodgepole pine (*P. contorta*). Not only are these species uniquely adapted to tolerate cold climate and nutrient-poor soils, but they may require them. The mountain pine beetle (*Dendroctonus ponderosae*) is a native species that feeds on the cambium of these pines. A form of natural disturbance in this region, these beetles irrupt every few decades and kill large tracts of subalpine forests. In recent decades, however, the forest die-off has been larger in area and more continuous than in the past, and many old-growth whitebark pine stands have suffered 70 to 90 percent mortality (Logan, Macfarlane, and Willcox 2010). The cause? Winter low temperatures have not been cold enough in recent years to slow the population growth rate of the beetles, as had been the case historically. Thus, a natural disturbance to which the pine trees were

FIGURE 1-1   Image of the Teton Range near Jackson, Wyoming. The ecosystems of the Rocky Mountains are framed by the harshness of the climate, raising questions on the effects of global warming. (Photo by Andrew J. Hansen.)

adequately adapted has been intensified by climate warming, putting high-elevation forests at risk.

Beyond leading to the recommendation of whitebark pine as a candidate threatened species, the forest die-off has effects that ripple across the ecosystem (chap. 15). The reduction in pine nuts, a major food source for grizzly bear (*Ursus arctos horribilis*) and other species, has led to the bears spending more time in lower-elevation habitats where they more frequently encounter humans, typically at the bears' expense. Loss of subalpine forest also increases the melt rate of mountain snow and reduces summer streamflows, which are critical to native trout populations, recreationalists, irrigation-fed agriculturalists, and the fast-growing local communities downstream from the mountains. Most of the whitebark pine stands are located in federally designated wilderness areas, where management options are limited by law.

Unfortunately, there are many other examples of unexpected responses to changing climate. Within rivers and streams in the region, loss of native Yellowstone cutthroat trout (*Oncorhynchus clarkii bouvieri*) is occurring through hybridization with exotic rainbow trout (*O. mykiss*), with rates of hybridization positively linked to warming stream temperatures (chap. 12). The conifer forests that dominate much of the east slope of the Rockies are projected by the end of the century to have climate suitable for desert scrub vegetation now found in the Wyoming basin (chap. 9). The North American elk (*Cervus elaphus*) in Yellowstone National Park may benefit from less snow in winter habitats, but reproduction may be impeded because summer warming of mountain grasslands has reduced the availability of green forage during the time that cow elk are recovering from nursing their offspring (Middleton et al. 2013).

In addition to these more subtle and indirect effects of climate change, there are direct and obvious effects. The iconic glaciers in Glacier National Park are melting, and the larger glaciers are forecast to disappear entirely by 2030 (US Geological Survey 2015). The frequency of severe fire has increased, and the extreme 1988 Yellowstone fires are projected to become the norm in future decades (chap. 10). Summer flows of rivers and streams have been declining and are projected to decline even more in the future (chaps. 7 and 12).

These examples from the Rocky Mountains illustrate how mountain ecosystems may be especially sensitive to climate change. Temperature, precipitation, and solar radiation levels vary with elevation and aspect in mountains. Many species are adapted to narrow ranges of climate in these systems. Under climate warming, species may be able to track suitable habitats by shifting to higher elevations. However, land area decreases at higher elevations and suitable climate conditions may eventually "move off the tops of the mountains," leaving species that depend on alpine conditions stranded (chaps. 6, 9, 10, and 15). Moreover, management options are constrained by law in the national parks, wilderness areas, and roadless areas that dominate land allocation at these higher elevations. For example, most of the whitebark pine stands are located in federally designated wilderness areas, where management options are limited by law (chaps. 10 and 15).

In contrast to the Rocky Mountains, signs of response to climate change are much less obvious in the Appalachian Mountains in the eastern United States (chaps. 5, 7, 8, and 11). The Appalachians are a veritable garden of Eden compared to the Rockies. The warm, humid climate, summer rains, and fertile soils result in the Appalachians being cloaked in forest. These forests are some of the fastest growing and most diverse in plant species in North America (Whittaker 1956). The effects of past climate change are less obvious here than in the Rockies. There has been some forest die-off of subalpine forests in the Great Smoky Mountains, but this is primarily due to air pollution and the introduction of exotic forest pests (chap. 8). The differences between the Rockies and the Appalachians in rates and reactions to climate change illustrate that local study is needed to understand and manage wildlands under global change.

Although the potential effects of climate change are both interesting and worrisome in some locations, our knowledge of the rates of climate change, the tolerances of species to these changes, and the potential changes in ecosystem services to humans is embryonic. For any given unit of land, such as a national park or national forest, certain fundamental questions have not yet been addressed:

- How much has climate changed over the past century, and how is it projected to change in the future?
- Has there been a trend in climate change above the natural variability?
- Which of the directional climate changes are significant ecologically?
- What ecosystem processes and species are most vulnerable to projected climate change and in which places are they most vulnerable?
- For vulnerable species, which adaptation and management options are feasible and likely to be effective?

Also poorly understood are means of managing wildland ecosystems to make them more resilient to climate change. Both the science and the management are challenged by the very nature of human-induced climate change (chap. 3). It is manifest over time periods (decades) that are long relative to resource management and scientific study horizons and even relative to the career spans of scientists and managers. It is occurring across regional to continental-sized areas that greatly exceed the spatial domains of individual national forests and national parks, necessitating interagency collaboration. Disentangling the signal of human-induced climate change from the pronounced natural variation is difficult and creates doubt in some sectors of society as to whether humans are altering global climate. Land use intensification around wildlands constrains management options. In combination, these factors result in climate change being a major challenge to resource managers. Agency policies are not yet well defined. Methods of linking climate science with management are underdeveloped. And few case studies exist of implementing management actions to mitigate the effects of climate change.

Fortunately, scientific and natural resource agencies and organizations in the United States have launched a plethora of initiatives, programs, and studies in recent years to bolster the capacity to respond and to increase knowledge on the science and management of climate change (chaps. 2,

3, and 13) (Halofsky, Peterson, and Marcinkowski 2015). For example, both the US Department of the Interior (DOI) and the US Department of Agriculture have initiated various programs to meet these management challenges. The National Park Service Inventory and Monitoring Program (NPS I&M) was created in 2000 to provide a framework for scientifically sound information on the status and trends of conditions in the national parks (Fancy, Gross, and Carter 2009).

Based partially on the success of the NPS I&M, in 2009 the DOI launched the creation of landscape conservation cooperatives (LCCs) across networks of the federal lands (US DOI 2009). The goal of the LCCs is to craft practical, landscape-level strategies for managing climate change impacts, with emphasis on (1) ecological systems and function, (2) strengthened observational systems, (3) model-based projections, (4) species-habitat linkages, (5) risk assessment, and (6) adaptive management. Related funding agencies, such as the National Science Foundation and the National Aeronautics and Space Administration (NASA) Earth Science Program, have created initiatives to support research and application of climate change science.

The DOI recently adapted an existing framework for linking science and management to cope with climate change (Glick, Stein, and Edelson 2011; Stein et al. 2014). The key elements of the framework are to (1) identify conservation targets, (2) assess vulnerability to climate change, (3) identify management options, and (4) implement management options (chap. 2). However, there are several challenges to implementing this framework (chap. 3), and few demonstrations of the approach exist to date (Janowiak et al. 2014; Halofsky, Peterson, and Marcinkowski 2015).

Aims of This Book

As stated at the opening of this chapter, a major challenge facing humankind is how to sustain both nature and ecosystem services in the remaining wildland ecosystems under climate and land use change. Toward this end, we seek in this book to develop and demonstrate means of bridging science and management to understand the rates and impacts of climate change in wildland ecosystems and to evaluate alternative strategies for managing to cope with these changes. We focus on two of the newly formed LCCs: the Great Northern LCC, which is centered on the northern Rocky Mountains, and the Appalachian LCC (fig. 1-2). Progress on merging climate science and management in these mountain ecosystems will hopefully provide a basis for subsequent applications in other LCCs. More specifically, our objectives are to:

- tell the story of change over the past century and potential change in the coming century for the Rockies and the Appalachians;
- evaluate the vulnerabilities of ecosystem processes and vegetation as a basis for prioritizing elements for management;
- develop and evaluate management alternatives for the most vulnerable elements and make recommendations for implementation;
- demonstrate the approach for climate adaptation planning that has been embraced by the US DOI;
- elucidate the lessons learned that may help these methods to be applied in other locations.

This book emerged from a five-year project called the Landscape and Climate Change Vulnerability Project funded by the NASA Earth Science Program, which seeks to use remote sensing products to inform climate change adaptation. The NASA Earth Science Program recognized the potential for the LCC program to make progress on the serious challenges that climate change poses to resource management and issued a call for proposals in 2010 to examine biological response to climate change in the context of newly forming LCCs. The past five-year period has been one of rapid progress in climate science, ecological forecasting, agency programs and policies on climate change, and management approaches. We hope to capture in this book the nature and excitement of this progress to best communicate the current "state of the art" of climate change adaptation in wildland ecosystems.

Members of the core project team have been at the center of this evolution in climate change adaptation. John E. Gross, with the National Park Service, was a leader in the development of the NPS I&M, which monitors change in the condition of natural resources across US national parks. During this project, he transitioned to becoming a lead scientist in the National Park Service Climate Change Response Program. In this role, he has contributed to pivotal policy documents, such as *Scanning the Conservation Horizon* (Glick, Stein, and Edelson 2011) and *Climate-Smart Conservation* (Stein et al. 2014). S. Thomas Olliff, as chief of resources for Yellowstone National Park, helped initiate the first climate change assessments across the Greater Yellowstone Ecosystem. He is currently expanding the assessment framework to other national parks as chief of landscape conservation

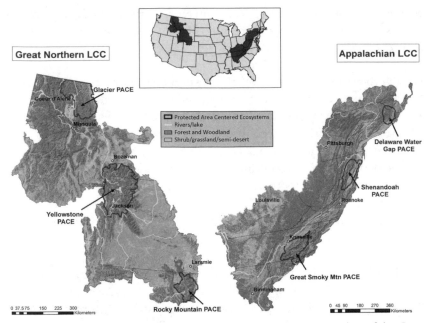

FIGURE 1-2   Our analyses focus on the US Rocky Mountains portion of the Great Northern Landscape Conservation Cooperative (LCC) (left) and the Appalachian LCC (right). Results were summarized across the LCCs and within national parks and their surrounding ecosystems (known as protected area centered ecosystems, or PACES). The PACES and the LCCs include vast wildlands as well as rapidly developing private lands. Managing across these public and private lands in the face of climate change is a special challenge in many wildland ecosystems.

and climate change for the National Park Service Intermountain Region and across the Great Northern LCC as its co-coordinator.

Forrest Melton and colleagues at NASA's Ames Research Center have developed a sophisticated computer simulation system to hindcast and forecast ecosystem processes under changing climate conditions. These modeling and analysis tools have been widely applied to natural resource issues, including wildland ecosystems (chap. 7), and currently to drought impact assessment in California. David M. Theobald, with Conservation Science Partners, has developed models of land use change across the United States and used the results to quantify connectivity of natural landscapes. His connectivity products are being widely used by LCCs and leading conservation organizations. Scott Goetz and Patrick Jantz of the Woods Hole Research Center have pioneered the use of remote sensing for regional-scale analysis of land use and biodiversity across the United States and globally, with a long history of applications in the Mid-Atlantic states.

William B. Monahan, formerly a lead NPS ecologist of the national I&M program and now working in a similar capacity with Forest Health Protection of the USDA Forest Service, is an expert in species distribution and climate modeling. He works closely with resource managers across the National Park Service system to integrate climate science into park management. Andrew J. Hansen, at Montana State University, has published widely on land use and climate effects on biodiversity, especially in the context of national parks and protected areas. Initially focusing on the Greater Yellowstone Ecosystem, he has expanded these studies to national and international applications. In addition to leading this current project, he is a science team leader for the North Central Climate Science Center.

Our agency collaborators have faced the challenge of deciding whether and how to include consideration of climate change in their daily activities. These collaborators (chap. 3) include natural resource specialists from each of the national parks in the project area and from the surrounding national forests and Bureau of Land Management lands. Agency scientists within the NPS I&M Rocky Mountain, Greater Yellowstone, and Eastern Rivers and Mountains networks have also contributed data, knowledge, and realism to the effort. These collaborators have been involved in the project from the outset and have played a vital role in the success of the project in "closing the science/management loop."

Overview of Chapters

The book is organized around the key steps in the *Scanning the Conservation Horizon* framework (table 1-1; Glick, Stein, and Edelson 2011) and the more detailed Climate-Smart Conservation framework that followed (Stein et al. 2014), both mentioned earlier in this chapter. Part 1 describes the overall approach. In chapter 2, lead author John E. Gross elaborates on the framework and how it was implemented in this project. Chapter 3, led by S. Thomas Olliff, explores the challenges resource managers face in confronting climate change and strategies used in the project to identify high-priority conservation needs.

Part 2 summarizes past and projected exposure to climate and land use change. Chapters 4 and 5, led by John E. Gross and Kevin Guay, quantify change in climate in the past century and projected for the com-

TABLE 1-1. Road map of the book chapters relative to the four steps in the *Scanning the Conservation Horizons* framework and by landscape conservation cooperative (LCC).

Part	Chapter	Identify Needs	Assess Vulnerability	Evaluate Management Options	Implement Management Options	Great Northern LCC	Appalachian LCC
Introduction	1	X	X	X	X	X	X
Approaches							
	2	X	X	X	X		
	3	X				X	X
Climate and Land Use Change							
	4		X			X	
	5		X				X
	6	X	X			X	
Ecological Consequences and Vulnerabilities							
	7		X			X	X
	8	X	X				X
	9	X	X			X	
	10	X	X			X	
	11	X	X				X
	12	X	X			X	X
Managing under Climate Change							
	13			X	X	X	X
	14			X	X	X	
	15			X	X	X	
	16	X	X	X		X	
Conclusion	17	X	X	X	X	X	X

ing century for the Great Northern LCC and the Appalachian LCC, respectively. Chapter 6, led by David M. Theobald, analyzes climate and land use change for landform units of interest to resource managers and relative to factors of adaptive capacity of those landforms within the Great Northern LCC.

Part 3 explores the ecological consequences of these changes in climate and land use. Chapters 7 through 12 evaluate the potential impacts of this climate change on ecosystem processes, such as primary productivity, and on tree species, plant communities, and native fish. These chapters also identify the species and communities that are most vulnerable to climate change.

Using climate science to inform management is the focus of part 4. In chapter 13, S. Thomas Olliff and agency collaborators examine approaches for developing and evaluating management alternatives for coping with climate change. The next two chapters focus on individual national parks and surrounding lands and aim to "tell the stories" of climate change and management in these parks. Focal parks include Rocky Mountain National Park (chap. 14) and Yellowstone/Grand Teton National Parks (chap. 15). Perhaps it is through the places that we know and love that we can best come to understand climate change and learn how to manage for healthy ecosystems. Chapter 16 takes a step back from climate adaptation planning for individual natural resources and asks how well the overall ecological integrity of the Greater Yellowstone Ecosystem has been sustained. It is at this full ecosystem scale that our success as stewards of wildlands can best be evaluated. The closing chapter draws together the main findings of the book and identifies the lessons learned that should be useful to applications in other places.

Intended Audience

The intended audience includes scientists and managers from federal programs, increasingly coordinated through LCCs, who are pioneering the incorporation of climate science into resource management. In this regard, we focused on a subset of the ecosystem processes, species, and resources that are of high importance in wildlands. We limited consideration to those response variables that we could hindcast, forecast, and analyze with rigorous scientific methods. By demonstrating the linkage of strong climate and ecological science and management for the topics for which we have expertise, we hope to help facilitate progress for other important components of ecosystems. The book is also intended for citizens and policy makers interested in climate change in the Rockies and the Appalachians. Ultimately, effective management of our federal lands is driven by the concern and input of our broader society, and we hope this book helps people better understand and appreciate the challenges climate change presents to these iconic mountain wildland ecosystems.

Acknowledgments

Linda B. Phillips prepared figure 1-2. Helpful comments on earlier drafts of the chapter were provided by David M. Theobald, S. Thomas Olliff, and William B. Monahan.

References

Fancy, S. G., J. E. Gross, and S. L. Carter. 2009. Monitoring the condition of natural resources in US national parks. *Environmental Monitoring and Assessment* 151:161–74.

Friedman, T. L. 2014. Stampeding black elephants. *New York Times*, November 22.

Glick, P., B. A. Stein, and N. Edelson, eds. 2011. *Scanning the Conservation Horizon: A Guide to Climate Change Vulnerability Assessment*. Washington, DC: National Wildlife Federation.

Halofsky, J. E., D. Peterson, and K. W. Marcinkowski. 2015. *Climate Change Adaptation in United States Federal Natural Resource Science and Management Agencies: A Synthesis*. USGCRP Climate Change Adaptation Interagency Working Group.

Hansen, A. J., N. Piekielek, C. Davis, J. Haas, D. Theobald, J. Gross, W. Monahan, T. Olliff, and S. Running. 2014. Exposure of U.S. national parks to land use and climate change 1900–2100. *Ecological Applications* 24 (3): 484–502.

Hansen, A. J., J. J. Rotella, M. L. Kraska, and D. Brown. 2000. Spatial patterns of primary productivity in the Greater Yellowstone Ecosystem. *Landscape Ecology* 15:505–22.

Janowiak, M. K., C. W. Swanston, L. M. Nagel, L. A. Brandt, P. R. Butler, S. D. Handler, P. Danielle Shannon, L. R. Iverson, S. N. Matthews, A. Prasad, and M. P. Peters. 2014. A practical approach for translating climate change adaptation principles into forest management actions. *Journal of Forestry* 112 (5): 424–33. doi: http://dx.doi.org/10.5849/jof.13-094.

Kalisz, P. J., and H. B. Wood. 1995. Native and exotic earthworms in wildland ecosystems. In *Earthworm Ecology and Biogeography in North America*, edited by P. F. Hendrix, 117–24. Boca Raton, FL: CRC Press.

Logan, J. A., W. W. Macfarlane, and L. Willcox. 2010. Whitebark pine vulnerability to climate-driven mountain pine beetle disturbance in the Greater Yellowstone Ecosystem. *Ecological Applications* 20:895–902.

Middleton, A. D., M. J. Kauffman, D. McWhirter, J. G. Cook, R. C. Cook, A. A. Nelson, M. D. Jimenez, and R. W. Klaver. 2013. Animal migration amid shifting patterns of phenology and predation: Lessons from a Yellowstone elk herd. *Ecology* 94 (6): 1245–56.

Millennium Ecosystem Assessment. 2005. *Ecosystems and Human Well-Being: Synthesis*. Washington, DC: Island Press.

Radeloff, V. C., et al. 2010. Housing growth in and near United States protected areas limits their conservation value. *Proceedings of the National Academy of Sciences of the United States of America* 107 (2): 940–45.

Stein, B. A., P. Glick, N. Edelson, and A. Staudt, eds. 2014. *Climate-Smart Conservation: Putting Adaptation Principles into Practice*. Washington, DC: National Wildlife Federation.

Theobald, D. M., and W. H. Romme. 2007. Expansion of the US wildland-urban interface. *Landscape and Urban Planning* 83 (4): 340–54.

US Department of the Interior. 2009. Addressing the impacts of climate change on America's water, land, and other natural and cultural resources. Secretarial Order 3289. Washington, DC: US Department of the Interior.

US Geological Survey, Northern Rocky Mountains Science Center. 2015. Retreat of glaciers in Glacier National Park. http://nrmsc.usgs.gov/research/glacier_retreat .htm.

Whittaker, R. H. 1956. Vegetation of the Great Smoky Mountains. *Ecological Monographs* 26 (1): 1–80.

Wilkie, D. S., et al. 2006. Parks and people: Assessing the human welfare effects of establishing protected areas for biodiversity conservation. *Conservation Biology* 20 (1): 247–49.

Wittemyer, G., et al. 2008. Accelerated human population growth at protected area edges. *Science* 321:123–26.

PART 1

Approaches for Climate Adaptation Planning

How can science and management be brought together to keep ecosystems healthy under climate and land use change? We open the book with an overview of the methods that have been proposed and a focus on a conceptual approach that has been widely embraced by federal resource agencies. The Climate-Smart Conservation framework expands on the traditional adaptive management approach that originated in the 1970s, in which management actions are done in the context of experiments such that resource managers implement management actions, monitor their effectiveness, and iteratively modify the approach to improve effectiveness. The Climate-Smart Conservation framework adds consideration of climate impacts and vulnerabilities to this "learn as you manage" approach.

Chapter 2 describes the steps in the Climate-Smart Conservation framework and elaborates on how we implemented the approach in our Landscape Climate Change Vulnerability Project (LCCVP). It provides details on the study areas of interest to the project and the collaborating agencies. The gulf between concept and application can be vast. Chapter 3 is an honest assessment of the numerous impediments to actually executing the Climate-Smart Conservation framework across US federal lands. The key challenges of bridging scientist and resource management cultures and mind-sets in the context of the complexities and uncertainties of climate change are elucidated. We identify our key agency partners and summarize some of the approaches we used in the LCCVP to begin to overcome these challenges.

In total, part 1 provides a road map for the book in moving from broad concepts and theory to rolling up sleeves and getting on with the business of managing ecosystems under climate change and land use change.

Chapter 2

Effectively Linking Climate Science and Management

John E. Gross and S. Thomas Olliff

Many agencies and managers recognize the adverse effects of climate change, but they are struggling to find information that helps them understand, plan for, and respond to existing and projected changes. The science of climate change adaptation is still a young and rapidly evolving field. Although there has been explosive growth in information on climate changes, much of the existing information is focused on general issues and at very broad scales that are more relevant to informing policy than to helping place-based resource managers. So, while the effects of climate change are now very well documented, there are still few examples of how to actually do climate adaptation planning for conservation at the scales most relevant to on-the-ground managers.

Our work on climate adaptation addressed the goals articulated in chapter 1 but, ultimately, real success will be measured by the transfer of knowledge from the scientific community to managers and the application of this knowledge to management decisions. Thus, our work had a very strong management focus and was organized around activities that reflect the latest developments in climate adaptation (Stein et al. 2014). Agencies' responses to climate change fall into the broad categories of adaptation and mitigation. Adaptation is the active response to a changing climate. Activities that constitute adaptation include on-the-ground actions as well as changes in planning, interpretation/education, flexible decision making, allocation of budgets to research and staffing, and a host of other strategic and operational activities (National Research Council 2010). Mitigation constitutes activities to reduce the magnitude of future changes, primarily by reducing emissions of greenhouse gases. Mitigation contributes to adaptation by reducing the magnitude of climate changes and necessary adaptive responses. There may be conservation co-benefits to mitigation actions (e.g., carbon sequestration via altered fire regimes), but mitigation and nature conservation activities need to be evaluated against their specific goals. Our work focused on adaptation, with the recognition that lands are managed to meet many goals, including climate mitigation and the provision of multiple ecosystem services.

Many existing and familiar tools and concepts will effectively address climate adaptation, but these may need to be applied in new ways or places and in the context of goals that facilitate transitions rather than stasis. Most managers simply don't have time to discover, learn, and apply a large and rapidly expanding suite of new tools and techniques. For the foreseeable future, the most effective climate adaptation will rely on developing partnerships in which scientists and science translators make available the relevant science—some of which already exists and much of which is in process—and collaborate with land managers to address resource issues in a climate-informed way. The future of managing natural resources will depend more and more on working with partners at ecosystem and landscape scales to address issues that transcend traditional administrative and disciplinary boundaries. Understanding and developing effective partnerships is thus the focus of chapter 3. First, in this chapter, we describe the overall approach and conceptual foundations of our work. Climate adaptation is a work in progress, and it's not perfect. Our experiences and those of others are contributing to better on-the-ground climate adaptation.

Adaptation Framework

Climate adaptation involves all of the steps in traditional "good planning" as well as some new steps. Stein and Glick (2011) presented a simple framework for climate change adaptation based on existing conservation planning frameworks (fig. 2-1). A key strength of this simple conceptual framework is the clarity with which it articulates key steps for climate adaptation without getting distracted by the detailed work required at each step. The four steps are to (1) identify conservation targets, (2) assess vulnerability to climate change, (3) identify management options, and (4) implement management options.

The simple framework in figure 2-1 captures the essence of the adaptation process and stresses the outcomes of each step that feed into

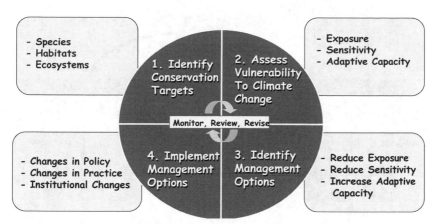

FIGURE 2-1 Four fundamental steps in a simple conceptual model of planning and implementing climate adaptation, emphasizing the products of each step that feed into the next step. In practice, each step in this diagram consolidates a number of activities. (Modified from Stein and Glick 2011.)

following activities. However, each of the four steps is composed of actions and activities that can pose serious challenges and require substantial time, resources, and effort to accomplish. To more fully illustrate and plan for the processes and key feedbacks involved in real-world adaptation projects, we followed a more detailed adaptation framework that embellishes the simple framework by breaking the four basic steps described in figure 2-1 into discrete activities and that explicates the monitoring and evaluation necessary for adaptive management (fig. 2-2; National Park Service 2010; Cross et al. 2013; Stein et al. 2014).

We focused primarily on this detailed framework—known as the Climate-Smart Conservation framework (Stein et al. 2014)—because the additional detail facilitated project management, and because the framework was developed with multiple partners, including land management agencies, at the table. Training designed around the Climate-Smart Conservation framework and supported by the US Fish and Wildlife Service National Conservation Training Center helps to generalize and promote the approach used by the Landscape Climate Change Vulnerability Project (LCCVP) team. The Climate-Smart Conservation framework builds from and embellishes on the four-step framework in *Scanning the Conservation Horizon* (Stein and Glick 2011) and expands it to include seven steps.

Key activities and outcomes of the seven steps in the Climate-Smart Conservation framework shown in figure 2-2 are as follows:

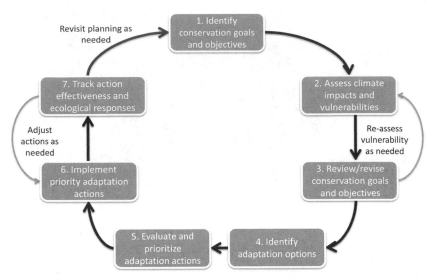

FIGURE 2-2 The Climate-Smart Conservation framework for designing and implementing climate adaptation for natural resources. (Modified from Stein et al. 2014.)

Step 1. *Identify conservation goals and objectives*. Assemble the initial team, and ensure that the project has a clear purpose. Examine the relevance of existing conservation goals and conservation targets (targets can be species/communities, ecological systems, biomes, or other resources). Agree on the geographic area, time frame, and other aspects of the overall design. Ensure that the overall effort is realistic and scaled to the detail needed to support relevant decisions, staff time, expertise, and funding.

Step 2. *Assess climate impacts and vulnerabilities*. Evaluate the impacts and results from vulnerability assessments, and identify key vulnerabilities. Vulnerability assessments identify what is at risk and why—while considering both climate and nonclimate factors. Climate change vulnerability is typically broken down into exposure, sensitivity, and adaptive capacity. Exposure accounts for physical conditions or changes in conditions (temperature, precipitation, freezing, and so forth) experienced by the conservation target. Exposure is primarily extrinsic, and it is often estimated using the results of climate models. Sensitivity represents the degree to which a target will be affected by, or responsive to, stimuli such as climate. Sensitivity is primarily composed of intrinsic characteristics, such as physiological sensitivity to heat or cold, pH, or humidity. Adaptive capacity refers to the ability of the target to adapt to changes and can be modified by management actions that, for example, alter habitat connectivity or external stressors.

Step 3. *Revisit (review/revise) conservation goals and objectives*. Based on results and insights from steps 1 and 2, conservation goals and objectives should be reviewed to ensure that they are still relevant and realistic. For example, the vulnerability assessment may suggest that it will be extremely difficult or impossible to sustain stable populations over the current range of a species.

Step 4. *Identify adaptation options*. Climate-informed goals and vulnerability assessments provide a basis for identifying potential management responses. In many situations, current practices will be inadequate in the future, and it is important to be creative and consider new practices. Techniques for identifying options include expert elicitation, scenario planning, and focus groups (West and Julius 2014).

Step 5. *Evaluate and prioritize adaptation actions*. With a broad range of potential responses identified in step 4, step 5 narrows the options. Evaluation of options often requires careful evaluation of ecological models and consideration of costs, practicality, risks, and public support. To avoid any inhibition of the creative process and to help ensure that all options are on the table, the evaluation and selection of adaptation options should almost always be distinct from the process of identifying potential options.

Step 6. *Implement priority adaptation actions*. Implementation is the natural consequence of selecting options, but often barriers must be overcome. These include regulatory compliance (e.g., with the National Environmental Policy Act), funding, resistance to changing management practices, or uncertainty in the outcome.

Step 7. *Track action effectiveness and ecological responses*. When implementation is being planned, monitoring and evaluation should be integral to the design. Currently, there are few examples of the implementation of climate adaptation and even fewer results from the evaluation of climate adaptation actions. Adaptation poses special challenges for monitoring (Bours, McGinn, and Pringle 2014).

The bulk of this volume describes how LCCVP developed and demonstrates practical ways to accomplish all steps in the framework up to implementation. We expect that the results of this project will lead to implementation of desirable adaptation options by a broad range of partners, including the National Park Service, the US Forest Service, and state and other public and private land managers. For federal land managers, implementation of some options will trigger lengthy review and compliance processes (e.g., National Environmental Policy Act and public outreach).

Study Area and Approach

Our work focused on the Rocky Mountains ecoregion of the Great Northern Landscape Conservation Cooperative (http://greatnorthernlcc.org/) and the Appalachian Landscape Conservation Cooperative (http://applcc.org/; fig. 1-2 in chapter 1). In addition to the landscape conservation cooperatives (LCCs), the project addressed a variety of other places at varying spatial scales that are relevant to conservation. Most commonly, these included the formal boundary of the national park units and the surrounding protected area centered ecosystems, or PACEs (Hansen et al. 2011). These areas of analysis provided ecological and management-relevant case studies for vulnerability assessment and management applications. Parks within these areas include Grand Teton, Yellowstone, and Rocky Mountain National Parks in the Great Northern LCC, and Delaware Water Gap National Recreation Area and Shenandoah and Great Smoky Mountains National Parks in the Appalachian LCC. All of the park PACEs include lands managed by the US Forest Service; additional parts of the PACEs are managed by the US Bureau of Land Management, the US Fish and Wildlife Service, states, private citizens, and other jurisdictions.

Analyses directed to managers of specific sites (i.e., parks, ecosystems) generally used the PACE as the area of analysis (Hansen et al. 2011). Few, if any, protected area boundaries within the United States were originally delineated specifically to sustain all of the native biodiversity that exists in the area. To help managers and scientists identify habitats—inside and outside a park's boundaries—required to sustain the integrity of a park's biodiversity, we developed and applied a methodology that accounted for species area requirements, key habitats, disturbances, and edge effects. These delineations of PACEs have proven useful for various analyses of park conditions, historical trends, and projections of future threats to parks (Hansen et al. 2011, 2014; Davis and Hansen 2011; Amberg et al. 2012; Piekielek and Hansen 2012).

Because the scope of our studies encompassed a broad geographical extent and considered both historical and projected observations, we relied on remotely sensed data and derived products (Gross et al. 2006, 2011). These were combined with on-the-ground observations of species occurrences, soils, runoff, and other data in statistical and ecological simulation models. We relied on publically available data from the National Aeronautics and Space Administration (NASA) and the models, data, computational facilities, and associated infrastructure available through the NASA Earth Exchange (NEX) at the NASA Ames Research Center (Nemani et al. 2009).

Approach of the Landscape Climate Change Vulnerability Project

From the beginning we recognized that climate adaptation would be a continuous, ongoing process with short- and long-term goals and objectives as well as options that applied at small to very large areas. The science-oriented component of our project thus required an integrative approach for assessing past changes and the current status of system drivers and states along with an ability to forecast a range of potential future climate scenarios and the resulting trends of key conservation targets and ecosystem properties.

To achieve this integration, we used the overall project design summarized in figure 2-3, which identifies land use and climate changes as dominant, broad-scale drivers of change. In effect, we used a complementary set of coarse- and fine-filter indicators (Noon, McKelvey, and Dickson 2009; Gross and Noon 2015). Our coarse-filter indicators characterized the environment at the landscape scale using vegetation and land cover maps derived from remotely sensed data, gridded and station-based climate data, maps of roads, and calculated indicators, such as naturalness and degree of human modification (chap. 6; Theobald 2013). The coarse-filter indicators we measured were defined independently of any particular species, and they attempted to capture key attributes and habitat requirements of entire species' communities. A basic assumption for using coarse-filter indicators is that they can provide an independent means to estimate and predict the distribution of species and conservation targets in the landscape. When coarse-filter indicators accurately account for the distribution of conservation targets, they can be an effective means to support decisions on management to prevent loss of species (Margules and Pressey 2000), to maintain landscape connectivity (Brost and Beier 2012), and, for LCCVP, to provide an assessment of landscape-level adaptive capacity (chap. 6).

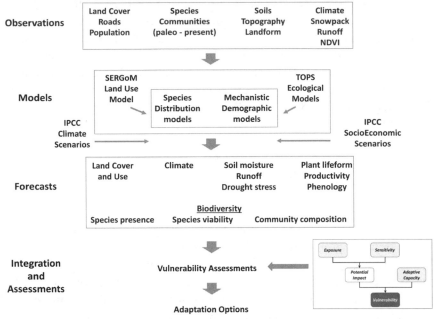

FIGURE 2-3   Simplified diagram of the ecological modeling approach that served as a foundation and framework for the project. This figure illustrates the flow of key input data (top) to the primary modeling environments (SERGoM and TOPS) and then to the synthesis and assessment products (bottom).

For more detailed assessments and to identify actionable, site-specific management options, we supplemented the coarse-filter indicators with fine-filter targets and analyses. The fine-filter targets focused on individual species, specific sites, and other ecological characteristics that required detailed site- and species-specific information. Fine-filter analyses include detailed assessments of forest tree species and fish (part 3 in this volume).

Ecological Models and Forecasting

Because most effects of climate changes will be expressed in the future, ecological models are central to planning for climate adaptation. Our approach used a complementary set of models that forecasted changes in the highest-level drivers: land use and land cover change, and climate-driven ecosystem dynamics. At the broadest level, many of our analyses relied on high-resolution (0.5 mile, 30 arc-second [800 meter]) downscaled climate data developed on NEX in collaboration with NASA Ames (Thrasher et al. 2013). Future changes in housing density, a key aspect of rural land use around protected areas, were forecast using SERGoM (Spatially Explicit Regional Growth Model; Theobald 2005), a spatially explicit model that estimates the density and location of development. Ecosystem dynamics were forecasted using models managed and executed within the NASA TOPS (Terrestrial Observation and Prediction System) ecosystems modeling environment (Nemani et al. 2009). The ecosystem models forecast future changes in vegetation productivity, distribution of broad vegetation classes, changes in seasonality of growth and senescence, runoff, and snow dynamics.

Results from LCCVP model experiments provided quantitative estimates of current ecosystem states that were compared with on-the-ground measurement to calibrate and verify model performance. Many of these results are presented in part 3. In general, LCCVP simulations used both a best-case reduced emissions pathway and a "business as usual" future greenhouse gas emissions pathway—representative concentration pathways 4.5 and 8.5, respectively. Although it appears unlikely that greenhouse gas emissions will be reduced in the immediate future, the range of future climate trajectories provides insights to the potential magnitude of ecological impacts that we can expect (chap. 7).

Vulnerability Assessment

Climate change vulnerability assessments are the key result of step 2 in the adaption framework (fig. 2-2). The three components of vulnerability (exposure, sensitivity, and adaptive capacity) were evaluated at ecological levels from species to biomes (table 2-1). Key components of exposure are the magnitude of change in climate (chaps. 4 and 5) and land use (chap. 6). These are critical, broad-scale constraints on ecological processes and biodiversity. Sensitivity of ecosystem processes was evaluated via changes in ecosystem processes, the breadth of climates that determine species distributions, and the response of broader vegetation types to climate changes. Adaptive capacity has proved to be particularly challenging to define and capture with ecological indicators (Stein, Glick, and Hoffman 2011; Nicotra et al. 2015). Chapters in part 3 focus on vulnerability of species and ecosystems, and they consider adaptive capacity in different ways. Chapter 6, for example, evaluates adaptive capacity as a function of exposure to hu-

TABLE 2-1. Components of vulnerability and general approach and data for evaluating the components at three levels of ecological organization.

Component of Vulnerability	Species/ Communities	Ecological System	Biomes
Exposure	Climate (NEX-DCP30[1], WorldClim[2]) and land use (SERGoM[3]) projections	Climate (NEX-DCP30, WorldClim) and land use (SERGoM) projections	Climate[4] and land use (SERGoM) projections
Sensitivity	Habitat suitability modeling	Habitat suitability modeling; TOPS[5] projections	Biome BGC projections[5]; controls of NPP; ecosystem model responsiveness
Adaptive Capacity	Species and habitat traits	Landscape physiography; ecosystem modifications; connectivity; protection	Diversity at ecological system level; conservation context

[1] Thrasher et al. 2013.
[2] Hijmans et al. 2005.
[3] Spatially Explicit Regional Growth Model; Theobald 2005.
[4] Rehfeldt et al. 2006.
[5] Terrestrial Observation and Prediction System; Nemani et al. 2009.

man modifications and physiographic diversity. Chapter 11 uses dispersal ability, propagule pressure, and fragmentation to assess adaptive capacity of forest trees. These studies are examples of new ways to generate and evaluate information needed for climate change adaptation.

Review Conservation Goals and Evaluate Management Options

A main objective of meetings early in the LCCVP project was to identify important conservation goals and objectives that guided us in the selection of resources for further study and assessment. With preliminary or final results of analyses of climate changes, species distribution models, and vulnerability assessments, we needed to revisit resource management goals and use our results to collaboratively identify potential management options. To identify and articulate generic adaptation actions and strategies, we reviewed existing management plans (e.g., Greater Yellowstone Coordinating Committee 2011) and developed alternatives based on published literature, current management practices, and projected impacts to ecological systems and focal species based on this project (e.g., West and Julius 2014; Schmitz et al. 2015). Other adaptation options emerged as insights from collaborations between scientists and managers (chap. 13).

Key Partners

Climate impacts are blind to jurisdictional boundaries, and effective adaptation requires a greater degree of collaboration among managers, agencies, and scientists than do many other management activities. LCCVP addresses a very broad and complex topic, across a vast geographical extent. This is the nature of many important climate adaptation projects, and effective and mutually beneficial partnerships are critical to the success of this activity and similar ones. As a result, the work reported in this volume was possible only because many partners generously shared their time, expertise, and insights. We thank all of our many partners.

Climate adaptation studies are fundamentally place based, and many LCCVP studies used national parks as an identifiable focal location, with the caveat that we used the park PACE as a starting point for defining the boundary for ecological analyses. The area of a park PACE always extended well beyond the park boundary (Hansen et al. 2011; Davis and Hansen 2011). NPS staff were key collaborators throughout the project, identifying high-priority resources, providing insights and feedback, and critically helping identify practical management options that respect agency goals, regulatory constraints, and the myriad practical considerations of managing complex ecosystems. Many parks in both the Great Northern LCC and the Appalachian LCC are adjacent to national forests, and US Forest Service scientists and managers have contributed throughout the project. Many high-priority resources—and thus LCCVP studies—rely on detailed analyses of forest trees and their response to climate changes. US Forest Service personnel are experts on these topics, and they provided data and advice and have been collaborators and coauthors. The LCCs were designed to facilitate landscape-scale conservation, and they have been integral to our overall efforts. Other noteworthy partners have been the Greater Yellowstone Coordinating Committee's Whitebark Pine Subcommittee, which was essential to our work on five-needle pines. The US Geological Survey's North Central Climate Science Center has worked closely in developing climate data products, supporting related climate adaptation studies, and sharing insights and discussions throughout this study.

Many of the key scientists were introduced in chapter 1. The LCCVP team is large, diverse, and distributed across agencies and institutions, from NASA's Ames Research Center in California to the Woods Hole Research Center in Massachusetts. Most of the key contributing scientists are authors on one or more chapters in this book.

Data and Information Transfer

Research-management partnerships to address important climate adaptation issues will invariably take a long time, engage many diverse audiences, and generate a huge variety of products. We found that effective communication and information transfer required the project team to develop and present a multitude of products that vary in length, technical detail, and format to meet specific and different needs (Gross et al. 2011). For a complex project, there's no "one-size-fits-all" communication product. A full array of documents—from peer-review publications to short, public-focused brochures—facilitated communication. The most effective communication requires the right information, delivered to the right people, in the right format, at the right time.

Table 2-2 summarizes six major categories of products used or produced by LCCVP. This table illustrates that product management for projects such as ours is an increasingly difficult challenge because products range from slide-based presentations, brochures, publications, and code to run models and statistical analyses to complex ecological model results and enormous raw data sets and imagery. There is no single host through which discovery, provision, and archiving services can be provided for all of these products, in part simply because of the huge volume of computer storage needed to archive climate and simulation model outputs. To ensure persistence and the long-term utility of the most important results, we adopted a multifaceted strategy for disseminating information and archiving results and products.

Our options for data and product management were driven largely by those who we most wanted to use the results from the project. Resource managers were our primary partners and audience, and the science–management team collaborated in the development of presentations, resource briefs, reports, and publications that were designed to support existing and future management planning processes. Most of these products are suitable for archiving in the National Park Service Information System (IRMA).

TABLE 2-2. Categories of products and strategies for their management to ensure discovery, access, and persistence.

Type of Product	Intended Audience	Disposition
Resource and climate briefs, presentations	Managers, engaged public; primarily nontechnical stakeholders	NPS IRMA/data.gov[1], project websites
Technical reports and publications	Scientists, resource managers, consultants, some decision makers	NPS IRMA/data.gov, peer-reviewed publications
Administrative and project reports, concept papers	Project staff, funders, auditors	NPS IRMA/data.gov; project website
Derived spatial data sets (habitat maps, connectivity, housing density)	Scientists, spatial analysts	Various; NPS IRMA/data.gov, North Central Climate Science Center, USGS GeoDataPortal, Great Northern LCC
Climate data	Scientists	Accessed from original sources; not stored by this project
Simulation model results (e.g., primary productivity, leaf area index, soil moisture, runoff)	Scientists	NASA Ames Research Center

[1] IRMA (Integrated Resource Management Applications; https://irma.nps.gov/App/Portal/Home) is a primary NPS application for long-term management of digital information. Services include extensive searching (discovery), direct access to materials, and long-term archiving. IRMA records are harvested by data.gov, a primary data source for the US government, providing a degree of visibility and discoverability.

Records in IRMA are harvested by US Department of the Interior (DOI) information systems and are discoverable on the DOI public websites (www.doi.gov/data and www.data.gov). Storage of large data sets, such as climate data and simulation model results, was facilitated by partnerships with other agencies, most notably NASA and the US Geological Survey (table 2-2).

Conclusion

Climate change poses one of the greatest and most complex challenges that natural resource managers have ever faced. Climate adaptation will be an ongoing process to respond to climate challenges, and it will require effective partnerships between scientists and managers. In this chapter, we described a range of processes, activities, analyses, interactions, and complexities that characterize broad-scale, real-world conservation projects such as climate adaptation. The framework we described explicitly shows how the tools of climate change adaptation—vulnerability assessments, climate hindcasting and projections, ecological models, and adaption options—fit together to help inform decisions and build a foundation for scientists and managers to work together on climate adaptation.

References

Amberg, S., K. Kilkus, S. Gardner, J. E. Gross, M. Wood, and B. Drazkowski. 2012. *Badlands National Park: Climate Change Vulnerability Assessment*. Natural Resource Report NPS/BADL/NRR-2012/505. Fort Collins, CO: National Park Service.

Bours, D., C. McGinn, and P. Pringle. 2014. Twelve reasons why climate change adaptation M&E is challenging. Phnom Penh, Cambodia: SEA Change CoP; Oxford: UKCIP.

Brost, B. M., and P. Beier. 2012. Use of land facets to design linkages for climate change. *Ecological Applications* 22:87–103.

Cross, M. S., P. D. McCarthy, G. Garfin, D. Gori, and C. A. F. Enquist. 2013. Accelerating adaptation of natural resource management to address climate change. *Conservation Biology* 27:4–13.

Daly, C., M. Halbleib, J. I. Smith, W. P. Gibson, M. K. Doggett, G. H. Taylor, J. Curtis, and P. P. Pasteris. 2008. Physiographically sensitive mapping of climatological temperature and precipitation across the conterminous United States. *International Journal of Climatology* 28:2031–64.

Davis, C. R., and A. J. Hansen. 2011. Trajectories in land use change around U.S. national parks and their challenges and opportunities for management. *Ecological Applications* 21:3299–3316.

Greater Yellowstone Coordinating Committee, Whitebark Pine Subcommittee. 2011. Whitebark Pine strategy for the Greater Yellowstone Area. Report.

Gross, J. E., A. J. Hansen, S. J. Goetz, D. M. Theobald, F. M. Melton, N. B. Piekielek, and R. R. Nemani. 2011. Remote sensing for inventory and monitoring of U.S. national parks. In *Remote Sensing of Protected Lands*, edited by Y. Q. Yang, 29–56. Boca Raton, FL: Taylor & Francis.

Gross, J. E., R. R. Nemani, W. Turner, and F. Melton. 2006. Remote sensing for the national parks. *Park Science* 24:30–36.

Gross, J. E., and B. R. Noon. 2015. Application of surrogates and indicators to monitoring natural resources. In *Surrogates and Indicators in Ecology, Conservation and Environmental Management*, edited by D. B. Lindenmayer, J. C. Pierson, and P. Barton, 169–78. Melbourne: CSIRO Publishing; London: CRC Press.

Hansen, A. J., C. Davis, N. B. Piekielek, J. E. Gross, D. M. Theobald, S. J. Goetz, F. Melton, and R. DeFries. 2011. Delineating the ecosystems containing protected areas for monitoring and management. *BioScience* 61:263–73.

Hansen, A. J., N. B. Piekielek, C. Davis, J. R. Haas, D. M. Theobald, J. E. Gross, W. B. Monahan, T. Olliff, and S. W. Running. 2014. Exposure of US national parks to land use and climate change 1900–2100. *Ecological Applications* 24:484–502.

Hijmans, R. J., S. E. Cameron, J. L. Parra, P. G. Jones, and A. Jarvis. 2005. Very high resolution interpolated climate surfaces for global land areas. *International Journal of Climatology* 25:1965–78.

IPCC (Intergovernmental Panel on Climate Change). 2014. *Fifth Assessment Report, Working Group II Glossary*.

Margules, C. R., and R. L. Pressey. 2000. Systematic conservation planning. *Nature* 405:243–53.

Monahan, W., and N. Fisichell. 2014. Climate exposure of US national parks in a new era of change. *PLOS ONE* 9 (7): e101302. doi: 10.1371/journal.pone .0101302.

National Park Service (NPS). 2010. National Park Service Climate Change Response Strategy. Fort Collins, CO: NPS Climate Change Response Program.

National Research Council (NRC). 2010. *Adapting to the Impacts of Climate Change*. Washington, DC: National Academies Press.

Nemani, R., H. Hashimoto, P. Votava, F. Melton, W. L. Wang, A. Michaelis, L. Mutch, C. Milesi, S. Hiatt, and M. White. 2009. Monitoring and forecasting ecosystem dynamics using the Terrestrial Observation and Prediction System (TOPS). *Remote Sensing of Environment* 113:1497–1509.

Nicotra, A. B., E. A. Beever, A. L. Robertson, G. E. Hofmann, and J. O'Leary. 2015. Assessing the components of adaptive capacity to improve conservation and management efforts under global change. *Conservation Biology*. 29: 1268–78.

Noon, B. R., K. S. McKelvey, and B. G. Dickson. 2009. Multispecies conservation planning on U.S. federal lands. In *Models for Planning Wildlife Conservation in Large Landscapes*, edited by J. J. Millspaugh and F. R. Thompson III, 51–84. San Diego: Elsevier Science.

Oyler, J. W., A. Ballantyne, K. Jencso, M. Sweet, and S. W. Running. 2014. Creating a topoclimatic daily air temperature dataset for the conterminous United States using homogenized station data and remotely sensed land skin temperature. *International Journal of Climatology Online*. doi: 10.10002/joc.4127.

Piekielek, N. B., and A. J. Hansen. 2012. Extent of fragmentation of coarse-scale habitats in and around U.S. national parks. *Biological Conservation* 155:13–22.

Rehfeldt, G. E., N. L. Crookston, M. V. Warwell, and J. S. Evans. 2006. Empirical analyses of plant-climate relationships for the western United States. *International Journal of Plant Sciences* 167:1123–50.

Schmitz, O. J., J. J. Lawler, P. Beier, C. Groves, G. Knight, D. A. Boyce Jr, J. Bulluck, K. M. Johnston, J. L. Klein, K. Muller, D. J. Pierce, W. J. Singleton, J. R. Strittholt, D. M. Theobald, S. C. Trombulak, and A. E. Trainor. 2015. Conserving biodiversity: Practical guidance about climate change adaptation approaches in support of land-use planning. *Natural Areas Journal* 35:190–203.

Stein, B. A., and P. Glick. 2011. Introduction to *Scanning the Conservation Horizon: A Guide to Climate Change Vulnerability Assessment*, edited by P. Glick, B. A. Stein, and N. Edelson, 6–18. Washington, DC: National Wildlife Federation.

Stein, B. A., P. Glick, N. Edelson, and A. Staudt, eds. 2014. *Climate-Smart Conservation: Putting Adaptation Principles into Practice*. Washington, DC: National Wildlife Federation.

Stein, B. A., P. Glick, and J. Hoffman. 2011. Vulnerability assessment basics. In *Scanning the Conservation Horizon: A Guide to Climate Change Vulnerability Assessment*, edited by P. Glick, B. A. Stein, and N. Edelson, 19–38. Washington, DC: National Wildlife Federation.

Theobald, D. M. 2005. Landscape patterns of exurban growth in the USA from 1980 to 2020. *Ecology and Society* 10:32. http://www.ecologyandsociety.org/vol10/iss1/art32/.

Theobald, D. M. 2013. A general model to quantify ecological integrity for landscape assessments and US application. *Landscape Ecology* 20:1859–74.

Thrasher, B., J. Xiong, W. Wang, F. Melton, A. Michaelis, and R. Nemani. 2013. Downscaled climate projections suitable for resource management. *Eos, Transactions American Geophysical Union* 94:321–23.

West, J. M., and S. H. Julius. 2014. Choosing your path: Evaluating and selecting adaptation options. In *Climate-Smart Conservation: Putting Adaptation Principles into Practice*, edited by B. Stein, P. Glick, N. Edelson, and A. Staudt, 141–52. Washington, DC: National Wildlife Federation.

Chapter 3

Challenges and Approaches for Integrating Climate Science into Federal Land Management

S. Thomas Olliff and Andrew J. Hansen

Although private lands are prevalent, federally managed public lands—those managed by the US Forest Service, the National Park Service (NPS), the US Fish and Wildlife Service (USFWS), and the Bureau of Land Management (BLM)—are uniquely positioned to coordinate climate adaptation across their lands and with other landowners and stewards. The federal government manages almost 30 percent of the land in the United States, nearly 650 million acres, including 401 units of the National Park System and 155 national forests and grasslands (GAO 2013).

Presidential Executive Order 13514 (October 5, 2009) directed federal agencies to develop adaptation approaches, and E.O. 13635 (November 2013) calls on federal agencies to work with states, tribes, and local governments to improve preparedness for the impacts of a changing climate. Department of the Interior Secretarial Order 3289 (September 14, 2009) established department-level programs that include the BLM, USFWS, and NPS. These governmental agencies have developed agency-specific climate change strategies (Gonzalez 2011; Shafer 2014). The president has issued a Climate Action Plan (Executive Office of the President 2013) and a Priority Agenda for Enhancing the Climate Resilience of America's Natural Resources (Council on Climate Preparedness and Resilience 2014). Climate adaptation directives are clear, yet managers working at the site level are challenged to implement these directives and few adaptation actions have been undertaken. This chapter describes the approach taken in the

Landscape Climate Change Vulnerability Project (LCCVP) toward helping managers understand the science and to frame and implement adaptation actions.

Challenges of Integrating the Science into the Management

There is a general consensus that science is critical to making successful resource management decisions (Williams et al. 2007; White, Garrott, and Olliff 2009), but incorporating the best available science in land management decisions—whether through agency planning, National Environmental Policy Act decision documents, cross-jurisdictional strategies, or on-the-ground projects—can be difficult for federal land managers. In a survey of land managers in the northern Rockies, managers ranked the following five barriers to using research: (1) the lack of funding to implement research findings, (2) the lack in research documents of clear management implications, (3) poor communication between science and management agencies, (4) limited budget and travel, and (5) conflicts between scientific recommendations and high-level political priorities (Northern Rockies Fire Sciences Network 2011).

Along with the typical challenges of integrating science into management, climate change has some unique challenges:

- The concepts and language are new and not well understood.
- The science is new, and different, from what managers are used to (chap. 13).
- High uncertainty leads to management indecision.
- There is little understanding of organizing frameworks to help managers make sense of the new science and new climate change tools.
- Climate change operates at such large scales that we have to build organizational bridges to work across boundaries (chaps. 14 and 15).
- Climate change is considered critical but not urgent.
- Climate change involves social complexity with fragmented stakeholders.

We discuss each of these challenges below.

Understanding Concepts

The amount of rigorous scientific inquiry into climate change is staggering—for example, a 2013 report found 11,944 manuscripts that were published from 1991 to 2013 (Cooke et al. 2013). As the scientific field has grown, it has developed its own concepts and terminology. However, most federal land managers did not become engaged in climate change prior to 2007, when the Intergovernmental Panel on Climate Change's 4th Assessment Report was issued (IPCC 2007). The earliest agency strategies regarding climate change were not issued until 2009. Even in 2014, many land managers were not conversant with climate change concepts and language. Those that are can still be classified as innovators or early adopters (Rogers 2003).

Managers commonly do not understand basic terminology, such as *general circulation model*, *downscaled climate model*, *vulnerability assessment*, *exposure*, *sensitivity*, *adaptive capacity*, and *adaptation options*. In a survey of NPS Intermountain Region employees, 90 percent of respondents correctly identified definitions of key terms—such as the greenhouse effect and mitigation of and adaptation to climate change—but far fewer (38 to 82 percent) correctly matched seven examples of actions with the terms *mitigation* and *adaptation* (Garfin et al. 2011). Current managers generally need to retrofit their skill sets to understand the concepts, language, and acronyms of climate change.

The Science of Climate Change Is New to Managers

The tools of climate change—scenario planning, vulnerability assessment, and development of adaptation options—are relatively new and not well known to land managers. While climate change science has been an active area of research for more than 40 years, the terminology is unfamiliar to many land managers whose responsibilities have only recently expanded to include climate change adaptation planning. As discussed earlier, the terminology and the concepts are new. Just as important, the science itself is new. Many of today's federal land managers worked early in their careers as biologists or technicians—that is, they spent their days in the field collecting data. These empirical data were then analyzed and used in reports and management documents.

Climate change science also uses empirical data but incorporates these data into vast models. General circulation models (GCMs), which provide the fundamental data sets for assessing climate change at global scales, are based on mathematical models that simulate earth's atmosphere or ocean. These models are simulated at such course scales that statistical downscaling methods are often used to more accurately project future changes in climate at a particular, management-relevant location (Thrasher et al. 2013). And while both historic climate analyses and projected changes in vegetation are based on empirical data points, climate data interpolated across the landscape using models to fill in the gaps between weather stations and vegetation changes are predicted using techniques such as climate niche modeling (Chang, Hansen, and Piekielek 2014).

Most of these modeling techniques are not well understood by land managers. In fact, managers often misinterpret the results of these models or simply see them as "black boxes" whose inner workings are not understood and thus not trusted. The current cadre of land managers did not learn this type of science and how to use these types of models during their formal educations and so are in the position of retrofitting their knowledge of climate change. Lemieux et al. (2013) found that the impacts and management implications associated with climate change are so novel and complex that most adaptations occurring to date have tended to be incremental and ad hoc, and have largely ignored both the dynamic linkages within and between agencies and the multiscale effects (i.e., management actions that potentially increase impacts on others or reduce their capacity to adapt).

Difficulty in Decision Making

Managers have been reluctant to acknowledge uncertainty in environmental assessments and management strategies (Williams 2009). Some managers believe that the uncertainty around climate change science makes it fundamentally different from other issues faced by land managers (Stein et al. 2014). Managers are often frustrated by the lack of unambiguous results from science (White, Garrott, and Olliff 2009) and respond differently to uncertainty. For instance, some managers proceed as though there is no uncertainty, while others focus on better understanding the uncertainty. However, most seem to put off taking action until the level of certainty improves. As one management partner stated: "Scientists embrace uncertainty; managers use it as a reason for inaction." In addition, most management decisions have a three- to five-year time frame. A few planning efforts (e.g., forest planning by the Forest Service) consider time frames of up to thirty years. But climate change projections often focus on fifty- to one-hundred-year time frames.

Working across Boundaries

The scale of climate change impacts will far exceed the ability of any one park, forest, agency, or organization to effectively respond as a single entity.

Climate change will not affect all areas equally, and in some areas it will exacerbate existing stressors. Collaboration between scientists and managers from different organizations and agencies will greatly help guide the development of appropriate expectations for climate change adaptation actions. These integrated discussions will also assist in emphasizing locally tailored solutions within the context of larger-scale regional strategies. Working at the landscape scale, across ecosystems, and across jurisdictional and political boundaries is critical given the scope of climate change impacts (McKinney, Scarlett, and Kemmis 2010).

Currently, climate change adaptation efforts have not been well coordinated across government agency boundaries or at large scales (GAO 2009, 2011). To a great extent, managers themselves believe that their agencies are performing either "neutrally" or "poorly" on most factors influencing their organization's ability to respond to climate change, including having a clear mandate to respond to climate change, having appropriate policies in place to help mitigate climate change, and directing line officers to make climate change response a high priority (Lemieux et al. 2013).

New Science and New Tools

Although several frameworks have been developed and promoted for implementing climate change adaptation (Glick, Stein, and Edelson 2011; Cross et al. 2012; National Fish, Wildlife and Plants Climate Adaptation Partnership 2012; Stein et al. 2014; Schmitz et al. 2015), no systematic approach for managing resources under climate change has gained wide acceptance among public lands managers (GAO 2011; Hansen et al. 2013). Lemieux and Scott (2011) found that when senior agency decision makers evaluated the institutional feasibility of the fifty-six most desirable adaptation options identified by an expert science panel, they found only two to be implementable, due largely to fiscal and internal capacity limitations.

Lemieux et al. (2013) compared the perceptions by federal and state land and natural resource managers of agency performance on adaptive capacity in two US regions (northern Colorado and southwestern South Dakota) and found that perceived importance was greater than performance on most adaptive capacity factors. To a great extent, managers surveyed perceived that their agencies were performing either neutrally or poorly on most factors influencing adaptive capacity; in other words, managers perceive that their respective agencies must take more action to improve their performance on climate change–related management issues. Most respondents to a managers' survey in the northern Rockies had not identified adaptation options or management actions that will help priority resources cope with climate change (Hansen et al. 2013).

Adaptation Actions

As one manager told us: "Climate change is something we think about at 5:30 pm when we are done for the day." There is a tendency for resource managers to focus primarily on short-term, required activities, leaving less time for addressing longer-term issues such as climate change (GAO 2011). Public land managers have a full plate, and their workload is increasing while the workforce is shrinking. For example, in the last decade, the Forest Service has, in many cases, combined two national forests when a forest supervisor retires (such as the Gallatin-Custer), rather than hiring a new supervisor. And among many land managers, there is still a feeling that climate change "is going to happen," rather than an understanding it has happened and will continue to accelerate—thus, most managers are managing for "business as usual" (Hansen et al. 2013).

Social Complexity and Fragmented Stakeholders

Public land managers make decisions based on fidelity to the law, best available science, and best public interest (National Park Service 2011)—yet the public interest is difficult to determine when climate change is the subject. One study found that only 42 percent of Americans are certain that global warming is happening and that it is human caused (the Alarmed and Concerned); another 25 percent (the Cautious) believe that climate change is real but are not certain (many of these do not know the cause); and 32 percent are either Disengaged (have given the issue little thought), Doubtful (uncertain), or Dismissive (certain that global warming is not happening) (Leiserowitz et al. 2012). Land managers themselves may not believe that global warming is a real phenomenon or that it is human caused; in addition, managers must respond to Congress, the Executive Branch, and the public. It is perhaps not surprising that a manager recently requested that a slide depicting the finding of the Intergovernmental Panel on Climate Change that "climate change is caused by humans" be removed from an introductory slide show aimed at introducing public land agency employees to climate change—or that another manager asked, "Why are we even talking about climate change; half of Congress does not even believe in it?"

The LCCVP's Collaborative Approach to Climate Adaptation

The LCCVP team is using the Climate-Smart Conservation framework (chap. 2) to develop and apply decision support tools that use National Aeronautics and Space Administration (NASA) data and other data and models to assess the vulnerability of ecosystems and species to climate and land use change and to evaluate management options. Objectives include the following:

- Quantify trends in ecological processes and ecological system types from past to present and under projected future climate and land use scenarios using NASA data and other data and models across two landscape conservation cooperatives (LCCs).
- Assess the vulnerability of ecological processes and ecological system types to climate and land use change by quantifying exposure, sensitivity, adaptive capacity, and uncertainty in and around focal national parks within LCCs.
- Evaluate management options for the more vulnerable ecosystem processes and types within these focal parks.
- Design multiscale management approaches for vulnerable elements to illustrate adaptation strategies under climate and land use change.
- Facilitate technology transfer of data, methods, and models to LCCs and federal agencies to allow the decision support tools to be applied more broadly.

To be most effective in developing tools and techniques that can be used in land management, the team developed the principles, practices, and products described below.

Developing Science Products

Managers make numerous daily decisions, from the mundane to the momentous. The most far reaching decisions—including land management plans that affect multiple threatened and endangered species, land designation, treatment options, visitor access, and preservation actions—embrace time frames decades into the future and can take years to develop. Increasingly, these types of plans are mandated to have a sound scientific basis (Public Law No. 105-391) and even require evaluation of climate change impacts (USDA Forest Service 2012). It is imperative in a decision space that mandates science that managers understand the nuances of the science they are using as inputs into decision making. Managers can better understand scientific models and studies if they are integrated early in the scientific process.

As noted earlier in this chapter, communications between scientists and managers are not always easy (Northern Rockies Fire Sciences Network 2011). Managers will make a decision with available information and little or no input from scientists or scientific studies. Scientists who want to make their information available to managers often must bear the burden of information transfer, with the caveat that good communication is (a) face-to-face (at least when developing relationships) and (b) repeated (one study found that information has to be repeated six times to become integrated into a person's thought process) (Edgelow 2005).

The LCCVP team found that managers need to be integrated into the science team from the beginning, including helping to shape proposals submitted for management-relevant science; attending regular workshops and team meetings to understand models, track analysis progress, and understand preliminary results; and coauthoring reports, briefs, and scientific manuscripts to add power to recommendations for management relevance. This collaborative approach to climate adaptation planning is depicted in figure 3-1. Not all managers, of course, have the time for or interest in this close interaction with scientists, but the innovators and early adopters commonly feel left out of the process without this type of engagement. These managers are often the most important for scientists to reach because they are the ones who will spread information to their colleagues.

Managers and scientists can work together in an iterative fashion, each contributing from his or her unique strengths. Managers are experts at evaluating their decision support needs, including staff education, budgets, policy, and available management options; scientists are experts at evaluating their analysis capabilities. Given this information, managers and scientists can jointly identify conservation targets, goals, and objectives. Scientists can then conduct analysis, including hindcasting and forecasting climate exposure and sensitivity of conservation targets, which managers can use in assessing the vulnerability of these resources and choosing adaptation actions to implement. The approach for vulnerability assessment used in the LCCVP is discussed in chapter 2.

As managers review and evaluate these analyses, they gain a deeper understanding of analysis techniques, including limits of the modeling and uncertainty in the analyzed outcome. In climate science, managers and scientists must work together to assess the vulnerability of conservation targets; managers, who often have years of experience including field ob-

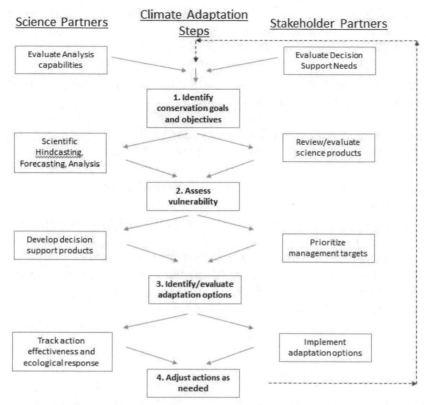

FIGURE 3-1   Depiction of the collaborative approach to climate adaptation planning in the Landscape Climate Change Vulnerability Project, noting the role of science partners (left column), the role of stakeholder partners (right column), and the interactions with each through the four fundamental steps of planning and implementing climate change adaptation. (From Glick, Stein, and Edelson 2011.)

servations, may have the best information on a resource's adaptive capacity. Scientists can develop decision support tools that help managers to identify and evaluate adaptation options and to prioritize management actions in response to current or projected climate change. Managers implement management actions, after which scientists can monitor to track the effectiveness of and ecological response to the management. Finally, in an adaptive management context, managers and scientists work together to understand how to adjust management actions as needed. The case studies contained in later chapters—managing whitebark pine in the Greater Yellowstone Area (chap. 15); analyzing limber pine in Rocky Mountain

National Park (chap. 14); and prioritizing populations of Yellowstone cutthroat trout for restoration throughout the species range (chap. 12)— illustrate these science–management interactions.

In addition to developing strong relationships with individual managers, scientists can provide managers with the tools and knowledge to adapt to climate change by doing the following:

- encouraging, developing, and supporting training
- providing science that is understandable and useful at management-decision scales
- supporting mechanisms to help managers work across boundaries and jurisdictions
- identifying early adopters and getting them involved even at the proposal writing stage of a project
- helping managers discover and understand new science
- developing and promoting organizing frameworks that help managers understand and use climate change analyses and tools
- developing effective communications tools

Raising Awareness of Climate Science

Different managers have varying interest in, and time for, learning the new concepts, terminology, science, acronyms, and tools useful in developing climate change adaptation strategies. The Ladder of Engagement, developed by EcoAdapt (Hansen and Hoffman 2011), is useful when working with a wide spectrum of managers to determine how engaged a person is with climate change. The ladder comprises six rungs: awareness, assessment, planning, implementation, integration, and sharing. Many federal land managers are just beginning to become aware of global warming and its impacts on natural resources; moving up the rungs of the ladder, fewer and fewer managers are knowledgeable and committed. In fact, very few federal land managers are currently implementing climate change adaptation projects (Halofsky, Peterson, and Marcinkowski 2015).

Scientists can facilitate increased knowledge by (a) engaging in formal training courses, such as those conducted by the National Conservation Training Center on Climate-Smart Conservation and Vulnerability Assessments; (b) developing or contributing to climate change workshops that engage managers in scenario planning, selecting conservation targets, and identifying adaptation options; and (c) developing presentations on past and projected future climate exposure, resource sensitivity, projected impacts to resources, and potential management actions that could help to mitigate projected impacts.

Developing a Communications Approach

Effective science communication only starts with a peer-reviewed article. Beyond publishing in the scientific literature, we work to write reports directed toward land managers; articles in popular magazines and periodicals; short, two- to four-page briefs highlighting the most important information; and simple sound bites that are brief, easy to digest, and easy to remember (see table 2-2 in chap. 2).

Edgelow (2005) emphasizes that effective communication is most often face-to-face. NPS managers who were surveyed confirmed this: they would prefer to learn about climate change through small groups, in-person lecture series, or classes of about ten to fifteen people, whereas if cost is an issue they would prefer to learn about climate change through user-friendly websites with clear graphics, links to background materials, and an ongoing discussion forum (Garfin et al. 2011). The LCCVP team held several meetings and workshops with land managers from Yellowstone and Grand Teton national parks, the US Forest Service and the Bureau of Land Management from the Greater Yellowstone area; and managers from Rocky Mountain National Park, Great Smoky Mountains National Park, Delaware Water Gap, and the Blue Ridge Parkway (table 3-1). The initial meetings were general in nature, introducing climate change framework, climate and ecological modeling, and the concepts of vulnerability assessments. As the team developed more sophisticated products and results, subsequent meetings, workshops, and conferences delved deeper into those results as well as management applications, feasibility, and limitations. These repeated face-to-face meetings resulted in shared learning among scientists and managers. Publications and other written material reinforced the information shared at face-to-face venues.

Useful and Important Products

A framework for engaging scientists and stakeholders in decision support around the four basic steps of climate change adaptation is depicted in

TABLE 3-1. Workshops with collaborators and presentations to understand information needs and to deliver science products.

Organization	Key Collaborators	Date	Needs We Can Address/ Information Delivered
Greater Yellowstone Coordinating Committee Whitebark Pine Subcommittee	Virginia Kelly, Karl Buermeyer, Dan Reinhart, Nancy Bokino, Kristin Legg	April 2012	Effectiveness of "GYCC WBP Strategy" under future climate
Rocky Mountain NP	Ben Bobowski, Judy Visty, Jeff Connor, John Mack, Larry Gamble, Jim Cheatham, Mary-Kay Watry, Nate Williamson	November 2012	Climate, land use, ecosystem interactions Limber pine Collaborative management among agencies
Yellowstone NP	Dave Hallac, Ann Rodman, P. J. White, Roy Renkin	November 2012 January 2013	Whitebark pine Grassland phenology YNP climate change program direction: monitoring, vulnerable resources, management options
Great Smoky Mountains NP, Shenandoah NP, Appalachian Highlands I&M	Jim Renfro, Jeff Troutman, Tom Remaley, Jim Schaberl, Paul Super, Jeb Wofford	November 2012	Vegetation communities (6 across elevation range) PACE methods Land use legacy in parks
Delaware Water Gap NRA	Rich Evans, Mathew Marshall, Leslie Moorlock	November 2012	Hemlock vegetation community Land use/hydrology
Grand Teton NP	Sue Consolo Murphy, Kelly McClosky, Cathy Melander	May 2013	Snowpack Whitebark pine
Yellowstone NP	Dave Hallac, Ann Rodman, P. J. White, Roy Renkin	February 2014	Preliminary results: whitebark pine distribution and GYA vegetation distribution
National Congress on Conservation Biology, LCCVP Symposiums 7 and 20	Ben Bobowski, Virginia Kelly, Doug Smith	July 2014	Preliminary results: whitebark pine distribution and GYA and APLCC vegetation distribution; whitebark pine management scenarios

Table 3-1. (*Continued*)

Organization	Key Collaborators	Date	Needs We Can Address/ Information Delivered
Crossing Boundaries: Science, Management, & Conservation in the Greater Yellowstone, the 12th Biennial Science Conference	Sue Consolo Murphy, Dave Hallac, Dan Reinhart, Virginia Kelly, Kristin Legg, Dan Reinhart	October 2014	Preliminary results: climate hindcasting and forecasting; vegetation distribution modeling in GYA; vegetation management scenarios
GYCC WBP Subcommittee	Virginia Kelly, Karl Buermeyer, Dan Reinhart, Nancy Bokino, Kristin Legg, Kelly McClosky	October 2014	Whitebark pine distribution based on climate envelope modeling
Greater Yellowstone Area Land Managers Adaptation Workshop	29 federal land managers and partners from NPS, USFS, and BLM	April 2015	Final results: climate exposure and GYA vegetation changes; feedback on adaptive capacity, adaptation options, and feasibility

Note: GYCC WBP = Greater Yellowstone Coordinating Committee whitebark pine; I&M = National Park Service Inventory and Monitoring Program; LCCVP = Landscape Climate Change Vulnerability Program; NP = national park; NRA = national recreation area.

figure 3.1. The framework recognizes that regarding any particular issue, such as climate change, resource managers may make decisions on a broad array of agency activities. Most obvious among these are decisions on active and passive management, but also relevant are decisions on deployment of monitoring schemes, priorities for research, content of training programs for staff, interpretative programs for visitors, budgets, and agency policies. The types of scientific products or the means for communicating these products may vary with decision types and should be designed accordingly. Efforts to quantify the effectiveness of decision support provide a feedback loop for redefining the client needs, science products, and communication vehicles to better enable effective decision making. When asked to rate the quality of climate change information provided by the NPS, over half of NPS staff (51.2 percent) reported that the information was of average quality and more than a quarter reported that the information was of below average (24.7 percent) or extremely poor (7.9 percent) quality (Garfin et al. 2011).

Early feedback from surveys indicated that managers were primarily interested in the LCCVP team developing downscaled climate models, how-to guides, and user-friendly tools, and testing realistic approaches to developing climate change adaptation actions (chap. 2). In addition to peer-reviewed articles, the LCCVP team has produced key data sets and modeling tools, including (a) the CMIP5 NASA Earth Exchange Downscaled Climate Projections (Thrasher et al. 2013) that are already being used by NPS through the Inventory and Monitoring (I&M) and Climate Change Response programs; (b) Standard Operating Procedures for using data sets and modeling tools; (c) decision support tools hosted in cooperation with the NPS I&M Program and the North Central Climate Science Center; and (d) climate primers for the Greater Yellowstone and Rocky Mountain National Park areas (chap. 2). Products from this project are also being used in agency policy documents, such as the *National Park Service Intermountain Region Climate Change Strategy and Action Plan* (Olliff et al. in review).

Facilitate Interaction Among Land Managers

Land management in the United States has typically been fragmented by the fact that so many agencies and organizations—each exercising jurisdiction in different geographies or with different species, and each with unique laws, mandates, policies, and goals—manage natural resources that shift locations, such as mammals or birds. Bison that have migrated out of Yellowstone National Park, for example, might be on land managed by the Forest Service, fall under the management jurisdiction of the State of Montana Department of Livestock, carry a disease that is under the jurisdiction of the US Animal and Plant Health Inspection Service, and be subject to a hunt supervised by the State of Montana Department of Fish, Wildlife, and Parks—all the while being advocated for and against by various interest groups, nongovernmental organizations, and individual citizens from across the nation.

In many areas, particularly in the northern Rockies, cross-boundary cooperation and coordination is already strong. Federal land managers in the Yellowstone area have cooperated under the Greater Yellowstone Coordinating Committee for over forty years. Since 1990, federal, state, and university scientists and provincial managers have worked across boundaries near Glacier National Park under the umbrella of the Crown Managers Partnership. State-led partnerships that include nongovernmental organizations and federal partners operate in the Columbia Basin in Washington

(Washington Connected) and in Wyoming (Wyoming Landscape Conservation Initiative). The High Divide Collaborative, led by area land trusts but involving federal and state managers as well as private landowners, works across boundaries from Yellowstone National Park to the Central Idaho Wilderness.

New partnerships have emerged in recent years to promote cross-boundary cooperation in response to climate change and other landscape-scale system stressors. LCCs and climate science centers, established in 2009 by Department of the Interior Secretarial Order 3289, are intended to provide the latest science to land managers and conservation partners and to work with federal, state, tribal, and local governments, private landowners, and nongovernmental organizations to "develop landscape-level strategies for understanding and responding to climate change impacts" and to help managers sustain the continent's natural and cultural resources.

Creating and Promoting the Framework

The Climate-Smart Conservation framework introduced in chapter 2 is an organizing framework to help managers and scientists understand where and how such tools as vulnerability assessments, adaptation options, selection of conservation targets, assessments of uncertainty, analysis of feasibility and risks, and effectiveness monitoring fit together. We worked with managers from Yellowstone and Grand Teton national parks to introduce this framework and complete portions of step 1 (identifying conservation targets) in a July 2012 workshop. The workshop organizers reviewed previous efforts in the Greater Yellowstone Ecosystem as a baseline for workshop participants to rank high-priority resources. These previous efforts ranged in scale from the park to the ecosystem to the landscape level and included the following:

- Yellowstone National Park Vital Signs Report (Yellowstone Center for Resources 2011)
- *Vital Signs Monitoring Plan for the Greater Yellowstone Network: Phase III Report* (Jean et al. 2005)
- Greater Yellowstone Science Agenda for Climate Change, Land Use, and Invasive Species, based on a November 2009 workshop with more than one hundred managers and scientists (Olliff et al. 2010)
- *Great Northern Landscape Conservation Cooperative Strategic Conservation Framework* (Chambers et al. 2012)

- NPS I&M High Elevation Climate Workshop May 2010 report (Bingham et al. 2010)

Based on these documents, we identified fifty-six species, ecological processes, ecosystems, and ecological stressors to be considered and ranked in the process. Using wildlife as an example, many of the species that managers currently spend most money and time to monitor and manage (e.g., wolves, grizzly bears, elk) were also species that were less vulnerable to climate change impacts (compared to highly vulnerable species, such as bats or amphibians). Other chapters examine exposure, sensitivity, and vulnerability of other conservation targets identified by managers, including vegetation, fish, and a keystone tree species—whitebark pine.

Conclusion

Federally managed public lands offer an unparalleled opportunity worldwide to act cohesively in response to rapid, human-induced climate change. Currently, that opportunity is not being realized because of several limiting factors:

- Most of these concepts and most of this science are new—many federal land managers are not familiar with them.
- Most park managers simply don't have time to track these tools and science down, learn to use it, and incorporate it into management—this process must add value without adding (too much) work.
- The science we need is or will be made available—unit managers do not have to conduct most of it themselves.
- Given the scale of current and projected future stressors to our resources, the future of managing resources on federal lands will depend more on working with partners at the ecosystem and landscape scale—the scale of the impacts.

The LCCVP team has developed a collaborative approach between scientists and managers to overcome many of these obstacles. Although this first iteration isn't perfect, we will incorporate adaptive management guidelines to continue to monitor, review, and revise these principles and products in the ongoing challenge of developing effective, cross-boundary, multijurisdictional, large-scale solutions to the problems posed by climate change.

Acknowledgments

The authors would like to thank the staff and collaborators of the Great Northern LCC (Yvette Converse, Sean Finn, Matt Heller, Mary McFadzen, Anne Carlson, John Pierce, Molly Cross, Vita Wright, Linh Hoang, and Michael Whitefield), staff from the NPS Climate Change Response Program (Leigh Welling, Cat Hawkins-Hoffman, Nick Fisichelli, Gregor Schuurman, Angie Richman, and Patrick Gonzales) and the NPS Greater Yellowstone I&M Program (Kristin Legg and David Thoma), the GYCC Climate Change Subcommittee (Virginia Kelly, Scott Barndt, Karri Cari, and Pam Bode), and NPS IMR climate change coordinator Pam Benjamin and associate regional director Tammy Whittington for contributing their ideas and experience to this chapter.

References

Bingham, B., M. Britten, L. Garrett, P. Latham, and K. Legg. 2010. *Enhanced Monitoring to Better Address Rapid Climate Change in High-Elevation Parks: A Multinetwork Strategy*. Natural Resource Report NPS/IMR/NRR—2011/285. Fort Collins, CO: National Park Service.

Chambers, N., G. Tabor, Y. Converse, T. Olliff, S. Finn, R. Sojda, and S. Bischke. 2013. The Great Northern Landscape Conservation Cooperative Strategic Conservation Framework. http://greatnorthernlcc.org/sites/default/files/documents/gnlcc_framework_final_small.pdf.

Chang, T., A. J. Hansen, and N. Piekielek. 2014. Patterns and variability of projected bioclimatic habitat for *Pinus albicaulis* in the Greater Yellowstone area. *PLOS ONE* 9 (11): e111669. doi: 10.1371/journal.pone.0111669.

Cooke, J., D. Nuccitelli, S. A. Green, M. Richardson, B. Winkler, R. Painting, R. Way, P. Jacobs, and A. Skuce. 2013. Quantifying the consensus on anthropogenic global warming in the scientific literature. *Environmental Research Letters* 8 024024. doi: 10.1088/1748-9326/8/2/024024.

Council on Climate Preparedness and Resilience, Climate and Natural Resources Working Group. 2014. *Priority Agenda for Enhancing the Climate Resilience of America's Natural Resources*. https://www.whitehouse.gov/sites/default/files/docs/enhancing_climate_resilience_of_americas_natural_resources.pdf.

Cross, M. S., E. S. Zavaleta, D. Bachelet, M. L. Brooks, C. A. F. Enquist, E. Fleishman, L. J. Graumlich, C. R. Groves, L. Hannah, L. Hansen, G. Hayward, M. Koopman, J. J. Lawler, J. Malcolm, J. Nordgren, B. Petersen, E. L. Rowland, D. Scott, S. L. Shafer, M. R. Shaw, and G. M. Tabor. 2012. The Adaptation for Conservation Targets (ACT) framework: A tool for incorporating climate change into natural resource management. *Environmental Management* 50:341–51.

Edgelow, C. 2005. *Communicating Organizational Change*. Edmonton, Alberta: Lost Creek Press.

Executive Office of the President. 2013. *The President's Climate Action Plan*. Washington, DC: The White House, June. https://www.whitehouse.gov/sites/default/files/image/president27sclimateactionplan.pdf.

GAO (US Government Accountability Office). 2009. *Climate Change Adaptation: Strategic Federal Planning Could Help Government Officials Make More Informed Decisions*. GAO-10-113. Washington, DC: GAO.

GAO (US Government Accountability Office). 2011. *Climate Change: Improvements Needed to Clarify National Priorities and Better Align Them with Federal Funding Decisions*. GAO-11-317. Washington, DC: GAO.

GAO (US Government Accountability Office). 2013. *Climate Change: Various Efforts Are Under Way at Key Natural Resource Management Agencies*. GAO-13-253. Washington, DC: GAO.

Garfin, G., H. Hartmann, M. Crescioni-Benitez, T. Ely, J. Keck, J. Kendrick, K. Legg, J. Wise, L. Graumlich, and J. Overpeck. 2011. *Climate Change Training Needs Assessment for the National Park Service Intermountain Region*. Tucson: University of Arizona.

Glick, P., B. A. Stein, and N. Edelson, eds. 2011. *Scanning the Conservation Horizon: A Guide to Climate Change Vulnerability Assessment*. Draft. Washington, DC: National Wildlife Federation.

Gonzalez, P. 2011. Science for natural resource management under climate change. *Issues in Science and Technology* 27 (4): 65–74.

Halofsky, J. E., D. Peterson, and K.W. Marcinkowski. 2015. Climate Change Adaptation in United States Federal Natural Resource Science and Management Agencies: A Synthesis. USGCRP Climate Change Adaptation Interagency Working Group.

Hansen, A., J. S. Goetz, F. Melton, B. Monahan, J. Gross, T. Olliff, and D. Theobald. 2013. NASA Applied Sciences Initial Assessment Report: Using NASA Resources to Inform Climate and Land Use Adaptation: Ecological Forecasting, Vulnerability Assessment, and Evaluation of Management Options across Two US DOI Landscape Conservation Cooperatives. http://www.montana.edu/lccvp/index.html.

Hansen, L. J., and J. R. Hoffman. 2011. *Climate Savvy: Adapting Conservation and Resource Management to a Changing World*. Bainbridge Island, WA: EcoAdapt.

IPCC (Intergovernmental Panel on Climate Change). 2007. *Climate Change 2007: Impacts, Adaptation, and Vulnerability*. Cambridge: Cambridge University Press.

Jean, C., A. M. Schrag, R. E. Bennetts, R. Daley, E. A. Crowe, and S. O'Ney. 2005. *Vital Signs Monitoring Plan for the Greater Yellowstone Network*. Bozeman, MT: National Park Service, Greater Yellowstone Network.

Lemieux, C. J., and D. J. Scott. 2011. Changing climate, challenging choices: Identifying and evaluating climate change adaptation options for protected areas management in Ontario, Canada. *Environmental Management* 48 (4): 675–90.

Lemieux, C. J., J. L. Thompson, J. Dawson, and R. M. Schuster. 2013. Natural resource manager perceptions of agency performance on climate change. *Environmental Management* 114:178–89.

Leiserowitz, A., E. Maibach, C. Roser-Renouf, G. Feinberg, and P. Howe. 2012. *Global Warming's Six Americas in September 2012*. New Haven, CT: Yale Project on Climate Change Communication and George Mason University, Center for Climate Change Communication. http://environment.yale.edu/climate-communication/files/Six-Americas-September-2012.pdf.

McKinney, M., L. Scarlett, and D. Kemmis. 2010. *Large Landscape Conservation: A Strategic Framework for Policy and Action*. Cambridge, MA: Lincoln Institute of Land Policy. http://www.lincolninst.edu.

National Fish, Wildlife and Plants Climate Adaptation Partnership. 2012. *National Fish, Wildlife and Plants Climate Adaptation Strategy*. Washington, DC: Association of Fish and Wildlife Agencies, Council on Environmental Quality, Great Lakes Indian Fish and Wildlife Commission, National Oceanic and Atmospheric Administration, and US Fish and Wildlife Service. http://www.wildlifeadaptationstrategy.gov/pdf/NFWPCAS-Final.pdf.

National Park Service. 2011. *A Call to Action: Preparing for a Second Century of Stewardship and Engagement*. August 25. http://www.nps.gov/calltoaction/PDF/C2A_2014.pdf.

Northern Rockies Fire Sciences Network. 2011. Obstacles to using research: A survey of fire and fuels managers, resource managers, and line managers in the Northern Rockies. http://nrfirescience.org/your-input/needsassessment/survey/obstacles.

Olliff, S. T., P. Benjamin, L. Chan, J. Cowley, N. Fisichelli, W. Monahan, L. Meyer, and T. Whittington. In review. *National Park Service Intermountain Region Climate Change Strategy and Action Plan*. Natural Resource Report NPS/IMRO/NRR—2015. Fort Collins, CO: National Park Service.

Olliff, T., G. Plumb, J. Kershner, C. Whitlock, A. Hansen, M. Cross, and S. Bischke. 2010. A science agenda for the Greater Yellowstone area: Responding to landscape impacts from climate change, land use change, and invasive species. *Yellowstone Science* 18 (2).

Rogers, E. M. 2003. *Diffusion of Innovation*. New York: Simon and Schuster.

Schmitz, O. J., J. J. Lawler, P. Beier, C. Groves, G. Knight, D. A. Boyce Jr., J. Bulluck, K. M. Johnston, M. L. Klein, K. Muller, D. J. Pierce, W. R. Singleton, J. R. Strittholt, D. M. Theobald, S. C. Trombulak, and A. Trainor. 2015. Conserving biodiversity: Practical guidance about climate change adaptation approaches in support of land-use planning. *Natural Areas Journal* 35 (1): 190–203.

Shafer, C. L. 2014. From non-static vignettes to unprecedented change: The U.S. National Park System, climate impacts and animal dispersal. *Environmental Science & Policy* 40:26–35.

Stein, B. A., P. Glick, N. Edelson, and A. Staudt, eds. 2014. *Climate-Smart Conservation: Putting Adaptation Principles into Practice*. Washington, DC: National Wildlife Federation.

Thrasher, B., J. Xiong, W. Wang, F. Melton, A. Michaelis, and R. Nemani. 2013. Downscaled climate projections suitable for resource management. *Eos, Transactions American Geophysical Union* 94 (37): 321–23.

USDA Forest Service. 2012. National Forest System Land Management Planning: Final rule and record of decision. *Federal Register* 77 (68).

White, P. J., R. A. Garrott, and S. T. Olliff. 2009. Science in Yellowstone: Contributions, limitations, and recommendations. In *The Ecology of Large Mammals in Central Yellowstone: Sixteen Years of Integrated Field Studies*, edited by R. A. Garrott, P. J. White, and F. G. R. Watson, 671–88. San Diego, CA: Elsevier Academic Press.

Williams, B. K., R. C. Szaro, and C. D. Shapiro. 2009. *Adaptive Management: The U.S. Department of the Interior Technical Guide*. Washington, DC: Adaptive Management Working Group, US Department of the Interior.

Yellowstone Center for Resources. 2011. *Yellowstone National Park Vital Signs Report*. http://www.nps.gov/yell/learn/management/upload/vitalsigns2-2.pdf.

PART 2

Climate and Land Use Change

The two major ways that humans influence wildland ecosystems are (1) by changing climate through greenhouse gases and (2) by changing habitats through conversion to different land cover types and through land uses that vary in intensity. In the context of vulnerability assessment, these factors are considered elements of exposure that elicit responses from ecological systems. Climate and land use each influence ecosystems in unique ways. They also interact, often in ways that make effective management even more challenging.

This section of the book uses historic data and computer forecasts to summarize past and projected climate across the two study areas for the period 1900–2100. Climate can be represented in many ways. Chapters 4, 5, and 6 frame climate in terms that are most relevant to ecosystems and the species they support. In addition to annual averages of temperature and precipitation, seasonal averages and variability are presented as are extreme warm or cold events. These data are summarized across landscape conservation cooperatives for broad context but, importantly, are also analyzed at finer scales within the ecosystems centered on national parks and on habitat types within those ecosystems. These finer scales are most relevant to management, with the goal of telling the stories of climate change that are compelling to the stewards of these lands. Summaries are produced over the past century to show the estimated degree of climate change in recent decades. Similarly, projections for the future are offered to allow us to visualize the range of plausible future conditions in the coming decades.

Although predicting the future might be called a fool's errand, using the best scientific methods to project plausible future conditions is an essential tool to motivate both managers and the public to anticipate the future, as well as providing credible information to express scenarios of possible future conditions. These projections differ for a variety of reasons, and this section of the book attempts to communicate the uncertainty of anticipated future conditions to facilitate management decisions that are robust in the face of this uncertainty. Although a two-hundred-year period is very long by human scales, it is quite short in terms of ecological and evolutionary scales. Thus, chapters 4 and 5 summarize climate change and major vegetation response since the last glacial period around fourteen thousand years ago. A sobering conclusion is that the projected change by 2100—just eighty-five years from now—is equal to or greater than that which occurred over the past fourteen centuries.

Concern about land use intensification has been at the forefront of conservation biology for several decades. The authors of this book's chapters have done extensive primary analyses of land use patterns and change at local to national levels. Rather than emphasize the primary analyses, we largely cite previous work on land use change in this book. Chapter 6 integrates land use classes and intensities into an index of human development. More specifically, the inverse of human development is used as an index of landscape integrity, with high values indicating lower exposure to intense land use. The chapter summarizes patterns of landscape integrity and climate change across the Great Northern Landscape Conservation Cooperative as elements of exposure in analyses of vulnerability of landforms to global change. In this regard, chapter 6 offers a coarse-filter approach to evaluating vulnerability, which nicely segues to chapters in part 3 that examine the specific ecological consequences of these changes.

Chapter 4

Analyses of Historical and Projected Climates to Support Climate Adaptation in the Northern Rocky Mountains

John E. Gross, Michael Tercek, Kevin Guay,
Marian Talbert, Tony Chang, Ann Rodman,
David Thoma, Patrick Jantz,
and Jeffrey T. Morisette

Most of the western United States is experiencing the effects of rapid and directional climate change (Garfin et al. 2013). These effects, along with forecasts of profound changes in the future, provide strong motivation for resource managers to learn about and prepare for future changes. Climate adaptation plans are based on an understanding of historic climate variation and their effects on ecosystems and on forecasts of future climate trends. Frameworks for climate adaptation thus universally identify the importance of a summary of historical, current, and projected climates (Glick, Stein, and Edelson 2011; Cross et al. 2013; Stein et al. 2014). Trends in physical climate variables are usually the basis for evaluating the exposure component in vulnerability assessments. Thus, this chapter focuses on step 2 of the Climate-Smart Conservation framework (chap. 2): vulnerability assessment. We present analyses of historical and current observations of temperature, precipitation, and other key climate measurements to provide context and a baseline for interpreting the ecological impacts of projected climate changes.

This chapter is limited to analyses of trends and patterns in a small subset of key climate variables. Interested readers will find additional information in the papers cited here and in the very informative analyses of Rocky Mountain climate patterns by Kittel et al. (2002), McWethy et al. (2010), and Rice, Tredennick, and Joyce (2012).

Geography and Climate

The Rocky Mountain region of the Great Northern Landscape Conservation Cooperative (Great Northern LCC) is an area of great geographical and
climatic variation (fig. 1-2 in chap. 1). The broader region encompassing
Glacier, Yellowstone, and Rocky Mountain national parks spans latitudes
from 40 to 49 degrees north, with elevations from 3,100 to more than
14,000 feet (950 to 4,300 meters). In many places, the mountains form a
significant barrier to moisture flowing from the Pacific Ocean, resulting in
greater precipitation on the westward side of the mountain ranges (Davey,
Redmond, and Simeral 2006, 2007). The high mountains in Rocky Mountain National Park (Rocky Mountain NP) can also form a barrier to moisture flowing from the east, which can collide with peaks of the Front Range
of Colorado and result in rare but tremendous precipitation events that
have resulted in loss of life and property (Western Water Assessment 2013).
Kittel et al. (2002) evaluated climates across this vast area and divided the
Great Northern LCC into three climatological regions, with Rocky Mountain, Yellowstone, and Glacier national parks in the southern, central, and
northern regions, respectively. These climatological regions share characteristics common to most mountain regions, such as significant elevational
gradients in precipitation and temperature, but the regions differ in their
relationships to regional-scale climate drivers.

Geography and elevation strongly influence climate throughout the
Great Northern LCC. In general, temperatures decline toward the north,
but within the Great Northern LCC parks this geographical pattern is tem-
pered by the generally lower elevations in Glacier and Yellowstone national
parks as compared to Rocky Mountain NP. Decreases in temperatures with
increasing elevation (lapse rate) vary with aspect, relative humidity, and local topography. Lapse rates are reported to be 2.4 to 3.6 degrees F (0.3 to 2
degrees C) per 1,000 feet elevation across the Great Northern LCC region,
with lapse rates generally increasing with latitude but with substantial local
variation (Wolfe 1992; Kittel et al. 2002; Al-Chokhachy et al. 2013). As
temperatures rapidly increase, these lapse rates are a partial determinant
of the rate at which species may need to migrate up in elevation to remain
within a climatically favorable zone.

Broad-Scale Pattern, Fine-Scale Complexity

Broad-scale patterns in climate change direction and magnitude provide a
context for interpreting the finer-scale variability that is important to man-

agement. While warming or increased precipitation are consistent changes at regional scales and over long time frames, there can be considerable variation at finer spatial scales (Dobrowski 2011).

Fine-scale variation in climate can be an extremely important determinant of both historical and future species distribution patterns, but it is still poorly understood. For example, temperatures in western Montana generally track northern hemisphere temperature cycles such that warm northern hemisphere winters translate to warm winters in western Montana. However, some individual climate stations (i.e., local areas) may not track regional trends because local variation in climate results from nonlinear controls exerted by global, regional, or local effects that vary in strength on decadal or even annual scales (Pederson et al. 2011a; Dobrowksi et al. 2009). Regional trends linked to global change include hydrological effects, such as changes in glacier and permanent snowfield mass balance (Pederson et al. 2004), forest disease (Logan, MacFarland, and Wilcox 2010), and fire frequency (Westerling et al. 2011). However, some regional changes are not seen at local scales. Areas locally resistant to regional-scale climate changes can act as climate refugia, allowing the preservation of species and communities that might otherwise perish as the broader region becomes climatically unsuitable.

Interactions between complex topography and edaphic factors alter the effects of climate on biota, making some locations favorable and others unfavorable under similar temperature and precipitation regimes (Stephenson 1990). Evidence for climate refugia comes from records of fossil pollen that show species persisted on landscapes through periods of inhospitable climates and were then able to recolonize when a favorable climate returned (Whitlock and Bartlein 1993; Gavin et al. 2014). Topographic complexity, particularly elevation and solar heat load, are important sources of local climatic variation that can result in climate refugia (McCune 2007; Dobrowski 2011). Similarly, soil properties that can help mitigate climatic water deficits during drought and beetle epidemics were described for the southwestern United States (Breshears et al. 2009; Peterman et al. 2012) and Sierra (Millar et al. 2012). In these studies, water deficit interactions with tree physiology were the primary determinants of mortality at the species and stands levels. These studies demonstrate that climate refugia exist in discrete geographic locations and suggest that understanding the relationship between water and energy is critically important for understanding the spatiotemporal aspects of climate and land surface interactions that affect ecosystem processes.

Paleoclimate Context

Analyses of paleoclimate records of the Great Northern LCC showed that a key driver of long-term climate change was the natural variability of the earth's position relative to the sun. At the last glacial maximum, 26,500 BP, the Cordilleran ice sheet covered most of the northern Rocky Mountains. Subsequent warming from increasing summer insolation caused melting of these ice sheets and exposed much of the land surface we see today (Clark et al. 2009). At the last glacial period, 14,000 BP, the climate was initially colder and wetter than at present. The climate inferred for Rocky Mountain NP was 3.6 to 8 degrees F (2 to 5 degrees C) cooler and had 2.8 to 6.3 inches (7 to 16 centimeters) per year more precipitation than today (Reasoner and Jodry 2000; Fall 1997; Leonard 1989). This cold/wet climate associated with surface exposure due to glacial retreats allowed the establishment of tundra communities throughout the Rocky Mountains. About 11,000 BP, widespread establishment of subalpine forests containing spruce (*Picea*), fir (*Abies*), and pine (*Pinus*) were apparent at Yellowstone, Glacier, and Rocky Mountain national parks (Whitlock and Bartlein 1993; Pierce 1979; Fall 1997). Associated with these colder temperatures, upper-treeline elevations for these forests were lower than present—in some cases, 780 to 2,300 feet (300 to 700 meters) below contemporary treelines (Fall 1997).

Conditions warmed from 9,500 to 4,000 BP, and glaciers reached their minimum extent (Millspaugh, Whitlock, and Bartlein 2000). By 4,000 BP, the region was characterized by temperatures 1.8 to 3.6 degrees F (1 to 2 degrees C) warmer than present, with greater variability in moisture regime as some regions became drier and some wetter depending on the geography and topography (Millspaugh, Whitlock, and Bartlein 2000). For instance, within Yellowstone NP, there was a division of warm, dry summer conditions in the southern region and warm, wet summer conditions in the northern region (Whitlock and Bartlein 1993). The general warming trend resulted in an increase in the elevation of the upper treeline, and the spatial variation in precipitation changes contributed to vegetation community response patterns that persist today. In the drier situations, montane species, such as Douglas-fir (*Pseudotsuga menziesii*) and lodgepole pine (*Pinus contorta*), invaded the lower limits of the subalpine (Whitlock and Bartlein 1993; Brunelle et al. 2005). In the wetter situations, subalpine species extended their range and moved down in elevation (Fall 1997). This phase ended around 5,000–2,000 BP, when conditions within the Great Northern LCC cooled and moisture patterns shifted opposite to what they were during the 9,500–4,000 BP period (i.e., dry shifted to wetter, and wet shifted to drier).

Within the past 2,000 years, two distinct climate anomalies occurred, the Medieval Warm Period and the Little Ice Age. The Medieval Warm Period (1,300–900 BP) resulted in warm episodes for many areas of the world, with responses of glacier recession, increased tree growth, and immigration of native peoples to formerly inhospitable lands (Soloman 2007; Whitlock 2002). Within the Great Northern LCC, glaciated areas stagnated or retreated, and treelines moved up in elevation, indicating regionwide increases in temperatures (Whitlock 2002). In the southern Rocky Mountains, regional climate during the Medieval Warm Period involved complex precipitation patterns. The growing season increased with the warmer conditions, but some subregions experienced increased winter precipitation and summer monsoonal moisture, while other subregions experienced more frequent droughts (Petersen 1994; Dean 1994).

The Little Ice Age followed the Medieval Warm Period about 700–150 BP, when glaciers expanded throughout the Great Northern LCC. Glaciers extended farther down valley, resulting in advances comparable to periods from the late Pleistocene leading to the prominent moraines visible today (Osborn and Luckman 1988). The most severe phase began about 1800–1880, with climate approaching late Pleistocene conditions and resulting in summer cooling of 1 to 3.6 degrees F (0.6 to 2 degrees C) in alpine regions and extreme wet summers and winter snowpack (Gray, Graumlich, and Betancourt 2007; Pederson et al. 2006; Brunstein 1996). In Glacier NP, the Little Ice Age marked the coldest and wettest period in the past seven hundred years (Pederson et al. 2006). The severity of these extreme cold/wet conditions extended to the southern Great Northern LCC region but diminished steadily with decreasing latitude (Gray, Graumlich, and Betancourt 2007; Whitlock 2002).

At the end of the Little Ice Age, temperatures increased and moisture decreased, leading to current climate conditions. Since the end of the Little Ice Age, the area of glaciers has diminished 30 to 70 percent and modern vegetation communities have become established, with treeline advancing back into today's alpine environments (Hall and Fagre 2003; Hessl and Baker 1997). Paleoecological records show that the Great Northern LCC vegetation has responded to many previous climate regimes. Plant associations have been dismantled and reformed several times, and treeline has moved both up and down an elevational gradient over the past fourteen thousand years. Vegetation communities are clearly responsive to climate changes; plant assemblages and distributions will likely continue to shift as a result of current and future climate trends.

Contemporary Trends in Key Climate Variables

Our analyses considered both the jurisdictional boundaries of parks and the broader areas critical to preserving the biodiversity of the focal parks. We call these broader regions protected area centered ecosystems (PACEs; described in chap. 2) (Hansen et al. 2011). PACEs for Rocky Mountain, Yellowstone, and Glacier national parks include substantial areas at lower elevations than the parks themselves. These lower-elevation areas are generally warmer and drier than the parks, and the seasonality of precipitation often differs with elevation (Davey et al. 2006, 2007). Climate differences are particularly large between the areas within Rocky Mountain NP and its PACE. Temperatures averaged across the entire PACE tend to be 3 to 4 degrees F (1.7 to 2.2 degrees C) higher than those of the park. These differences reflect the large elevation gradients in the vicinity of Rocky Mountain NP, which are greater than those of Glacier and Yellowstone NP. Analyses for Yellowstone NP include the contiguous areas of Yellowstone and Grand Teton national parks and the John D. Rockefeller Jr. Memorial Parkway.

We evaluated historical temperature trends from TopoWx, a 30-arc-second (about 0.5 mile; [800 meter] resolution) daily gridded climate data set for the period of 1948–2012, aggregated to monthly intervals (Oyler et al. 2014). Algorithms for TopoWx correct systematic errors in high-elevation climate stations associated with changes in instrumentation (Oyler et al. 2015). TopoWx does not include precipitation, and we estimated historical precipitation from PRISM 30 arc-second gridded data (Daly et al. 2008). Monthly projected climate variables were evaluated from the National Aeronautics and Space Administration (NASA) Earth Exchange Downscaled Climate Projections (NEX-DCP30) data for the United States (Thrasher et al. 2013). NEX-DCP30 climate projections are 30-arc-second gridded, bias corrected, and spatially downscaled (Thrasher et al. 2013). Ensembles include output from thirty-four global climate models and four representative concentration pathways (RCPs; Moss et al. 2010), consistent with the approach recommended by Mote et al. (2011). We estimated future daily climate variables from three global climate models that performed well for the northwestern United States (CanESM2, CCSM4, and CNRM-CM5; Rupp et al. 2013).

For analyses of projected climates, we focused on RCPs 4.5 and 8.5 (Moss et al. 2010). RCP 4.5 is consistent with a rapid stabilization in greenhouse gas emissions (primarily carbon dioxide) to a level that achieves an anthropogenic climate forcing of 4.5 Watts per square meter in the year 2100. Stabilization consistent with RCP 4.5 will require globally coordinated reductions in greenhouse gas emissions that are much more aggressive than current practices. RCP 8.5 is consistent with increases in greenhouse gas emissions at a rate similar to the present. RCP 4.5 is estimated to result in global warming of approximately 2.5 degrees F (1.4 degrees C) by 2050 and approximately 3.3 degrees F (1.8 degrees C) by approximately 2100. RCP 8.5 is estimated to result in global warming of approximately 3.6 degrees F (2.0 degrees C) by 2050 and about 6.7 degrees F (3.7 degrees C) by 2100 (IPCC 2013). Projected rates of warming are highly variable at regional to local scales.

Temperature

All three parks and the surrounding regions have already experienced significant increases in temperature (table 4-1; fig. 4-1; Hansen et al. 2014; Monahan and Fisichelli 2014). Average annual minimum temperatures have increased more rapidly than maximum temperatures, and statistically significant temperature trends over the historic period (1948–2012) can mostly be attributed to increases that have occurred since 1970.

All projections for the future include sustained and very rapid increases in temperature until at least the middle of the century (figs. 4-2 and 4-3; table 4-1). If global emissions of greenhouse gases are curbed and stabilize (i.e., RCP 4.5), temperatures will eventually stabilize 3 to 5 degrees F (1.7 to 2.8 degrees C) higher than the 2000 era (table 4-1). If the current rate of increase in greenhouse gas emissions is sustained (RCP 8.5), temperatures by the end of the century are projected to exceed anything experienced in the past ten thousand years (fig. 4-2; Marcott et al. 2013). Because of the long life of some greenhouse gases, we are now committed to experience temperatures well beyond those of the last century (Solomon et al. 2009).

Projected temperature increases based on ensemble global climate model results are roughly similar across all three parks (table 4-1). For RCP 4.5, the ensemble average projected rate of increase is about 4.5 degrees F (2.5 degrees C) per century, with most of this increase occurring prior to 2050. For RCP 8.5, the average rate of increase is about 9.5 degrees F (5.3 degrees C) per century, with a slightly greater rate of increase in the second half of the century.

TABLE 4-1.   Observed and projected average precipitation (inches/year), maximum temperature (Tmax, degree F), and minimum temperature (Tmin; degree F) for the Glacier NP, Rocky Mountain NP, and Yellowstone NP PACEs.

	1950–1959	2000–2009	2050–2059		2090–2100	
			RCP 4.5	RCP 8.5	RCP 4.5	RCP 8.5
Glacier PACE						
Precipitation	44.5	38.6	5% (−32, 37)	5% (−30, 34)	7% (−31, 39)	11% (−27, 44)
Tmax	46.7	49.0	3.2 (0.3, 6.1)	4.6 (1.4, 8.0)	4.3 (1.4, 7.4)	9.2 (5.8, 12.9)
Tmin	27.1	28.9	3.6 (1.3, 6.1)	5.1 (2.4, 8.0)	4.5 (2.0, 7.2)	9.8 (6.7, 13.0)
Rocky Mountain PACE						
Precipitation	24.1	24.7	4% (−32, 32)	7% (−29, 24)	6% (−31, 36)	9% (−28, 39)
Tmax	49.3	50.3	3.2 (0.8, 5.6)	4.5 (1.9, 7.1)	4.4 (1.8, 6.9)	9.6 (6.6, 12.8)
Tmin	25.5	26.8	3.2 (1.3, 5.2)	4.7 (2.7, 6.8)	4.2 (2.2, 6.3)	9.4 (6.8, 11.9)
Yellowstone PACE						
Precipitation	29.8	29.4	8% (−29, 38)	8% (−26, 36)	8% (−29, 39)	16% (−23, 47)
Tmax	45.2	46.8	3.1 (0.4, 5.8)	4.6 (1.6, 7.6)	4.4 (1.5, 7.1)	9.7 (6.2, 13.2)
Tmin	23.4	24.8	3.5 (1.3, 5.9)	5.1 (2.6, 7.8)	4.6 (2.2, 7.1)	10.2 (7.1, 13.2)

Note: Values for 1950–1959 and 2000–2009 are observations. Values for future periods are differences from 2000–2009 in degree F (Tmin, Tmax) or percent (precipitation). Numbers in parentheses are 25th and 75th percentiles of observations from the set of thirty-four models.

Source: Historical temperature data from TopoWX; historical precipitation from PRISM; projections are downscaled NEX-DCP30 data. Values are ensemble averages of monthly data for RCP 4.5 and RCP 8.5.

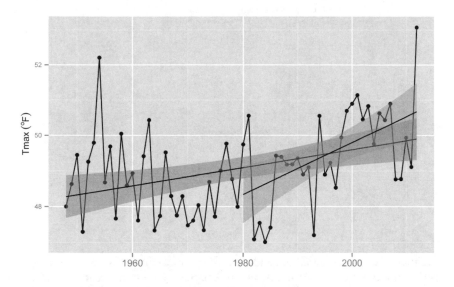

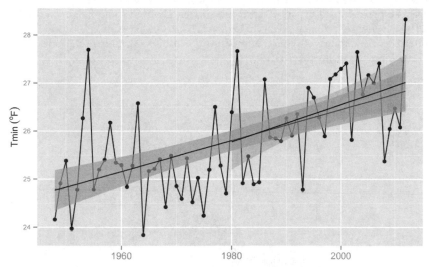

FIGURE 4-1 Trends in annual average maximum (Tmax) and minimum temperature (Tmin) for the Rocky Mountain National Park PACE. Lines are linear regressions for the periods 1948–2012 and 1980–2012. All trends are statistically significant.

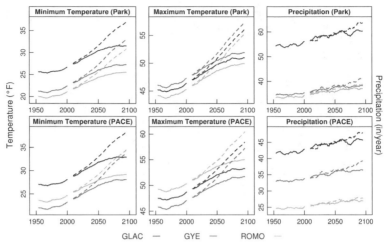

FIGURE 4-2 Annual minimum and maximum temperatures for the areas within park and PACE boundaries and two representative concentration pathways (RCP 8.5 shown with a dashed line and RCP 4.5 shown with a solid line) for 1950–2100. Data for the 1950–2006 period represents modeled "observations"; data for 2010–2100 are ensemble annual averages. Key: GLAC = Glacier National Park, GYE = Greater Yellowstone Ecosystem, and ROMO = Rocky Mountain National Park. (Data from Thrasher et al. 2013.)

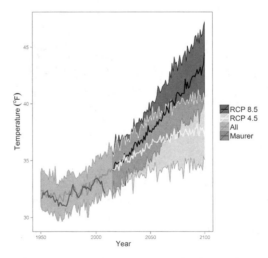

FIGURE 4-3 Projected average annual temperatures for the Rocky Mountain National Park PACE for a higher-emissions pathway (RCP 8.5) and a lower-emissions pathway (RCP 4.5) for an ensemble of global climate models. Shaded zones are ± 1 standard deviation. Results for other parks and PACEs are similar. "Maurer" in the key represents historical data (Maurer et al. 2002).

Projected changes in seasonal temperatures were similar between parks. For all parks, summer maximum temperatures (Tmax) are projected to increase most rapidly, with the smallest Tmax changes in winter. Differences in projected Tmax between summer and winter are large, and projected changes in Tmax are on the order of 2.7 to 3.8 degrees F (1.5 to 2 degrees C) greater in summer than winter by midcentury and 3.8 to 5.4 degrees F (2 to 3 degrees C) by 2100. By contrast, projected changes in minimum temperatures (Tmin) are greatest for winter in Glacier and Yellowstone national parks (an increase of 12.6 degrees F [7 degrees C] by 2100). For Rocky Mountain NP, spring Tmin are projected to change slightly less than other seasons.

Substantial increases in average temperatures are certainly concerning, but these mask much more impressive changes in ecologically relevant temperature thresholds estimated from daily climate projections. Summaries of temperature thresholds summarized in table 4-2 relate to plant growth, precipitation as snow versus rain, snowmelt, and runoff. Days below –8 degrees F (about –20 degrees C) is an index to temperatures low enough to cause significant winter mortality (and thus population control) of pine beetle (Bentz et al. 2010).

Projections of daily climate variables portend huge changes. The number of days below freezing is projected to decline by about 20 percent (e.g., from 200 to 160 days per year) by midcentury, and by 30 to 50 percent by 2100 (table 4-2). The number of days with temperatures above 90 degrees F (>32 degrees C) is projected to increase dramatically, particularly at lower elevations, where it could increase from less than a week per year to more than seven weeks (table 4-2). Earlier snowmelt, reduced snow cover in spring, and higher summer temperatures will drive changes in runoff timing, stream flow, soil moisture, fire, and forest pests.

Precipitation

Ensemble average projections for 2010–2100 suggest small changes in precipitation across the Great Northern LCC (table 4-1), but uncertainty and annual variation are high. Projections for the Rocky Mountain NP PACE favor a small increase in precipitation (on average, less than 10 percent). Averaged across models, annual precipitation for the Yellowstone NP PACE is projected to increase about 4.7 inches (120 millimeters) per year, or 16 percent, under RCP 8.5 by 2100. The average projected change in precipitation for the Glacier NP PACE is 2.7 inches (70 millimeters) per year, or

TABLE 4-2. Climate metrics calculated from daily climate variables for park PACEs, by vegetation type and decadal periods for RCP 8.5.

		Period			
Park/veg. Type	Metric*	1950– 1959	2000– 2009	2050– 2059	2090– 2099
Glacier PACE					
Grassland	above 90	4	7	21	50
	AGDD-32	4099	4601	5886	7408
	days < 32	208	192	150	111
	period > 28	107	127	154	195
	period > 32	72	91	123	161
	days < –8	8	7	5	2
Mixed conifer	above 90	5	8	22	49
	AGDD-32	3899	4389	5643	7159
	below 32	216	199	157	117
	period > 28	99	118	148	191
	period > 32	69	85	115	157
	days < –8	6	5	3	2
Spruce-fir	above 90	1	3	12	32
	AGDD-32	3356	3818	5002	6456
	below 32	234	216	175	135
	period > 28	86	104	133	172
	period > 32	56	72	102	142
	days < –8	9	8	5	3
Rocky Mountain PACE					
Alpine	above 90	0	0	0	~0
	AGDD-32	1864	2188	3052	4180
	below 32	289	271	234	197
	period > 28	67	85	116	144
	period > 32	38	54	87	121
	days < –8	13	11	4	2
Lodgepole pine	above 90	0	0	2	14
	AGDD-32	3241	3648	4695	5968
	below 32	247	232	199	164
	period > 28	98	111	138	165
	period > 32	68	84	116	142
	days < –8	10	8	3	2
Ponderosa pine	above 90	4	7	20	41
	AGDD-32	4629	5090	6285	7672
	below 32	210	197	164	132
	period > 28	125	135	161	184
	period > 32	101	113	139	165
	days < –8	6	5	2	1

TABLE 4-2. (*Continued*)

Park/veg. Type	Metric*	1950– 1959	2000– 2009	2050– 2059	2090– 2099
		Period			
Spruce-fir	above 90	0	0	0	4
	AGDD-32	2510	2880	3837	5036
	below 32	270	253	217	182
	period > 28	82	98	127	153
	period > 32	52	67	100	132
	days < –8	11	10	4	2
Yellowstone PACE					
Montane sage	above 90	1	4	15	34
	AGDD-32	3379	3854	4943	6317
	below 32	246	229	195	160
	period > 28	86	101	125	158
	period > 32	58	73	99	133
	days < –8	14	10	6	3
Alpine	above 90	0	0	0	7
	AGDD-32	2081	2465	3376	4575
	below 32	280	263	230	195
	period > 28	62	77	101	131
	period > 32	40	53	79	108
	days < –8	19	15	8	5
Lodgepole pine	above 90	0	1	8	27
	AGDD-32	2962	3422	4469	5800
	below 32	260	242	208	173
	period > 28	75	92	117	148
	period > 32	50	66	93	122
	days < –8	15	12	7	4
Spruce-fir	above 90	0	0	2	15
	AGDD–32	2471	2899	3863	5122
	below 32	271	254	221	185
	period > 28	67	85	109	139
	period > 32	44	59	85	114
	days < –8	18	14	7	4

*Metrics are as follows: above 90 = number of days per year above 90 degrees F; AGDD-32 = growing degree days with 32 degrees F growth threshold; days < 32 = number of days per year below 32 degrees F; period > 28 = consecutive days with Tmin above 28 degrees F; period > 32 = consecutive days with Tmin above 32 degrees F; days < –8 = days per year below –8 degrees F (–22 degrees C).

7 percent for RCP 4.5, and 4.2 inches (110 millimeters) per year, or 11 percent, for RCP 8.5 by 2100. Projected seasonal changes in precipitation are consistent across all three parks, with summer drying and wetter winters. Analyses of the Greater Yellowstone system project increased aridity because the effects of projected temperature increases on evaporation more than compensate for increased precipitation (Chang and Hansen 2015).

While there is considerable uncertainty about future trends in precipitation, higher temperatures in the recent past and future projections point very clearly toward an increasing portion of precipitation falling as rain rather than snow, particularly in the middle and lower elevations (Knowles, Dettinger, and Cayan 2006; Klos, Link, and Abatzoglou 2014). At broad scales in the western United States, projections are for a decrease of 34 to 56 percent in winter wet days conducive to snowfall in the southern and northern US Rockies. By the mid-twenty-first century, projected changes in the extent of snow-favorable temperatures will likely reduce the length of the snow-dominated season from November to March (five months) to December to February (three months) (table 4-2; Klos, Link, and Abatzoglou 2014).

Snow and Runoff

The mountain snowpack in the Rockies is hugely influential to both natural and human-dominated processes. Water stored in the snowpack is the predominant source of spring runoff and contributes about 75 percent of surface water (Service 2004). The timing and magnitude of runoff are critical to maintaining the integrity of riparian and riverine ecosystems (Poff 2010). A deep snowpack restricts access to forage by large herbivores, including elk and bison, and snow strongly influences large herbivore distribution, health, and reproduction. On-snow activities are a primary source of revenue for gateway communities for the long winter months, and the opening of roads in the spring is highly consequential to park operations and to surrounding communities. Because of the great importance of snow and water, there are established monitoring networks with good measurements of snow dynamics via SNOTEL sites, and excellent records of runoff from US Geological Survey and other stream gauging networks.

Throughout the Rockies, there are well-documented trends from the 1950s to the present toward a declining snowpack, earlier peak runoff, and more rapid melting of the winter snowpack (Mote et al. 2005; Clow 2010; Pederson et al. 2011a, 2011b). Pederson et al. (2011b) noted that the magnitude of the late-twentieth-century snowpack declines are almost unprecedented, and they attributed the declines mostly to rapid springtime warming. The strong influence of temperature on snowmelt is important because it may signal a shift in the 1980s to a dominating control of snow-pack dynamics by temperature rather than precipitation (Pederson et al. 2011a, 2011b).

Since 1980, snowmelt and spring flow timing has shifted on the order of two to three weeks earlier and snow water equivalent (at its peak and on April 1) has declined at a rate of 1.6 inches (4 centimeters) per decade (Clow 2010). Similarly, the number of snow-free days increased by an average of 14 days from 1969 to 2007 in the northern Rockies (Pederson et al. 2011a). These trends are embedded in records with high variability across years and sites, but the consistent trends in peak snow water equivalent, timing of runoff, and measurements of snow-free days unambiguously confirm these changes. Snowmelt is strongly influenced by temperatures above 32 degrees F (0 degrees C), especially in the spring, and both the historical and projected trends are toward more above-freezing days (table 4-2; Pederson et al. 2010).

Projected increases in temperature will obviously result in reduced snow accumulation, increasingly earlier spring melt and runoff, and re-duced snow cover (McKelvey et al. 2011; Diffenbaugh, Scherer, and Ash-faq 2013; Klos, Link, and Abatzoglou 2014). For projected temperature increases of 3.6 to 7.2 degrees F (2 to 4 degrees C), runoff could occur four to five weeks earlier (Stewart, Cayan, and Dettinger 2004). The effects of reduced snowpacks and more rapid snowmelt on streamflow, groundwater recharge, plant available moisture, and fire risk will likely be exacerbated by increased evapotranspiration caused by higher temperatures in the spring, summer, and fall.

Climate Data to Support Management

The Great Northern LCC is experiencing rapid and directional climate changes that are already affecting natural resources, recreational use of parks, and park operations. Projected increases in summer temperatures and relatively small changes in precipitation will likely result in substantial reductions in water available for plant growth (Cook, Ault, and Smerdon 2015) and increased wildfire (Westerling et al. 2006, 2011). Other observed changes in resources and disturbances will surely be exacerbated by climate changes, including outbreaks of forest insect pests (Bentz et al. 2010), changes in hydrological patterns (Clow 2010; Pederson et al. 2011a, 2011b), and changes in seasonality (Ault et al. 2011). Other chapters in this volume address climate-driven changes in plants, animals, and ecosystem processes.

Managers and scientists will continue needing a range of climate products that is nearly as large and diverse as the projected ecological impacts. Three major challenges to the use of climate data are (1) deciding on the most appropriate source of data, (2) identifying relevant variables and metrics, and (3) acquiring, analyzing, and reporting relevant information.

There is a huge range of data available to characterize the historic and future climate of a national park or other management unit. The appropriate choice of data source(s) and evaluation process will depend on the climate variable(s) of most interest and on the key questions or issues. Historical observations may come from weather stations operated by federal, state, local, or private organizations. Site-based measurements are the basis for derived data sets, including the interpolated gridded climate data evaluated in this chapter (i.e., PRISM, TopoWx). Most station or in situ data sources provide data on temperature and precipitation, but temperatures may be reported as daily minimum and maximum or as actual temperature at set intervals (typically 15 minutes to 1 hour). Gridded observations are typically summarized as minimum and maximum temperature at a daily or monthly time step. Precipitation may be rain only, snow, or both, and the interval between measurements varies. The most useful metric may be an average, median, or percentage departure from a reference period, or another statistical property.

Decisions on selection of data to assess future climates are further complicated by the variety of global climate models, downscaling methods, time and spatial scales, and methods to summarize and report results. The number and complexity of decisions on climate data and analyses emphasizes the advantages of interdisciplinary climate-ecological science teams that can work through many of these choices to present relatively straightforward summaries of climate for a given area and management issue.

Results in this chapter made extensive use of two sets of tools developed specifically to facilitate the use of climate data for natural resource applications as well as custom analyses that required project-specific coding in various computer languages. The tools can be used by many, while the custom coding we used will be inaccessible to most managers and many science teams. The first tool, Climate Analyzer (www.climateanalyzer.org), was developed by Michael Tercek in collaboration with the National Park Service. Most recently, Michael Tercek and Ann Rodman collaborated with other Yellowstone NP staff to identify and develop a very broad range of management-relevant summaries of climate, streamflow, and snow data, at scales from sites to areas, and from periods of days to decades. Data sources are primarily from stations (i.e., point locations), although many of the results are for an area and they are aggregates of multiple stations. Reports include statistical summaries in tables, graphs, and maps. Climate Analyzer currently provides data from historical observations.

In addition to Climate Analyzer, this chapter includes many results from the Graphics Catalog produced by the US Geological Survey's North Central Climate Science Center. The Graphics Catalog focuses on making projections from climate models accessible. To do so, modeled (gridded) historical data sources are used, but the focus is on using the information technology resources of the US Geological Survey and other data providers, such as NASA, to access enormous volumes of climate data and the computational resources needed to process and graphically summarize results from downscaled global climate models. A key challenge in developing the Graphics Catalog was identifying a manageable (i.e., relatively small) set of metrics and the most informative formats for reporting. Use of the Graphics Catalog to support the Landscape Climate Change Vulnerability Project, climate workshops, resource condition reports, and vulnerability assessments contributed to the evolving set of outputs.

Managers will always be most interested in the impacts of climate changes on resources, rather than temperature changes or other physical manifestations of climate. But analyses of the basic driving variables will remain important because it will never be possible to identify, much less forecast, changes to all important resources. The foundational analyses in this chapter illustrate the profound changes that are virtually certain to occur if greenhouse gas emissions are not rapidly curbed. The tools described at the end of this chapter and the many sources cited in this chapter provide a rich picture of the many changes forecast for the Rocky Mountains and the Great Northern LCC.

Conclusion

The composition and distribution of the Great Northern LCC's ecosystems are the result of broad-scale atmospheric circulation and geographical features. Glacial retreat and subsequent warming that occurred in the last 14,000 years BP are largely responsible for the current composition and distribution of these ecosystems. The ecosystems have already experienced the impacts of rapid climate changes, and models consistently project that temperatures will increase 4 to 10 degrees F (2.2 to 5.6 degrees C) in the next eighty-five years—about the same increase as since the last glacial period, 14,000 BP. The rate and magnitude of projected future climate changes, especially temperature increases, will have profound effects on vegetation, hydrology, and ecological processes, such as floods, fire, and pests.

Acknowledgments

We particularly thank Forrest Melton, Alberto Guzman, Rama Nemani, and Bridget Thrasher for providing access to and considerable assistance using the NEX climate data, analysis tools and code, and the NASA Ames High End Computing (HEC) system. This work would not have been possible without their support. Any use of trade, product, or firm names is for descriptive purposes only and does not imply endorsement by the US government.

References

Al-Chokhachy, R., J. Alder, S. Hostetler, R. Gresswell, and B. Shepard. 2013. Thermal controls of Yellowstone cutthroat trout and invasive fishes under climate change. *Global Change Biology* 19:3069–81.

Ault, T. R., A. K. Macalady, G. T. Pederson, J. L. Betancourt, and M. D. Schwartz. 2011. Northern hemisphere modes of variability and the timing of spring in western North America. *Journal of Climate* 24:4003–14.

Bentz, B. J., J. Regniere, C. J. Fettig, E. M. Hansen, J. L. Hayes, J. A. Hicke, R. G. Kelsey, J. F. Negron, and S. J. Seybold. 2010. Climate change and bark beetles of the western United States and Canada: Direct and indirect effects. *BioScience* 60:602–13.

Breshears, D. D., O. B. Myers, C. W. Meyer, F. J. Barnes, C. B. Zou, C. D. Allen, Nathan G. McDowell, and W. T. Pockman. 2009. Tree die-off in response to global change-type drought: Mortality insights from a decade of plant water potential measurements. *Frontiers in Ecology and the Environment* 7:185–89.

Brunelle, A., C. Whitlock, P. Bartlein, and K. Kipfmueller. 2005. Holocene fire and vegetation along environmental gradients in the Northern Rocky Mountains. *Quaternary Science Reviews* 24:2281–2300.

Brunstein, F. C. 1996. Climate significance of the bristlecone pine latewood frost-ring record at Almagre Mountain, Colorado, USA. *Artic and Alpine Research* 28:65–76.

Chang, T., and A. Hansen. 2015. Historic and projected climate change in the Greater Yellowstone Ecosystem. *Yellowstone Science* 23:14–19.

Clark, P. U., A. S. Dyke, J. D. Shakun, A. E. Carlson, J. Clark, B. Wohlfarth, J. X. Mitrovica, S. W. Hostetler, and A. M. McCabe. 2009. The last glacial maximum. *Science* 325:710–14.

Clow, D. W. 2010. Changes in the timing of snowmelt and streamflow in Colorado: A response to recent warming. *Journal of Climate* 23:2293–2306.

Cook, B. I., T. R. Ault, and J. E. Smerdon. 2015. Unprecedented 21st century drought risk in the American Southwest and Central Plains. *Science Advances* 1 (1): e1400082.

Cross, M. S., P. D. McCarthy, G. Garfin, D. Gori, and C. A. F. Enquist. 2013. Accelerating adaptation of natural resource management to address climate change. *Conservation Biology* 27:4–13.

Daly, C., M. Halbleib, J. I. Smith, W. P. Gibson, M. K. Doggett, G. H. Taylor, J. Curtis, and P. P. Pasteris. 2008. Physiographically sensitive mapping of climatological temperature and precipitation across the conterminous United States. *International Journal of Climatology* 28:2031–64.

Davey, C. A., K. T. Redmond, and D. B. Simeral. 2006. *Weather and Climate Inventory, National Park Service, Greater Yellowstone Network*. Natural Resource Technical Report NPS/GRYN/NRTR—2006/001. Fort Collins, CO: National Park Service.

Davey, C. A., K. T. Redmond, and D. B. Simeral. 2007. *Weather and Climate Inventory, National Park Service, Rocky Mountain Network*. Natural Resource Technical Report NPS/ROMN/NRTR—2007/036. Fort Collins, CO: National Park Service.

Dean, J. S. 1994. The medieval warm period on the southern Colorado Plateau. In *The Medieval Warm Period*, edited by M. K. Hughes and H. F. Diaz, 225–41. Springer Netherlands.

Diffenbaugh, N. S., M. Scherer, and M. Ashfaq. 2013. Response of snow-dependent hydrologic extremes to continued global warming. *Nature Climate Change* 3:379–84.

Dobrowski, S. Z. 2011. A climatic basis for microrefugia: The influence of terrain on climate. *Global Change Biology* 17:1022–35.

Dobrowski, S. Z., J. T. Abatzoglou, J. A. Greenberg, and S. G. Schladow. 2009. How much influence does landscape-scale physiography have on air temperature in a mountain environment? *Agricultural and Forest Meteorology* 149:1751–58.

Fall, P. L. 1997. Timberline fluctuations and late Quaternary paleoclimates in the Southern Rocky Mountains, Colorado. *Geological Society of America Bulletin* 109:1306–20.

Garfin, G., A. Jardine, R. Merideth, M. Black, and S. LeRoy, eds. 2013. *Assessment of Climate Change in the Southwest United States*. Washington, DC: Island Press.

Gavin, D. G., M. C. Fitzpatrick, P. F. Gugger, K. D. Heath, F. Rodríguez-Sánchez, S. Z. Dobrowski, et al. 2014. Climate refugia: Joint inference from fossil records, species distribution models and phylogeography. *New Phytologist* 204:37–54.

Glick, P., B. Stein, and N. Edelson. 2011. *Scanning the Conservation Horizon: A Guide to Climate Change Vulnerability Assessment.* Washington, DC: National Wildlife Federation.

Gray, S. T., L. J. Graumlich, and J. L. Betancourt. 2007. Annual precipitation in the Yellowstone National Park region since AD 1173. *Quaternary Research* 68:18–27.

Hall, M. P., and D. B. Fagre. 2003. Modeled climate-induced glacier change in Glacier National Park, 1850–2100. *BioScience* 53:131–40.

Hansen, A. J., C. Davis, N. B. Piekielek, J. E. Gross, D. M. Theobald, S. J. Goetz, F. Melton, and R. DeFries. 2011. Delineating the ecosystems containing protected areas for monitoring and management. *BioScience* 61:263–73.

Hansen, A. J., N. B. Piekielek, C. Davis, J. R. Haas, D. M. Theobald, J. E. Gross, W. B. Monahan, T. Olliff, and S. W. Running. 2014. Exposure of US national parks to land use and climate change 1900–2100. *Ecological Applications* 24:484–502.

Hessl, A. E., and W. L. Baker. 1997. Spruce and fir regeneration and climate in the forest-tundra ecotone of Rocky Mountain National Park, CO, USA. *Arctic and Alpine Research* 29:173–83.

IPCC (Intergovernmental Panel on Climate Change). 2013. Summary for policymakers. In *Climate Change 2013: The Physical Science Basis*, edited by T. F. Stocker, D. Qin, G.-K. Plattner, M. Tignor, S. K. Allen, J. Boschung, A. Nauels, Y. Xia, V. Bex, and P. M. Midgley, 1–30. Contribution of Working Group I to the Fifth Assessment Report of the Intergovernmental Panel on Climate Change. Cambridge: Cambridge University Press.

Kittel, T. G. F., P. E. Thornton, J. A. Royle, and T. N. Chase. 2002. Climates of the Rocky Mountains: Historical and future patterns. In *Rocky Mountain Futures*, edited by J. S. Baron, 59–82. Washington, DC: Island Press.

Klos, P. Z., T. E. Link, and J. T. Abatzoglou. 2014. Extent of the rain-snow transition zone in the western U.S. under historic and projected climate. *Geophysical Research Letters* 41:4560–68. doi: 10.1002/2014GL060500.

Knowles, N., M. D. Dettinger, and D. R. Cayan. 2006. Trends in snowfall versus rainfall in the western United States. *Journal of Climate* 19:4545–59.

Leonard, E. M. 1989. Climatic change in the Colorado Rocky Mountains: Estimates based on modern climate at late Pleistocene equilibrium lines. *Arctic and Alpine Research* 21:245–55.

Logan, J. A., W. W. MacFarland, and L. Wilcox. 2010. Whitebark pine vulnerability to climate-driven mountain pine beetle disturbance in the Greater Yellowstone Ecosystem. *Ecological Applications* 20:895–902.

Marcott, S. A., J. D. Shakun, P. U. Clark, and A. C. Mix. 2013. A reconstruction of regional and global temperature for the past 11,300 years. *Science* 339:1198–1201.

Maurer, E. P., A. W. Wood, J. C. Adam, D. P. Lettenmaier, and B. Nijssen. 2002. A long-term hydrologically based dataset of land surface fluxes and states for the conterminous United States. *Journal of Climate* 15:3237–51.

McCune, B. 2007. Improved estimates of incident radiation and heat load using non-parametric regression against topographic variables. *Journal of Vegetation Science* 18:751.

McKelvey, K. S., J. P. Copeland, M. K. Schwartz, J. S. Littell, K. B. Aubry, J. R. Squires, S. A. Parks, M. M. Elsner, and G. S. Mauger. 2011. Climate change predicted to shift wolverine distributions, connectivity, and dispersal corridors. *Ecological Applications* 21:2882–97.

McWethy, D. B., S. T. Gray, P. E. Higuera, J. S. Littell, G. T. Pederson, A. J. Ray, and C. Whitlock. 2010. *Climate and Terrestrial Ecosystem Change in the U.S. Rocky Mountains and Upper Columbia Basin: Historical and Future Perspectives for Natural Resource Management.* Natural Resource Report NPS/GRYN/ NRR—2010/260. Fort Collins, CO: National Park Service.

Millar, C. I., R. D. Westfall, D. L. Delany, M. J. Bokach, A. L. Flint, L. E. Flint, and J. Ridge. 2012. Forest mortality in high-elevation whitebark pine (*Pinus albicaulis*) forests of eastern California, USA: Influence of environmental context, bark beetles, climatic water deficit, and warming. *Canadian Journal of Forest Research* 42:749–65.

Millspaugh, S. H., C. Whitlock, and P. J. Bartlein. 2000. Variations in fire frequency and climate over the past 17,000 years in central Yellowstone National Park. *Geology* 28:211–14.

Monahan, W., and N. Fisichelli. 2014. Climate exposure of US national parks in a new era of change. *PLOS ONE* 9: e101302. doi: 10.1371/journal.pone .0101302.

Moss, R. H., J. A. Edmonds, K. A. Hibbard, M. R. Manning, S. K. Rose, D. P. Van Vuuren, et al. 2010. The next generation of scenarios for climate change research and assessment. *Nature* 463:747–56.

Mote, P. W., L. Brekke, P. B. Duffy, and E. Maurer. 2011. Guidelines for constructing climate scenarios. *Eos, Transactions American Geophysical Union* 92:257–58.

Mote, P. W., A. F. Hamlet, M. P. Clark, and D. P. Lettenmaier. 2005. Declining mountain snowpack in western north America. *Bulletin of the American Meteorological Society* 86:39–49.

Osborn, G., and B. H. Luckman. 1988. Holocene glacier fluctuations in the Canadian cordillera (Alberta and British Columbia). *Quarternary Science Reviews* 7:115–28.

Oyler, J. W., A. Ballantyne, K. Jencso, M. Sweet, and S. W. Running. 2014. Creating a topoclimatic daily air temperature dataset for the conterminous United States using homogenized station data and remotely sensed land skin temperature. *International Journal of Climatology Online.* doi: 10.10002/joc.4127.

Oyler, J. W., S. Z. Dobrowski, A. P. Ballantyne, A. E. Klene, and S. W. Running. 2015. Artificial amplification of warming trends across the mountains of the western United States. *Geophysical Research Letters* 42:153–61.

Pederson, G. T., D. B. Fagre, S. T. Gray, and L. J. Graumlich. 2004. Decadal-scale climate drivers for glacial dynamics in Glacier National Park, Montana, USA. *Geophysical Research Letters* 31:L12203. doi: 12210.11029/12004GL019770.

Pederson, G. T., L. J. Graumlich, D. B. Fagre, T. Kipfer, and C. C. Muhlfeld. 2010. A century of climate and ecosystem change in Western Montana: What do temperature trends portend? *Climatic Change* 98:133–54.

Pederson, G. T., S. T. Gray, T. Ault, W. Marsh, D. B. Fagre, A. G. Bunn, C. A. Woodhouse, and L. J. Graumlich. 2011a. Climatic controls on the snowmelt hydrology of the northern Rocky Mountains. *Journal of Climate* 24:1666–87.

Pederson, G. T., S. T. Gray, D. B. Fagre, and L. J. Graumlich. 2006. Long-duration drought variability and impacts on ecosystem services: A case study from Glacier National Park, Montana. *Earth Interactions* 10:1–28.

Pederson, G. T., S. T. Gray, C. A. Woodhouse, J. L. Betancourt, D. B. Fagre, J. S. Littell, E. Watson, B. H. Luckman, and L. J. Graumlich. 2011b. The unusual nature of recent snowpack declines in the North American cordillera. *Science* 333:332–35.

Peterman, W., R. H. Waring, T. Seager, and W. L. Pollock. 2012. Soil properties affect pinyon pine—juniper response to drought. *Ecohydrology Online*. doi: 10.1002/eco.1284.

Petersen, K. L. 1994. A warm and wet Little Climatic Optimum and a cold and dry Little Ice Age in the southern Rocky Mountains, USA. *Climatic Change* 26:243–69.

Pierce, K. L. 1979. History and Dynamics of Glaciation in the Northern Yellowstone National Park Area. *USGS Professional Paper* 29-F:1-90.

Poff, N. L., M. I. Pyne, B. P. Bledsoe, C. C. Cuhaciyan, and D. M. Carlisle. 2010. Developing linkages between species traits and multiscaled environmental variation to explore vulnerability of stream benthic communities to climate change. *Journal of the North American Benthological Society* 29:1441–58.

Reasoner, M. A., and M. A. Jodry. 2000. Rapid response of alpine timberline vegetation to the Younger Dryas climate oscillation in the Colorado Rocky Mountains, USA. *Geology* 28:51–54.

Rice, J., A. Tredennick, and L. A. Joyce. 2012. Climate Change on the Shoshone National Forest, Wyoming: A Synthesis of Past Climate, Climate Projections, and Ecosystem Implications. US *Forest Service RMRS-GTR-264.*

Rupp, D. E., J. T. Abatzoglou, K. C. Hegewisch, and P. W. Mote. 2013. Evaluation of CMIP5 20th century climate simulations for the Pacific Northwest USA. *Journal of Geophysical Research: Atmospheres* 118:10884–906.

Service, R. F. 2004. As the west goes dry. *Science* 303:1124–27.

Solomon, S. S. 2007. The last 2,000 years. In *Climate Change 2007: The Physical Science Basis*, edited by S. Solomon, D. Qin, M. Manning, Z. Chen, M. Marquis, K. B. Averyt, M. Tignor, and H. L. Miller, sec. 6.6, box 6.4. Contribution of Working Group I to the Fourth Assessment Report of the Intergovernmental Panel on Climate Change. Cambridge: Cambridge University Press, for the Intergovernmental Panel on Climate Change.

Solomon, S. S., G. K. Plattner, R. Knutti, and P. Friedlingstein. 2009. Irreversible climate change due to carbon dioxide emissions. *Proceedings of the National Academy of Sciences of the United States of America* 106:1704–09.

Stein, B. A., P. Glick, N. Edelson, and A. Staudt, eds. 2014. *Climate-Smart Conservation: Putting Adaptation Principles into Practice*. Washington, DC: National Wildlife Federation.

Stephenson, N. L. 1990. Climatic control of vegetation distribution: The role of water balance. *American Naturalist* 135:649–70.

Stewart, I., D. Cayan, and M. Dettinger. 2004. Changes in snowmelt runoff timing in western North America under a "business as usual" climate change scenario. *Climatic Change* 62:217–32.

Thrasher, B., J. Xiong, W. Wang, F. Melton, A. Michaelis, and R. Nemani. 2013. Downscaled climate projections suitable for resource management. *Eos, Transactions of the American Geophysical Union* 94:321–23.

Westerling, A. L., H. G. Hidalgo, D. R. Cayan, and T. W. Swetnam. 2006. Warming and earlier spring increase western US forest wildfire activity. *Science* 313:940–43.

Westerling, A. L., M. G. Turner, E. A. H. Smithwick, W. H. Romme, and M. G. Ryan. 2011. Continued warming could transform Greater Yellowstone fire regimes by mid-21st century. *Proceedings of the National Academy of Sciences of the United States of America* 108:13165–70.

Western Water Assessment. 2013. Severe flooding on the Colorado front range: A preliminary assessment. *CIRES Western Water Assessment*. Boulder: University of Colorado. http://wwa.colorado.edu/resources.

Whitlock, C. 2002. Paleoenvironmental history of the Rocky Mountain Region. In *Rocky Mountain Futures: An Ecological Perspective*, edited by J. Baron, 41–57. Covelo, CA: Island Press.

Whitlock, C., and P. J. Bartlein. 1993. Spatial variations of Holocene climatic change in the Yellowstone region. *Quaternary Research* 39:231–38.

Wolfe, J. A. 1992. An analysis of present-day terrestrial lapse rates in the western conterminous United States and their significance to paleoaltitudinal estimates. *U.S. Geological Survey Bulletin* 1964.

Chapter 5

Historical and Projected Climates as a Basis for Climate Change Exposure and Adaptation Potential across the Appalachian Landscape Conservation Cooperative

Kevin Guay, Patrick Jantz, John E. Gross,
Brendan M. Rogers, and Scott J. Goetz

Global temperatures have risen over the last few decades, and even the most conservative climate models project these trends to continue over the next eighty-five years (IPCC 2013). As climate changes, flora and fauna will be forced to adapt or migrate (Aitken et al. 2008). Many species have been able to adapt to past changes in climate, moving south during glacial periods and north during interglacial periods. However, anthropogenic climate change in most areas is occurring much faster than previous climatic shifts. Flora, in particular, may be unable to adapt or disperse quickly enough to track suitable climate conditions (Corlett and Wescott 2013). Understanding historical and projected future trends in temperature, precipitation, and other climate variables is important for evaluating the current context and likely consequences of climate changes in national parks, and in developing effective strategies for climate adaptation.

Physical climate variables are used to measure the potential exposure of management targets to climate changes. The term *climate change exposure* refers to the character, magnitude and rate of change in climatic conditions that a species or system is likely to experience (Klausmeyer et al. 2011). As such, it is of fundamental interest when assessing climate vulnerability. Many parks and protected areas in the Appalachian Landscape Conservation Cooperative (Appalachian LCC) are already experiencing abnormally high temperatures, and the strong likelihood that climates will continue to warm may necessitate a shift in management focus away from preserving historical conditions and toward managing for potential future conditions (Monahan and Fisichelli 2014).

Geography and Climate

The Appalachian LCC, our focal area for this assessment, includes much of the area from approximately 32 to 42 degrees of latitude north and 74 to 87 degrees of latitude west. This area encompasses a large portion of the Appalachian Mountains from Georgia to just south of the Adirondacks in New York state (fig. 5-1). Within this larger geographic unit, our analyses focus on three protected areas and their associated protected area centered ecosystems (PACEs; Hansen et al. 2011): Delaware Water Gap (DEWA) National Recreation Area, Shenandoah National Park (NP), and Great Smoky Mountains NP. The PACE for a park is the area needed to sustain the existing biodiversity in the park, and it represents an area where land cover and land use changes would be expected to impact park resources (fig. 5-1; Hansen et al. 2011). The PACE thus defines a more ecologically relevant area of analysis than that provided by most park administrative boundaries. Climate trends within our focal PACEs are a broad sample of climates across the Appalachian LCC and within units that share similar management constraints and opportunities.

Temperature in the Appalachian LCC follows gradients where temperature generally declines with latitude and elevation. Temperatures on the highest peaks can be more similar to conditions hundreds of kilometers

FIGURE 5-1 Outline of the Appalachian Landscape Conservation Cooperative (LCC) with boundaries for the Delaware Water Gap (DEWA), Shenandoah (SHEN), and Great Smoky Mountains (GRSM) protected areas and their respective protected area centered ecosystems (PACEs).

north than to those in the immediate surroundings. Much of the Appalachian LCC is marked by multidecadal temperature oscillations. Temperatures were relatively high in the first part of the twentieth century but then displayed a marked cooling midcentury followed by increases in recent decades (Pan et al. 2013). The duration and magnitude of these multidecadal trends vary across the eastern United States.

The southern portion of the Appalachian LCC is part of a larger area in the southern and central United States where temperatures have cooled over the last century, leading many to question whether observed trends are a result of internal climate variability related to sea surface temperatures (Robinson, Reudy, and Hansen 2002), external forcing from greenhouse gas emissions (Kunkel et al. 2006), aerosols (Portmann, Solomon, and Hegerl 2009; Leibensperger et al. 2012; Yu et al. 2014), land surface feedbacks (Pan et al. 2013), or some combination of the above. There is no consensus on the source of these temperature trends, a fact that limits our ability to assess how future trends might evolve. A question of particular interest for climate adaptation planning is whether temperature in the Southeast will "catch up" with the rest of the United States over the next few decades or continue to be an area of moderate to no warming.

Precipitation in the Appalachian LCC decreases with distance from coasts but can be significantly higher in the mountains; some of the highest precipitation amounts in the country are found in Great Smoky Mountains NP. Mean annual precipitation has changed relatively little in much of the eastern United States over the past one hundred years, although parts of the Northeast have experienced increases (NOAA 2015). Precipitation variability, or the frequency of high rainfall events in the Southeast, has increased (Li et al. 2011). This is in line with general predictions of increased precipitation intensity for many regions of the world. In the Southeast, increased precipitation variability has been attributed to the shifting position of the Bermuda High pressure system (Li et al. 2011). Both global general circulation modeling (Li, Li, and Deng 2013) and regional dynamic climate modeling (Gao et al. 2012) results suggest that the frequency of extreme events may continue to increase over the next century.

Past climate assessments for various geographies in the eastern United States have used a variety of climate data sets. For historical climate analysis, most have utilized gridded data sets created by interpolating meteorological station observations. The Parameter-elevation Regressions on Independent Slopes Model (PRISM; Daly et al. 2008) data set (0.5- and 2.5-mile [800-meter and 4-kilometer] resolution) is frequently used in the

United States, but others are available for larger regions (e.g., Worldclim and CRU). For assessing potential future climates, most groups have relied on relatively coarse projections on the order of tens or hundreds of miles, which dampen the considerable climate variability in the mountainous National Park Service units of the East. Our objective here is to combine historical PRISM data with high-resolution downscaled climate projections to conduct a climate exposure analysis for selected National Park Service units within the Appalachian LCC that are facing pressing climate-related management issues.

Contemporary Patterns and Future Trends in Key Climate Variables

We used PRISM data to estimate observed climate for the Appalachian LCC parks and PACEs from 1895 to 2010. PRISM data are generated using observations from climate stations across the conterminous United States, combined with data on elevation, aspect, and other topographical features that influence local climate (Daly et al. 2008). Our analyses used monthly PRISM gridded climate data at 0.5-mile resolution from 1895 to 2010 for precipitation and mean daily minimum and maximum temperatures. PRISM showed strong agreement with station data at Great Smoky Mountains NP at various elevations (fig. 5-2).

We projected temperature and precipitation using the NASA Earth Exchange Downscaled Climate Projections (NEX-DCP30) data set (Thrasher et al. 2013). NEX-DCP30 is a 30-arc-second (0.5-mile) downscaled product based on model output from the Coupled Model Intercomparison Project Phase 5 (CMIP5). NEX-DCP30 data are available at monthly time steps for both historical analysis (1950–2005) and future projections (2006–2100). We used the NEX-DCP30 ensemble statistics, a product calculated from thirty-three individual CMIP5 general circulation models. Patterns in the historical NEX-DCP30 data were similar to those observed with PRISM, which provides some confidence in their ability to capture the major climate features and trends in the eastern United States. For the future, we used model projections that were driven by representative concentration pathways (RCPs) (Moss et al. 2010) 4.5 and 8.5 (chap. 4). The RCP 4.5 scenario assumes that greenhouse gas emissions, mostly carbon dioxide, rapidly decline and are stabilized by about 2050. RCP 8.5 assumes that we continue to increase greenhouse gas emissions at a rate similar to present. The low and high scenarios (RCP 4.5 and 8.5, respectively)

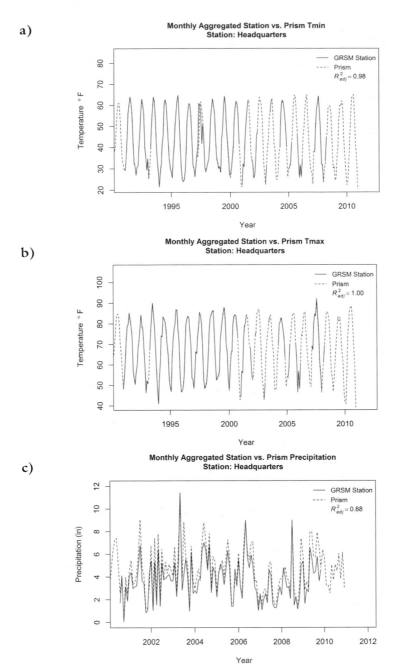

FIGURE 5-2   Monthly PRISM data and data from the Great Smoky Mountains National Park climate station located at the park headquarters for (a) minimum temperature, (b) maximum temperature, and (c) precipitation.

result in an anthropogenic climate forcing of 4.5 or 8.5 Watts per square meter by the year 2100.

The PRISM and NEX-DCP30 data were temporally aggregated from monthly to annual products. Since both PRISM and NEX-DCP30 exhibit considerable year-to-year variability, we used a moving thirty-year average to reduce this variability and more clearly show patterns at the decade and century scales. In addition, to evaluate changes in climate variables at higher temporal resolution, we generated daily NEX-DCP30 data (i.e., bias corrected and spatially downscaled as in Thrasher et al. 2013) using a tool created by Thrasher (personal communication). Output from the Beijing Normal University Earth System Model in the CMIP5 archive was used because it performed favorably when compared with daily climate station data from Great Smoky Mountains NP.

Temperature

Relative to global and continental averages (as well as the Great Northern LCC), temperatures in the Appalachian LCC remained relatively stable from 1895 to 2010. Over the observation period, monthly average minimum and maximum temperatures increased slightly at DEWA, did

TABLE 5-1. Trends in annual precipitation, mean monthly maximum temperatures, and mean monthly minimum temperatures for the three PACEs and parks.

	Domain Type	Delaware Water Gap (DEWA)	Great Smoky Mountains (GRSM)	Shenandoah (SHEN)
Precipitation (inches per year per decade)	PACE	0.559*	0.225	0.289
	Park	0.508*	0.229	0.438*
Maximum temperature (degrees F per decade)	PACE	0.044	−0.061	−0.023
	Park	0.042	−0.060	−0.034
Minimum temperature (degrees F per decade)	PACE	0.088*	−0.065	0.017
	Park	0.116*	−0.075*	0.025

Note: Trends were calculated as changes per decade using PRISM data from 1895 to 2010. Significant linear trends ($p < 0.5$) are denoted with an asterisk (*).

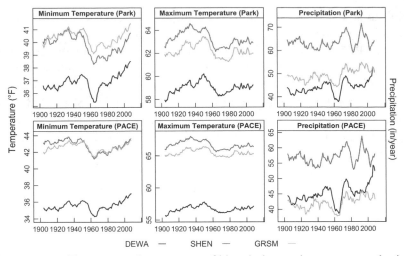

FIGURE 5-3   Ten-year moving averages of historical annual average annual minimum temperature, maximum temperature, and precipitation derived from PRISM data for Great Smoky Mountains (GRSM; light gray) and Shenandoah (SHEN; gray) national parks and the Delaware Water Gap National Recreation Area (DEWA; black) and their protected area centered ecosystems.

not change appreciably at Shenandoah NP, and decreased at Great Smoky Mountains NP (fig. 5-3; table 5-1). Trends during the historical interval were significant for warming in DEWA and for cooling in Great Smoky Mountains NP, but not for Shenandoah NP (table 5-1). This geographical pattern in the eastern United States has been widely reported, although the causes of the "southeastern warming hole" remain uncertain. One hypothesis concerns large-scale circulations in the Atlantic and Pacific oceans (Robinson, Reudy, and Hansen 2002; Kunkel et al. 2006). Alternatively, Yu et al. (2014) found a strong correlation between shortwave cloud forcing (facilitated by aerosols) and the decline in maximum temperature in the southeastern United States. Given that climate models may not capture the ultimate cause, it is unknown whether these factors will continue to dampen temperature increases in the Southeast relative to other regions during the twenty-first century.

Despite relatively constant temperatures from 1895 to 2010, all three parks showed a sharp decline in minimum temperature of roughly 2.5 to 3 degrees F (1.4 to 1.7 degrees C) between 1950 and 1970, which generally mimics global cooling during the same period. While temperatures in DEWA rebounded within a decade, temperatures in Shenandoah and

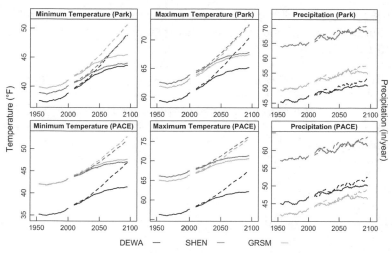

FIGURE 5-4   Historical and projected annual mean monthly minimum temperatures, mean monthly maximum temperatures, and precipitation from the NEX-DCP30 data set for Great Smoky Mountains (GRSM; light gray) and Shenandoah (SHEN; gray) national parks and the Delaware Water Gap National Recreation Area (DEWA; black) and their protected area centered ecosystems. Interannual variability is smoothed using a ten-year moving average. Projections are shown for the RCP 4.5 (solid line) and RCP 8.5 (dashed line) scenarios.

Great Smoky Mountains national parks took nearly fifty years to reach pre-1950 levels (fig. 5-3). Maximum temperatures showed less variability than minimum temperatures, but unlike minimum temperatures, maximum temperatures did not reach 1950 levels by 2010.

Temperatures in Shenandoah and Great Smoky Mountains national parks were relatively similar, while the temperatures in DEWA were about 2.5 degrees F lower. When comparing the parks and PACEs, the PACE was about 1.5 degrees F (0.8 degree C) warmer than the park for Shenandoah and Great Smoky Mountains, yet it was 1 degree F (0.6 degree C) cooler for DEWA. PACEs for mountain parks, such as Shenandoah and Great Smoky Mountains, are warmer because they add low-elevation areas outside the park boundary. DEWA is a low-elevation, river-based park, and the PACE areas outside the park boundary are mostly at higher elevation and thus cooler.

Future projections from RCP 4.5 and 8.5 display similar accelerations in warming until about 2040 (0.7 to 0.8 degrees F [0.4 degrees C] per decade averaged across the PACEs; fig. 5-4 and table 5-2). This is more

TABLE 5-2. NEX-DCP30 trends for annual precipitation, mean monthly maximum temperatures, and mean monthly minimum temperatures averaged across the PACEs.

	Historic	RCP 4.5		RCP 8.5	
	1956–2005	2006–2049	2050–2099	2006–2049	2050–2099
Precipitation (inches per year per decade)	0.341	0.213	0.040	0.529	0.560
Maximum temperature (degrees F per decade)	0.339	0.715	0.239	0.834	1.142
Minimum temperature (degrees F per decade)	0.324	0.692	0.239	0.829	1.172

Note: Trends are presented as changes per decade for historical (1956–2005) and two future time periods (2006–2049 and 2050–2099) using RCP 4.5 and 8.5 model ensembles.

than twice the warming rate seen during the previous fifty years, and it highlights that climate models do not project a continuation of the warming hole. After 2040, temperatures increased in RCP 8.5 considerably more than in RCP 4.5 (1.2 versus 0.3 degrees F per decade [0.7 versus 0.2 degrees C per decade]), reaching total increases of 4.4 degrees F (2.4 degrees C) in RCP 4.5 and 9.5 degrees F (5.3 degrees C) in RCP 8.5 by 2100, compared to 2006 across the three PACEs.

The number of days with temperatures below freezing (32 degrees F [0 degrees C]—i.e., frost days) per year is an important metric with respect to a species' exposure to climate change. For example, the number of frost days is related to length of the growing season, plant phenological processes, and life cycles of insects like the hemlock woolly adelgid (*Adelges tsugae*). There has been a small decline in the number of frost days per year in all three parks since 1950 (fig. 5-5). Over the next eighty-five years, annual frost days are projected to decrease by 25 to 75 days depending on RCP and PACE. DEWA has the most frost days per year (150) and is projected to decrease by the most (38 and 75 days for RCP 4.5 and 8.5, respectively).

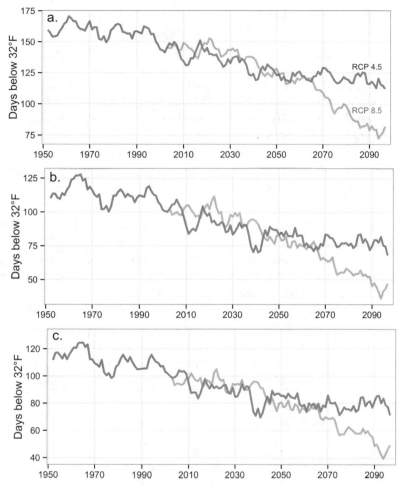

FIGURE 5-5   Projections of days below freezing (32 degrees F) for (a) Delaware Water Gap National Recreation Area, (b) Shenandoah National Park, and (c) Great Smoky Mountains National Park. Daily, downscaled NEX-DCP30 (BNU-ESM) data were used for both the historical and the future predictions (Thrasher et al. 2013). RCP 4.5 and RCP 8.5 scenarios are represented by dark and light gray lines, respectively. A five-year moving average was used to smooth year-to-year variability.

Precipitation

Precipitation in the Appalachian LCC has been quite variable over the last century. Yearly totals between 1895 and 2010 ranged from 38 to 52 inches (97 to 132 centimeters) per year in DEWA, 59 to 72 inches (150 to 183 centimeters) per year in Shenandoah NP, and 45 to 54 inches (114 to 137 centimeters) per year in Great Smoky Mountains NP (fig. 5-3). Precipitation increased slightly during the twentieth century for all three PACEs, although the increase was significant only for DEWA and the Shenandoah PACE (table 5-1), which generally confirms wetter conditions for the region (NOAA 2015). For DEWA and Great Smoky Mountains NP, this increase came largely as a step change around 1970. There was also a decrease in precipitation that coincided with regional cooling around 1960 for both DEWA and Great Smoky Mountains NP, although this was not seen at Shenandoah NP. Modeled historical precipitation using NEX-DCP30 aligned with the PRISM data for all three parks, showing small increases during the latter half of the twentieth century (fig. 5-4; table 5-2).

Precipitation is generally projected to continue to increase in a similar manner to how it did during the second half of the twentieth century (fig. 5-4; table 5-2). Interannual variability is also projected to increase, which several studies have linked with more intense droughts and floods (Easterling et al. 2000). Models forced with RCP 4.5 and 8.5 both project increases in future precipitation, with total increases of 1.2 inches (3.0 centimeters) per year in RCP 4.5 and 5.2 inches (13.2 centimeters) per year in the more extreme RCP 8.5 scenario by 2100. For ecosystem functions, an important question is whether these increases will be enough to compensate for longer, hotter summers with increased evaporative demand.

Climate Data to Support Management

A primary goal of national park management has been to maintain or restore ecosystems to historical conditions. Aspects of the climate in many parks have already exceeded the historical ranges of variability (Monahan and Fisichelli 2014). Climate models indicate that with current rates of increase in greenhouse gas emissions, climates in the future will move to a state that is completely outside that experienced in the past (Mora et al. 2013). Because of this, prospective management strategies are likely to be the most effective use of limited time and resources given expected directional changes in climate (Stein et al. 2014). Fundamental management questions related to climate change include the following: How much will the climate change? How do the magnitude and rate of projected climate change compare to historical climate variability? What are the major uncertainties?

Ensemble average projections derived from NEX-DCP30 indicate large increases in temperature and modest increases in precipitation over the next century in eastern US parks and PACEs (fig. 5-4). This is similar to previous work that projects increasing temperature trends across the eastern United States (Kunkel et al. 2006; Kumar et al. 2013). Temperature trends are projected to be most rapid in DEWA but are still within the projected range of variation of the other units. Although rapid, the projected rates of temperature change are on the order of estimates for past rapid climate change of 3.5 to 7 degrees F (1.9 to 3.9 degrees C) per century in the eastern United States associated with the Medieval Warm Period (Cronin et al. 2003; Willis and MacDonald 2011). However, absolute temperatures are projected to be substantially higher than any experienced since the last glacial maximum (Cronin et al. 2003), and likely higher than any interglacial period during the Pleistocene. While the NEX-DCP30 ensemble average projections display low interdecadal variability, this is a function of model averaging; future trends may therefore be marked by periods of stable temperatures or cooling that alternate with periods of warming that exceed mean trends, as has been seen in the past (Cronin et al. 2003).

The frost-free period has increased by twenty days in the northeastern United States since 1920 (McCabe, Betancourt, and Feng 2015), and our results using NEX-DCP30 projections show a large decreasing trend in the number of frost days throughout the next century. If frost days decline as projected, there could be large impacts on social and ecological systems. The length of the frost-free period controls spring vegetation green-up and senescence in the fall, with far-reaching effects from changes in vegetation productivity to changes in animal life cycles. For example, decreases in frost days can exacerbate the impact of forest pests if it increases the number of generations they can complete in an annual cycle.

Increased precipitation variability and more extreme storm events (Gao et al. 2012) suggest that flood risk could rise in all three parks considered here. More extreme storms and flooding can undercut roads, increase erosion, change aquatic habitats, and cause widespread tree mortality. DEWA experienced "100-year" floods in 2004, 2005, and 2006. These floods caused major infrastructure damage and toppled trees and branches onto trails, creating hazards to visitors (Richard Evans, personal communication). Moving infrastructure away from flood zones or taking steps to protect infrastructure from floods is a proactive approach that may minimize damage and costs from future storms.

Drought events are also projected to increase in frequency, which has important management implications. For example, the evaporative power of a warmer atmosphere coupled with periodic drought could increase plant water stress, impacting forests across the eastern United States, as has been seen in the West (Anderegg, Kane, and Anderegg 2012; Williams et al. 2013). In addition, drier fuels could promote more intense fires, as has also been observed in western parks (Easterling et al. 2000), leading to rapidly increasing and shifting demands on National Park Service management. Baseflow in streams may also be affected, reducing the quality and amount of aquatic habitat.

Because of the lack of analogs for the magnitude of projected warming and the increasing availability of higher-resolution climate projections, climate models are being used more and more frequently to inform management decisions. However, climate models have known limitations. Hawkins and Sutton (2009) identify three primary sources of uncertainty. Uncertainty related to internal climate variability includes natural fluctuations in the climate system that occur in the absence of external forcing. Model uncertainty arises from differences in how models represent the effects of greenhouse gas forcing on physical processes. Scenario uncertainty represents the inherent difficulty in projecting greenhouse gas forcing based on economic, demographic and policy processes. A holistic understanding of these uncertainty sources can help guide climate impact studies. Generally, climate model results will be most useful for lead times of several decades (roughly thirty to fifty years). Uncertainty for shorter lead times (i.e., ten to twenty years) tends to be dominated by internal variability, which can manifest as global pauses or accelerations in a given climate variable or as geographically localized trends that run counter to the global mean. Uncertainty for longer lead times (roughly eighty to one hundred years) tends to be dominated by scenario variability. In addition, climate models have better skill in representing some variables than others. Temperature projections tend to be more reliable than precipitation, and projections of mean values tend to be more reliable than projections of variability.

Recently, Thrasher et al. (2013) improved the management relevance of general circulation models that were part of the CMIP5 by downscaling the data to 30-arc-second (roughly 0.5-mile) resolution. Although an improvement over past models, 0.5-mile data still miss significant topoclimatic variability (Fridley 2009). Even fine-scale climate projections at 0.5 mile may therefore fail to identify or underestimate the extent of refugia that can provide suitable climatic conditions that allow plants and animals to persist in warming conditions. Such refugia may help explain why past warming events were not accompanied by large-scale extirpations (Willis and MacDonald 2011) and may be particularly important when formulating management plans over the next few decades. Despite short-comings, these models provide our best picture yet of how climate might evolve over the next few decades, although natural variability and unavoid-able uncertainty mean that forecasts will surely vary in some regards from the actual climate (Deser et al. 2012). In light of this, management strate-gies must maintain flexibility and adaptability to be able to respond to such changes.

Conclusion

Future temperature and precipitation conditions in the Appalachian LCC and the parks within it are predicted to depart substantially from past cli-mate, which will create new conditions to which flora and fauna will be exposed. They will need either to adapt or to migrate to areas with more suitable environments. Management will be challenged with protecting flora and fauna in relatively small parks and protected areas that may no longer harbor conditions under which many of the current species and vegetation associations have evolved and adapted to in recent times (cen-turies to millennia). To protect priority resources or maintain desired eco-logical functions, new management strategies may need to be considered, including assisted migration or establishing broad-scale connectivity by establishing protected area networks with federal, state, and local partners. Regular updates of climate forecasts and cross-checking with data from weather and ecological monitoring programs can help managers determine potential early onset of important ecosystem changes associated with cli-mate change.

Acknowledgments

This work was funded by the NASA Climate and Biological Response Program (10-BIOCLIM10-0034). We would like to thank Tina Cormier and Forrest Melton for helpful ideas and contributions throughout.

References

Aitken, S. N., S. Yeaman, J. A. Holliday, T. Wang, and S. Curtis-McLane. 2008. Adaptation, migration or extirpation: Climate change outcomes for tree popu-lations. *Evolutionary Applications* 1:95–111.

Anderegg, W. R. L., J. M. Kane, and L. D. L. Anderegg. 2012. Consequences of widespread tree mortality triggered by drought and temperature stress. *Nature Climate Change* 3:30–36.

Corlett, R., and D. Westcott. 2013. Will plant movements keep up with climate change? *Trends in Ecology & Evolution* 28:482–88.

Cronin, T. M., G. S. Dwyer, T. Kamiya, S. Schwede, and D. A. Willard. 2003. Medieval Warm Period, Little Ice Age and 20th century temperature variability from Chesapeake Bay. *Global and Planetary Change* 36:17–29.

Daly, C., M. Halbleib, J. I. Smith, W. P. Gibson, M. K. Doggett, G. H. Taylor, J. Curtis, and P. P. Pasteris. 2008. Physiographically sensitive mapping of climatological temperature and precipitation across the conterminous United States. *International Journal of Climatology* 28:2031–64.

Deser, C., R. Knutti, S. Solomon, and A. S. Phillips. 2012. Communication of the role of natural variability in future North American climate. *Nature Climate Change* 2:775–79.

Easterling, D. R., G. A. Meehl, C. Parmesan, S. A. Changnon, T. R. Karl, and L. O. Mearns. 2000. Climate extremes: Observations, modeling, and impacts. *Science* 289:2068–75.

Fridley, J. D. 2009. Downscaling climate over complex terrain: High finescale (< 1000 m) spatial variation of near-ground temperatures in a montane forested landscape (Great Smoky Mountains). *Journal of Applied Meteorology and Climatology* 48:1033–49.

Gao, Y., J. S. Fu, J. B. Drake, Y. Liu, and J.-F. Lamarque. 2012. Projected changes of extreme weather events in the eastern United States based on a high resolution climate modeling system. *Environmental Research Letters* 7:044025.

Hansen, A. J., C. R. Davis, N. Piekielek, J. Gross, D. M. Theobald, S. Goetz, F. Melton, and R. DeFries. 2011. Delineating the ecosystems containing protected areas for monitoring and management. *BioScience* 61:363–73.

Hawkins, E., and R. Sutton. 2009. The potential to narrow uncertainty in regional climate predictions. *Bulletin of the American Meteorological Society* 90:1095–1107.

IPCC. 2013. Summary for policymakers. In *Climate Change 2013: The Physical Science Basis*, edited by T. F. Stocker, D. Qin, G.-K. Plattner, M. Tignor, S. K. Allen, J. Boschung, A. Nauels, Y. Xia, V. Bex, and P. M. Midgley. Contribution of Working Group I to the Fifth Assessment Report of the Intergovernmental Panel on Climate Change. New York: Cambridge University Press.

Klausmeyer, K. R., M. R. Shaw, J. B. MacKenzie, and D. R. Cameron. 2011. Landscape-scale indicators of biodiversity's vulnerability to climate change. *Ecosphere* 2.

Kumar, S., J. Kinter, P. A. Dirmeyer, Z. Pan, and J. Adams. 2013. Multidecadal climate variability and the "warming hole" in North America: Results from CMIP5 twentieth- and twenty-first-century climate simulations. *Journal of Climate* 26:3511–27.

Kunkel, K. E., X. Z. Liang, J. Zhu, and Y. Lin. 2006. Can CGCMs simulate the twentieth-century "warming hole" in the central United States? *Journal of Climate* 19:4137–53.

Leibensperger, E. M., L. J. Mickley, D. J. Jacob, W. T. Chen, J. H. Seinfeld, A. Nenes, P. J. Adams, D. G. Streets, N. Kumar, and D. Rind. 2012. Climatic effects of 1950–2050 changes in US anthropogenic aerosols. Part 2: Climate response. *Atmospheric Chemistry and Physics* 12:3349–62.

Li, L., W. Li, and Y. Deng. 2013. Summer rainfall variability over the southeastern United States and its intensification in the 21st century as assessed by CMIP5 models. *Journal of Geophysical Research: Atmospheres* 118:340–54.

Li, W., L. Li, R. Fu, Y. Deng, and H. Wang. 2011. Changes to the North Atlantic subtropical high and its role in the intensification of summer rainfall variability in the southeastern United States. *Journal of Climate* 24:1499–1506.

McCabe, G. J., J. L. Betancourt, and S. Feng. 2015. Variability in the start, end, and length of frost-free periods across the conterminous United States during the past century. *International Journal of Climatology* 35 (15): 4673–80.

Monahan, W. B., and N. A. Fisichelli. 2014. Climate exposure of US national parks in a new era of change. *PLOS ONE* 9: e101302.

Mora, C., A. G. Frazier, R. J. Longman, R. S. Dacks, M. M. Walton, E. J. Tong, J. J. Sanchez, L. R. Kaiser, Y. O. Stender, J. M. Anderson, C. M. Ambrosino, I. Fernandez-Silva, L. M. Giuseffi, and T. W. Giambelluca. 2013. The projected timing of climate departure from recent variability. *Nature* 502: 183–87.

Moss, R. H., J. A. Edmonds, K. A. Hibbard, M. R. Manning, S. K. Rose, D. P. Van Vuren, et al. 2010. The next generation of scenarios for climate change research and assessment. *Nature* 463:747–56.

NOAA (National Oceanic and Atmospheric Administration). 2015. National Centers for Environmental Information. http://www.ncei.noaa.gov.

Pan, Z., X. Liu, S. Kumar, Z. Gao, and J. Kinter. 2013. Intermodel variability and mechanism attribution of central and southeastern U.S. anomalous cooling in the twentieth century as simulated by CMIP5 models. *Journal of Climate* 26:6215–37.

Portmann, R. W., S. Solomon, and G. C. Hegerl. 2009. Spatial and seasonal patterns in climate change, temperatures, and precipitation across the United States. *Proceedings of the National Academy of Sciences of the United States of America* 106:7324–29.

Robinson, W. A., R. Reudy, and J. E. Hansen. 2002. General circulation model simulations of recent cooling in the east-central United States. *Journal of Geophysical Research: Atmospheres* 107.

Stein, B. A., P. Glick, N. Edelson, and A. Staudt. 2014. *Climate-Smart Conservation: Putting Adaptation Principles into Practice.* Washington, DC: National Wildlife Federation.

Thrasher, B., J. Xiong, W. Wang, F. Melton, A. Michaelis, and R. Nemani. 2013. Downscaled climate projections suitable for resource management. *Eos, Transactions American Geophysical Union* 94:321.

Williams, A., C. Allen, A. Macalady, D. Griffin, C. Woodhouse, D. Meko, T. Swetnam, S. Rauscher, R. Seager, H. Grissino-Mayer, et al. 2013. Temperature as a potent driver of regional forest drought stress and tree mortality. *Nature Climate Change* 3:292–97.

Willis, K. J., and G. M. MacDonald. 2011. Long-term ecological records and their relevance to climate change predictions for a warmer world. *Annual Review of Ecology, Evolution, and Systematics* 42:267–87.

Yu, S., K. Alapaty, R. Mathur, J. Pleim, Y. Zhang, C. Nolte, B. Eder, K. Foley, and T. Nagashima. 2014. Attribution of the United States "warming hole": Aerosol indirect effect and precipitable water vapor. *Scientific Reports* 4:6929.

Chapter 6

Assessing Vulnerability to Land Use and Climate Change at Landscape Scales Using Landforms and Physiographic Diversity as Coarse-Filter Targets

David M. Theobald, William B. Monahan,
Dylan Harrison-Atlas, Andrew J. Hansen,
Patrick Jantz, John E. Gross, and S. Thomas Olliff

In this chapter, we examine how climate change will likely affect areas of the Great Northern Landscape Conservation Cooperative (Great Northern LCC), but rather than using a fine-filter approach that focuses on a particular species, as has been done in many of the other chapters (e.g., chaps. 9, 10, and 12), we have applied a coarse-filter approach with which we consider our conservation targets to be broader levels of biodiversity. A coarse-filter approach focuses not on an individual species but, rather, on the community that supports a species (Noss 1987) or even on the physical environments as "arenas" of biological activity (Hunter, Jacobson, and Webb 1988). More recently, coarse-filter conservation has been interpreted in a climate change context, in which coarse-filter strategies seek to conserve sites that are minimally affected by climate change (Tingley, Darling, and Wilcove 2014).

Identifying conservation targets associated with various levels of ecological organization is also consistent with the Great Northern LCC landscape integrity vision (Chambers et al. 2013; Finn et al. 2015), which employs a hierarchical organization of targets that builds on the fine-filter/coarse-filter approach (Hunter, Jacobson, and Webb 1988). The coarse-filter systems in the Great Northern LCC are called ecosystems and habitats and include six aquatic targets (riparian corridors, riverine, wetlands, alpine lakes, uplands, and pothole lakes) and five terrestrial types (alpine, subalpine, woodland, dry forests, and sage shrub/grasslands). Roughly sixteen species are identified as fine-filter conservation targets; these include, for example, cutthroat trout (*Oncorhynchus clarkii*), whitebark pine (*Pinus albicaulis*), and grizzly bear (*Ursus arctos horribilis*).

While management on federal lands has traditionally focused on individual species, agencies are challenged to conduct vulnerability assessments for all fine-filter targets, either because they lack the necessary time or resources or because the detailed data required are not available. Related to this, we recognize that agency planning has been slow to embrace coarse-filter landscape units, such as "landscape facets" (Beier and Brost 2010). However, managers increasingly are encouraged to "scale up" and manage within a landscape context (e.g., National Park Service 2011; Clement et al. 2014). There are important opportunities to inform resource management plans with coarse-filter targets that provide a better, seamless coverage for ecologically defined planning areas that we want to scale up to (e.g., National Park Service natural resource condition assessments, US Fish and Wildlife Service landscape conservation designs, US Forest Service watershed vulnerability assessments, Bureau of Land Management land management plans). Our intent is not to argue that coarse-filter landscape units are surrogates for biodiversity per se; rather, we advance them as a way to understand general ecological settings that are relevant to management of species and ecological processes.

More specifically, in this chapter we focus on the vulnerability of species that may be generally influenced by future climate change in the context of current land use and existing landforms and soils. We distinguish two ways that such characteristics can be interpreted as relevant to biodiversity, and we discuss their application to a few example species to illustrate our work. First, an individual landform type might itself be the conservation target under consideration, because there is a close association between the fine-filter conservation target and a landform type. One such example might be the "riparian corridors" identified in the Great Northern LCC strategic vision (Chambers et al. 2013; Finn et al. 2015), which is represented and mapped by the valley bottom landform type. Another example could be whitebark pine, which associates with upper slope and ridge landform types (Tomback, Arno, and Keane 2001). Second, the diversity of landforms and soil types (together called "physiography" or "land facets") can be considered to be the conservation target, such as identifying areas of high geodiversity that often are associated with high levels of biodiversity. In addition, certain generalist species, such as grizzly bear, can utilize a wide variety of habitats that are shaped by different physiographic features (Noss et al. 1996).

We assessed vulnerability as part of the Climate-Smart Conservation framework for these two types of coarse-filter conservation targets: feature based (e.g., riparian corridors associated with valley bottoms) and diversity based (e.g., grizzly bears associated with high physiographic diversity). To do this we quantified three *exposure* or *potential impact* variables—temperature change, climate change velocity, and biome velocity—and three *adaptive capacity* variables—the degree of human modification, landscape permeability, and physiographic diversity. Strictly speaking, potential impact is a function of exposure and sensitivity, but we did not explicitly estimate sensitivity for our feature- and diversity-based conservation targets. We calculated vulnerability as a function of potential impact and adaptive capacity: high vulnerability consists of high impact and low adaptive capacity, whereas low vulnerability consists of low impact and high adaptive capacity.

We believe this coarse-filter approach provides a practical way to assess vulnerability based on important ecological settings, which complements resource managers' knowledge and facilitates field-level decision making. In addition, managers recognize that we do not have the capacity, resources, or—in many instances—scientific knowledge required to formally assess the vulnerability of all valued fine-filter targets at landscape scales, so coarse-filter vulnerability assessments provide a crucial and practical first step toward identifying and "scaling up" management options. As such, this chapter addresses step 2 of the Climate-Smart Conservation framework: vulnerability assessment (chap. 2; Stein et al. 2014).

Methods

To examine how changes in climate will likely affect the vulnerability of our coarse-filter conservation targets, and how current patterns of land use may modify the capacity to adapt to those changes, we briefly review a classification of landforms and a calculation of physiographic diversity and describe three exposure and three adaptive capacity variables, on which we base our working estimates of landscape vulnerability to future climate change. We provide a summary of these variables both for the entire region and for more localized and stakeholder-relevant management geographies that we call eco-management zones. These zones include the following:

Columbia Plateau in Washington
Blue Mountains in Oregon
Crown of the Continent

Canadian Rocky Mountains
Greater Yellowstone Ecosystem
Idaho-Montana High Divide
Intermountain West Joint Venture Wetland Landscape
North Cascades and Pacific ranges
Okanagan Valley
Wyoming Landscape Conservation Initiative

Because these eco-management zones were defined by land managers and stakeholders as important geographies, some of which spanned the international boundary into Canada, we elected to calculate vulnerability across the full United States and Canada portions of the Great Northern LCC. We also included protected area centered ecosystem boundaries around national park units (Hansen et al. 2011).

We generated three physiographic maps: (1) detailed landform classes, (2) physiography that combines landform classes with lithology (i.e., basic edaphic properties about geologic parent material), and (3) multi-scale physiographic diversity. We used a classification of landforms developed specifically for climate adaptation applications, derived from hillslope position and dominant physical processes (see details in Theobald et al. 2015). Briefly, we used a multiscale topographic position index that provides a measure of relative topographic relief (Guisan, Weiss, and Weiss 1999; Dickson and Beier 2002) to delineate four hillslope positions: ridges/peaks (summits), upper slopes (shoulders), lower slopes (foot slopes), and valley bottoms (toe slopes). We then further differentiated each of these hillslopes as a function of solar orientation to reflect how ecological processes are influenced by solar insolation or shading. This was measured using incident radiation and heat load (McCune and Keon 2002) that combines slope, aspect, and latitude to predict the effects of potential direct insolation. We also identified features at the extremes of hillslope gradients, including very steep areas (i.e., "cliffs" > 50 degrees) and flat areas (i.e., areas < 2 degrees).

Exposure and Impact Variables

We calculated three exposure variables that are treated as potential impact in the coarse-filter assessment of vulnerability: temperature change (E_t), climate velocity (E_v), and climate velocity for habitat types (E_h; i.e., Rehfeldt's biomes). To estimate temperature change from the present to roughly

2070, we used current and future gridded estimates of annual mean temperature from WorldClim (Hijmans et al. 2005). Data were obtained at 30-arc-second spatial resolution and reprojected using an equal-area projection to 800 meters. Estimates of future temperature were based on the ensemble average of seventeen individual climate models available through the Coupled Model Intercomparison Project Phase 5, downscaled and calibrated (bias corrected) using WorldClim as the current (1950–2000) baseline. We considered a 2061–2080 future (referenced as 2070) and a "business as usual" representative concentration pathway (RCP) of 8.5 Watts per square meter (RCP 8.5).

To estimate climate velocity, we calculated the mean rate of change in temperature over time (future – baseline; degrees C/year) divided by the rate of temperature change over space (degrees C/kilometer) following Loarie et al. (2009). The resulting units for velocity are kilometers per year, which may be interpreted as the average distance per year that individual plants and animals must move in order to track or keep pace with the baseline temperatures experienced at a given location. To estimate climate velocity for habitat types, we calculated the movement velocity that a habitat type ("biomes" in Rehfeldt et al. 2012) would need to migrate to maintain constant climate conditions projected for 2060, following the method of forward distance velocity in Hamann et al. (2015). We employed forecasts from Rehfeldt et al. (2012) that portrayed consensus among three climate models (CGCM3, HadCM3, and CM2.1) for two emission scenarios (A2 and B1), which provides conservative results compared to climate velocity calculated for RCP 8.5.

Adaptive Capacity

We estimated adaptive capacity based on four variables that characterize the ability of plants and animals to adapt to climate change exposure and that relate explicitly to the terrestrial (A_g) and watershed-based (A_w) ecological processes highlighted in the Great Northern LCC landscape integrity vision (Chambers et al. 2013; Finn et al. 2015). The first two variables reflect the degree to which a location is surrounded by low levels of human modification (H), based on the assumption that if there is lower land use intensity (i.e., fewer roads, urban areas, agricultural areas, and so forth), then landscape permeability is higher, allowing movement and dispersal processes to occur more naturally (e.g., Watson, Iwamura, and Butt 2013). These variables are most broadly applicable to the "feature" or specific landform type per se. Here, adaptive capacity is the complement of human modification: $A = 1-H$. A second set of variables reflects the notion that areas with high physiographic diversity and low human modification (A_{gp}, A_{wp}) are important in facilitating species' responses to climate change (Dobrowski 2010), and so these variables are a useful characterization of natural landscape diversity that many generalists might be associated with (e.g., grizzly bear).

To estimate H, we represented the four main stressors and methods described in Theobald (2013) but adjusted the input data sets to account for the transboundary nature of the Great Northern LCC study area: urban and developed areas, transportation, agricultural areas, and resource extraction. We mapped urban and agricultural development using data from the North American Land Change Monitoring System (2013), assigning urban classes = 0.9, cropland agricultural = 0.5, and otherwise = 0.0. Housing density data were obtained from block-level data from the US Census of Housing 2010 and from the 2011 Census (Stats Canada 2014), and resource extraction areas were mapped using VIIRS nighttime lights for 2013 (1,500 feet, or 450 meters; Elvidge et al. 2013). The four individual stressors were combined using the "increasive function" (Theobald 2013) and resulted in values of H ranging from 0.0 (natural) to 1.0 (completely human modified).

We calculated a watershed-based estimate of adaptive capacity (A_w) by calculating the mean value of $1-H$ within twelve-, ten-, eight-, and six-digit hydrologic unit codes and then taking the mean value across scales (following Theobald 2013). Similarly, we calculated a terrestrial-based estimate of adaptive capacity, A_t, by calculating the mean value of $1-H$ within a series of moving windows scaled to be similar in size as twelve-, ten-, eight-, and six-digit hydrologic unit codes.

We estimated physiographic diversity by calculating the diversity of both landforms and parent material (lithology; Soller et al. 2009) using Shannon's equitability (E_H; 0 to 1), which is calculated by normalizing the Shannon-Weaver diversity index (S):

$$E_H = \frac{H}{H_{max}} = \frac{S}{Ln_S}$$

$$S = -\sum_{i=1}^{s} (p_i \, Ln \, p_i)$$

where p_i is the proportion of observations (cells) of type i in a given neighborhood and S is the number of geomorphological types. Therefore, we calculated two additional variables reflecting both physiographic diversity and human modification, where $A_p = E_H$; $A_{pt} = A_p \times A_t$; and $A_{pw} = A_p \times A_w$.

Vulnerability

Vulnerability is described as being a function of impact and adaptive capacity (Glick, Stein, and Edelson 2011), but guidance is limited on how to quantify this relationship (e.g., Klausmeyer et al. 2011). We calculated vulnerability V, which ranges from 0 (low) to 1 (high) as the product of exposure/impact variables E and the complement of adaptive capacity A, for all combinations: $V_{tg} = E_t \times (1-A_g)$; $V_{tw} = E_t \times (1-A_w)$; $V_{vg} = E_v \times (1-A_g)$; $V_{vw} = E_v \times (1-A_w)$; $V_{hg} = E_h \times (1-A_g)$; $V_{hw} = E_h \times (1-A_w)$.

We min-max normalized E_t (min = 2.45, max = 9.7), E_v (min = 0.008, max = 10.21), and E_h (min = 0, max = 65) using minimum and maximum values from North America (lacking any specific domain to refer to) to scale the exposure/impact values to a 0 → 1 variable so that they could be combined with adaptive capacity to calculate vulnerability. We also min-max normalized the adaptive capacity diversity values to reflect our example application to grizzly bear and so found minimum and maximum values within the historical grizzly bear range (Laliberte and Ripple 2004). Note that A_g and A_w already ranged from 0 → 1.

Results of Habitat Types and Landforms

The Great Northern LCC covers an expanse of 1.2 million square kilometers stretching across two countries and nine states/territories and is dominated by forest (46 percent) and shrublands (34.3 percent), with patches of grassland (10 percent), alpine (5.3 percent), and wetland/freshwater-dominated/other (4.5 percent) habitat types (table 6-1). Upper-slope and lower-slope landforms dominate the Great Northern LCC (42.7 percent and 41.6 percent, respectively), with 4.1 percent occurring in the peak/ridge classes and 11.6 percent in the valley bottom classes. Roughly three quarters of the landforms are in the neutral heat load classes, with 14.4 percent in warm areas and 10.8 percent in cool areas.

We also examined the relationship of fine-grained (100 feet, or 30 meters) landforms to the exposure variables and to the base data used for adaptive capacity $(1 - H)$. Although we expected to find little pattern in

TABLE 6-1. Summary of how landform types intersect with habitat types across the Great Northern Landscape Conservation Cooperative.

Landform Class	Habitat Type						Total	Percentage
	Alpine tundra	Forest	Shrub-land	Grassland	Water	Other		
Peak/ridge warm	551	1,657	2,324	282	0	176	4,988	0.4%
Peak/ridge	2,040	15,096	10,276	1,929	1.2	903	30,246	2.5%
Peak/ridge cool	894	1,925	719	130	0	365	4,033	0.3%
Mountain/ divide	3,475	2,468	1,428	208	0.1	1,001	8,580	0.7%
Cliff	1,249	412	154	29	0	233	2,077	0.2%
Upper slope warm	8,220	25,925	30,026	3,431	2.3	2,151	69,756	5.70%
Upper slope neutral	12,338	165,683	118,393	36,307	69.1	11,491	344,281	28.4%
Upper slope cool	8,130	34,177	6,141	877	1.4	2,163	51,490	4.2%
Upper slope flat	42	9,232	28,891	12,078	188.2	2,336	52,767	4.3%
Lower slope warm	5,515	24,678	25,683	2,810	3.9	1,653	60,344	5.0%
Lower slope neutral	10,912	163,196	105,562	33,658	107.7	12,126	325,562	26.8%
Lower slope cool	5,750	32,735	4,603	606	3.4	1,641	45,338	3.7%
Lower slope flat	42	11,160	34,328	16,266	328.7	11,762	73,888	6.1%
Valley	4,327	56,443	38,196	10,639	95.3	4,478	114,179	9.4%
Valley (narrow)	567	13,556	9,443	2,164	5.3	590	26,326	2.2%
Total (km²)	64,052	558,346	416,167	121,414	806.8	53,068	1,213,854	100%
Percentage	5.3%	46.0%	34.3%	10.0%	0.1%	4.4%		

the summaries because of the disparity between scales, we found reasonable trends relating the landform classes to the exposure variables (table 6-2)—that is, we found that warm landforms, regardless of hillslope position, will warm by slightly higher amounts than will neutral positions. Peak/ridge landforms will likely warm the most, and narrow valley bottoms the least. Patterns of climate velocity also showed reasonable trends, with flat landforms facing order of magnitude greater mean velocities. Finally, although the overall degree of human modification is fairly low throughout the Great Northern LCC, not surprisingly we found that flat landforms have a higher proportion of human modification compared to peaks and ridges. Expressed in terms of adaptive capacity (A_g), flat landforms had lower adaptive capacity (upper flat = 0.91, lower flat = 0.92) compared to peaks and ridges (above 0.97). The values of adaptive capacity that include physiographic diversity (A_{pg}, A_{pw}) show similar patterns where flat and neutral slopes tend to have lower physiographic adaptive capacity, but note that they are not directly comparable (in an absolute sense) to A_g and A_w because their $0 \rightarrow 1$ scaling is related to the entropy calculation as well as being normalized over a smaller domain (for historical grizzly bear distribution). Overall vulnerability scores are quite low, mostly due to the relatively low level of human modification in the Great Northern LCC, but trends of specific landform classes are consistent, showing lower-slope and warmer classes being more vulnerable than upper-slope and cooler classes.

Results of Impact, Adaptive Capacity, and Vulnerability

Patterns of potential future climate exposure in the Great Northern LCC are complex. Annual mean temperature is consistently warmer, ranging from 3.9 to 5.6 degrees C in 2070 relative to recent past conditions (1950–2000), with especially large changes being forecast for the southern portions of the LCC, particularly the lower elevations in Idaho and southwestern Wyoming (fig. 6-1a). Patterns of climate velocity, also calculated using annual mean temperature in 2070, suggest species will need to move longer distances in order to keep pace with (i.e., "track") changes in temperature (fig. 6-1b). Climate velocities are especially high in lower-elevation and relatively flat areas, such as southwestern Wyoming, the Columbia Plateau in Washington, and the Rocky Mountain Front in Alberta. Meanwhile, nearly one quarter of the Great Northern LCC will experience a shift in habitat type, and the mean climate velocity for habitats in the Great Northern LCC is estimated to be 0.75 mile per year (1.20 kilometers per

TABLE 6-2. A summary of three exposure/impact (t = temperature change, v = climate velocity, h = habitat climate velocity), four adaptive capacity variables (g = terrestrial-based, w = watershed-based), and two measures of vulnerability (physiographic diversity with terrestrial-based adaptive capacity = pg, with watershed-based adaptive capacity = pw) against landform types for the Great Northern Landscape Conservation Cooperative.

Landform Class	E_t	E_v	E_h	A_g	A_w	A_{pg}	A_{pw}	V_{pg}	V_{pw}
Peak/ridge warm	0.32	0.02	0.01	0.97	0.96	0.47	0.47	0.06	0.06
Peak/ridge	0.31	0.02	0.02	0.97	0.96	0.45	0.44	0.06	0.06
Peak/ridge cool	0.31	0.02	0.02	0.99	0.98	0.48	0.48	0.06	0.06
Mountain/divide	0.31	0.01	0.02	0.99	0.98	0.50	0.50	0.05	0.05
Cliff	0.31	0.01	0.02	0.99	0.98	0.50	0.50	0.05	0.05
Upper slope warm	0.32	0.01	0.01	0.97	0.97	0.47	0.46	0.06	0.06
Upper slope neutral	0.31	0.03	0.02	0.96	0.96	0.41	0.41	0.06	0.06
Upper slope cool	0.31	0.01	0.02	0.98	0.98	0.48	0.48	0.05	0.05
Upper slope flat	0.31	0.13	0.02	0.91	0.93	0.37	0.38	0.09	0.09
Lower slope warm	0.32	0.01	0.02	0.97	0.97	0.47	0.47	0.06	0.06
Lower slope neutral	0.31	0.03	0.02	0.96	0.96	0.42	0.42	0.06	0.06
Lower slope cool	0.31	0.01	0.03	0.98	0.98	0.48	0.48	0.05	0.05
Lower slope flat	0.32	0.12	0.02	0.92	0.95	0.39	0.4	0.09	0.09
Valley	0.31	0.04	0.02	0.96	0.96	0.43	0.43	0.07	0.07
Valley (narrow)	0.31	0.04	0.02	0.95	0.96	0.42	0.43	0.07	0.07

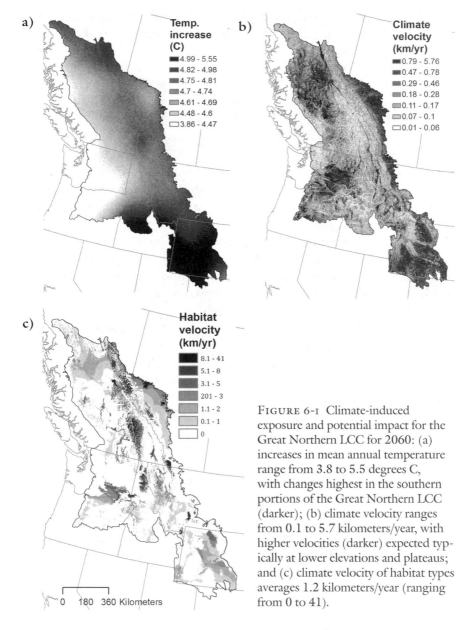

FIGURE 6-1 Climate-induced exposure and potential impact for the Great Northern LCC for 2060: (a) increases in mean annual temperature range from 3.8 to 5.5 degrees C, with changes highest in the southern portions of the Great Northern LCC (darker); (b) climate velocity ranges from 0.1 to 5.7 kilometers/year, with higher velocities (darker) expected typically at lower elevations and plateaus; and (c) climate velocity of habitat types averages 1.2 kilometers/year (ranging from 0 to 41).

year; min = 0, SD = 3.57, max = 41). The pattern of habitat velocity is more abrupt and varied, with rapid velocities required in a variety of current habitat interfaces throughout the region but including high-elevation areas in the Crown of the Continent area (fig. 6-1c).

Overall, the adaptive capacity in the Great Northern LCC is quite high (fig. 6-2), with a mean value of 0.96 and 0.97 for the terrestrial (A_t; fig. 6-2a) and watershed-based index (A_w; fig. 6-2b). This region boasts great physiographic diversity as well (figs. 6-2c and 6-2d), with 204 different combinations of landforms/lithology (out of 270 possible), which is reasonably well distributed throughout the Great Northern LCC, with an average equitability (E_H) value of 0.55. More localized places and watersheds with lower A do occur, however, predominantly in eastern Washington and southern parts of British Columbia and Alberta.

Focusing on the eco-management zones within the Great Northern LCC, a variety of opportunities exist for helping managers adapt to future changes in climate. Relating the three estimates of exposure with the three for adaptive capacity reveals varied and complex patterns (fig. 6-3). Overall, exposure of zones depends greatly on the variable (table 6-3). For example, the Greater Yellowstone Ecosystem will likely face high exposure to increasing temperatures, moderate to low exposure to climate velocity, and high exposure to habitat shifts, but it has high adaptive capacity. The Columbia Plateau in Washington will likely face relatively low exposure to temperature (but still increases > 4.25 degrees C), moderate exposure to climate velocity, and low exposure to habitat shifts. However, it has the lowest adaptive capacity measures due to higher levels of human land use but has moderately high adaptive capacity from high physiographic diversity. Dinosaur National Monument faces among the highest exposure to temperature increases and climate velocity.

Discussion

Overall, we found that exposure of the Great Northern LCC to climate change is high for all three variables we examined—with up to 5.5 degrees C warming, climate velocity rates up to 5 kilometers per year, and nearly 25 percent of habitats shifting to a different type. Not surprisingly, alpine areas will likely be strongly affected—more than 80 percent will face climate conditions conducive to conversion to forested vegetation. But other habitat types will likely be stressed at their current ecotones as well—roughly 17 percent of forested lands will experience a transition toward shrubland climate niche, and 16 percent of shrublands will become a grassland climate niche.

By eco-management zone (table 6-3), we found that Dinosaur National Park in northwestern Colorado and the Wyoming Landscape Conser-

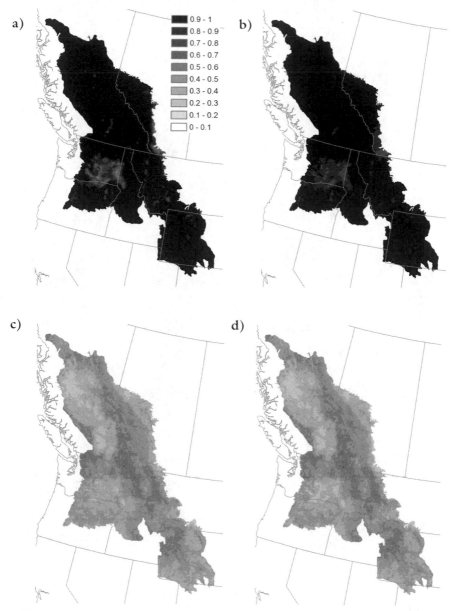

FIGURE 6-2 Adaptive capacity values in the Great Northern LCC reflecting types of movement, with dark values depicting higher adaptive capacity: (a) equal direction reflecting terrestrial fauna; (b) preferential within the hierarchy of watersheds; (c) in response to physiographic diversity composed of landforms and parent material (lithology) and terrestrial response; and (d) physiographic diversity and watershed response.

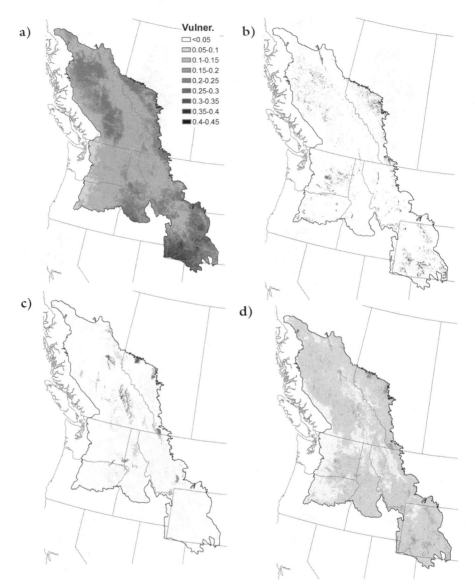

FIGURE 6-3 Overall vulnerability values for the Great Northern LCC assuming physiographic diversity and terrestrial adaptive capacity (note that watershed adaptive capacity patterns are generally similar), with exposure/impact variables normalized to the minimum and maximum value obtained in historic grizzly bear range for (a) temperature exposure; (b) impact from climate velocity; (c) climate velocity of habitat types; and (d) an integrated index of vulnerability by averaging (a) through (c). The legend in (a) also applies to (b), (c), and (d).

TABLE 6-3. For the stakeholder-defined eco-management zones in the Great Northern LCC, a summary of the mean exposure/impact value for each zone.

Eco-management Zone (*PACE)	Abbreviation	Temperature Exposure		Climate Velocity Impact		Habitat Velocity Impact	
		Terrestrial	Watershed	Terrestrial	Watershed	Terrestrial	Watershed
Columbia Basin in Washington	CBW	0.1373	0.1340	0.0384	0.0371	0.0112	0.0108
Bighorn Canyon*	BICA	0.1940	0.1927	0.0333	0.0329	0.0097	0.0095
Blue Mountains	BM	0.1433	0.1446	0.0121	0.0122	0.0044	0.0045
Crown of the Continent	COC	0.1547	0.1540	0.0241	0.0234	0.0044	0.0045
Crater Lake*	CRLA	0.1036	0.1056	0.0104	0.0106	0.0115	0.0119
Canadian Rocky Mountains	CRM	0.1494	0.1505	0.0086	0.0087	0	0
Craters of the Moon*	CRMO	0.1805	0.1799	0.0105	0.0105	0.0200	0.0202
Dinosaur National Monument*	DINO	0.2301	0.2302	0.0344	0.0344	0	0
Glacier*	GLAC	0.1394	0.1407	0.0121	0.0121	0.0008	0.0008
Great Northern LCC	GNLCC	0.1609	0.1608	0.0244	0.0242	0.0093	0.0097
Greater Yellowstone Ecosystem	GYE	0.1672	0.1681	0.0168	0.0168	0.0105	0.0105
Intermountain High Divide	IMHD	0.1590	0.1598	0.0121	0.0121	0.0065	0.0065
Intermountain West Joint Venture	IMWJV	0.1603	0.1576	0.0607	0.0593	0.0057	0.0057
Lake Roosevelt*	LARO	0.1357	0.1344	0.0252	0.0246	0.0101	0.0100
Mount Rainier*	MORA	0.1059	0.1091	0.0055	0.0057	0.0005	0.0005
North Cascades Pacific Range	NCPR	0.1021	0.0986	0.0039	0.0037	0	0
North Cascades	NOCA	0.1069	0.1064	0.0038	0.0037	0.0002	0.0002
Okanagan	OKA	0.1518	0.1521	0.0117	0.0117	0.0008	0.0008
Wyoming Landscape Conservation Initiative	WLCI	0.2015	0.2018	0.0428	0.0428	0.0066	0.0065
Yellowstone*	YELL	0.1518	0.1541	0.0127	0.0129	0.0060	0.0060
						0.0049	0.005

vation Initiative in southwestern Wyoming had the highest temperature exposure values (> 0.2), while most zones in northern Washington had the lowest (< 0.11), including Crater Lake, Mount Rainier, North Cascades Pacific Range, and North Cascades National Park. The impact of climate velocity is very high in the Intermountain West Joint Venture and the Wyoming Landscape Conservation Initiative (> 0.04), and high in the Arid Lands Initiative (Columbia Plateau region), Dinosaur and Bighorn Canyon park units (> 0.03). Interestingly, the highest habitat velocity was found in the Canadian Rocky Mountains region.

The adaptive capacity of the Great Northern LCC is high, as much of its landscapes are relatively intact and there is great physiographic diversity that will likely facilitate the adaptation of natural features to these changes. However, the magnitude of potential climate exposure in the Great Northern LCC remains fairly profound—with increases in temperature predicted to be 3.9 to 5.6 degrees C warmer. Recent work (Oyler et al. 2014) has identified a subtle bias in estimates of temperature at high elevation of up to 0.5 degrees C for elevations above about 6,600 feet (about 2,000 meters). Consequently, exposure for mountain/ridge landforms might be particularly sensitive to this bias and therefore deserves caution when interpreting, but overall exposure patterns should hold.

Also, the pattern of potential climate exposure in the Great Northern LCC is quite complex. For example, climate velocity measures the distance per year that a species would need to move to "track" changes in climate so that the species retains the climatic conditions experienced in a given pixel. Some vagile species (e.g., migratory birds) would likely be able to adapt readily, but the plant species that make up their habitat might not—and certainly many species will likely be unable to disperse at these rates (Chen et al. 2011). Also, nearly one quarter of the Great Northern LCC will likely face significant enough climatic changes to result in a complete shift of a general "biome" or habitat type.

In this chapter, we have emphasized a coarse-filter approach to conservation using landforms and physiographic diversity as relevant targets in the face of climate change and described potential effects on all landscapes as an initial data exploration and illustration. Our findings also would provide the basis to refine analysis to examine the effects on a narrower conservation target (e.g., a specific species). This would require knowledge from field biologists and land managers to be combined with landscape-level information to further isolate important ecological processes that will drive likely future patterns. For example, honing in on specific landform features could provide the landscape context for some specific habitat features, such as riparian corridors associated with valley bottoms. Interestingly, the valley bottom (narrow) landform type was associated with all habitat types (table 6-1), but predominantly with forest, shrubland, and grassland. The locations where this landform coincides with general habitat types (defined by current vegetation cover) provide useful information about key plant communities that support a wealth of biodiversity. Another example would be to characterize the landscape context for a conservation target that is more diversity oriented, such as grizzly bears, which are associated with high physiographic diversity and low human modification.

We emphasize that although we provide summary descriptions for entire eco-management zones, there is clearly much variability that can be addressed by examining the patterns provided in the map figures. An important illustration of our work is to better "downscale" climate change information to make it more applicable to land management—that is, our exposure variables were derived from the finest available data sets based on mesoclimate scale (800 meters), yet they still remain decoupled from the finer-scale topographic variation seen in landforms and physiography. In and around the Greater Yellowstone Ecosystem, the spatial patterning of major habitats is evident in the distribution of landforms (figs. 6-4a and 6-4b). However, a standard measure of exposure—future change in annual mean temperature—shows little to no correspondence with either the biome distributions evident in the imagery, or the landforms. Nevertheless, we know from a combination of first principles and empirical work (Dobrowski 2010) that steep, north-facing slopes can be buffered against climatic changes—decoupled from large-scale synoptic conditions—and serve as microrefugia. This will allow managers who know their systems to conduct more specific, targeted analyses for a given species that could be aligned with a specific landform or landforms to provide management relevant information.

We also foresee other uses of the landforms-based analysis—for example, as a way to focus sampling design for a particular conservation target that makes use of known affinities for fine-grained associations related to physiography, such as additional data collection for whitebark pine species distribution modeling. Finally, a minor but important aspect of this work has been the development of transboundary data sets, recognizing that the Great Northern LCC occurs in Canada as well as United States. This required extra effort to locate, acquire, and synthesize various data sets (e.g., on human land use in particular). This need also applies to extending the ecosystem boundaries surrounding various protected areas (Hansen et al. 2011).

We also discovered a number of data impediments that should be addressed for future refinements, including the lack of a standardized map

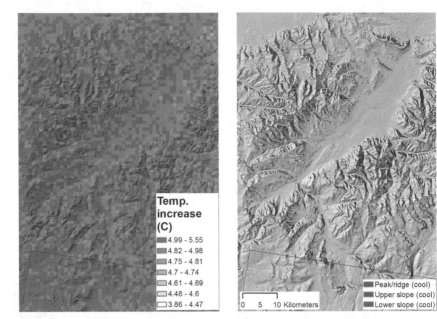

FIGURE 6-4   An illustration of "downscaling" relatively coarse climate data with fine-grained landform features within the Greater Yellowstone Ecosystem (following the Yellowstone River out of the park), focusing on all "cool" landform features: (a) mesoscale climate exposure data on temperature increase, and (b) landform features that occur in "cool" heat load index locations.

of "habitats," consistent high-resolution (30 meters) land cover data, and detailed maps on resource extraction activities and infrastructure (pipelines, power lines, and so forth). Another type of data gap that we are challenged to track and map is specific on-the-ground actions that managers and partners have taken that would affect their interpretation of climate adaptation (e.g., fire management activities, installation of wildlife crossing structures, translocation activities, or even decisions made to not engage in some management activity). We need to better link the data sets and types of analyses that we conduct in the vulnerability cycle step with those management steps that come later in the process, to complete the "cycle."

Conclusion

In this chapter, we applied a coarse-filter approach to examine the exposure, impact, and vulnerability of different geographies of interest to Great

Northern LCC managers and stakeholders. Again, we have emphasized that detailed maps of landforms and physiographic diversity can be useful to managers to understand the general ecological settings that are relevant to the management of species and ecological processes. We conclude with key summary messages:

- Careful classification of landforms is a useful way to identify, at a high resolution (about 10 to 30 meters), those landscape features that provide ecological context, especially for conservation targets with specific habitat needs.
- Application of a vulnerability assessment can be fine-tuned to specific conservation targets by characterizing exposure, impact, and adaptive capacity variables using relevant spatial domains (e.g., historical grizzly bear range).
- Specific areas identified by our vulnerability analysis to have relatively high impact/vulnerability include the Wyoming Landscape Conservation Initiative, Dinosaur National Monument, and Bighorn Canyon National Recreation Area.
- The magnitude of potential climate exposure in the Great Northern LCC remains fairly profound—with increases in temperature predicted to be 3.9 to 5.6 degrees C warmer—although the adaptive capacity of the Great Northern LCC is high.

References

Beier, P., and B. Brost. 2010. Use of land facets to plan for climate change: Conserving the arenas, not the actors. *Conservation Biology* 24:701–10.

Chambers, N., G. Tabor, Y. Converse, T. Olliff, S. Finn, R. Sojda, and S. Bischke. 2013. The Great Northern Landscape Conservation Cooperative Strategic Conservation Framework. Unpublished report.

Chen, C., J. K. Hill, R. Ohlemuller, D. B. Roy, and C. D. Thomas. 2011. Rapid range shifts of species associated with high levels of climate warming. *Science* 333:1024-26.

Clement, J. P., A. Belin, M. J. Bean, T. A. Boling, and J. R. Lyons. 2014. *A Strategy for Improving the Mitigation Policies and Practices of the Department of the Interior*. A report to the Secretary of the Interior from the Energy and Climate Change Task Force, Washington, DC.

Dickson, B. G., and P. Beier. 2002. Home-range and habitat selection by adult cougars in southern California. *Journal of Wildlife Management* 66:1235–45.

Dobrowski, S. Z. 2010. A climatic basis for microrefugia: The influence of terrain on climate. *Global Change Biology*. doi: 10.1111/j.13652486.2010.02263.x.

Elvidge, C. D., K. Baugh, M. Zhizhin, and F. C. Hsu. 2013. Why VIIRS data are superior to DMSP for mapping nighttime lights. *Proceedings of the Asia-Pacific Advanced Network* 35:62–69.

Finn, S., Y. Converse, T. Olliff, M. Heller, R. Sojda, E. Beever, S. Pierluissi, J. Watkins, N. Chambers, and S. Bischke. 2015. *Great Northern Landscape Conservation Cooperative Science Plan, 2015–2019.* Unpublished report.

Glick P., B. A. Stein, and N. A. Edelson, eds. 2011. *Scanning the Conservation Horizon: A Guide to Climate Change Vulnerability Assessment.* Washington, DC: National Wildlife Federation.

Guisan, A., S. B. Weiss, and A. D. Weiss. 1999. GLM versus CCA spatial modeling of plant species distribution. *Plant Ecology* 143:107–22.

Hamann, A., D. R. Roberts, Q. E. Barber, C. Carroll, and S. E. Nielsen. 2015. Velocity of climate change algorithms for guiding conservation and management. *Global Change Biology* 31:997–1004.

Hansen, A. J., C. R. Davis, N. Piekielek, J. Gross, D. M. Theobald, S. Goetz, F. Melton, and R. DeFries. 2011. Delineating the ecosystems containing protected areas for monitoring and management. *BioScience* 61 (5): 363–73.

Hijmans, R. J., S. E. Cameron, J. L. Parra, P. G. Jones, and A. Jarvis. 2005. Very high resolution interpolated climate surfaces for global land areas. *International Journal of Climatology* 25:1965–78.

Hunter, M. L., G. L. Jacobson, and T. Webb. 1988. Paleoecology and the coarse-filter approach to maintaining biological diversity. *Conservation Biology* 2:375–85.

Klausmeyer, K. R., M. R. Shaw, J. B. MacKenzie, and D. R. Cameron. 2011. Landscape-scale indicators of biodiversity's vulnerability to climate change. *Ecosphere* 2 (8): 88.

Laliberte, A. S., and W. J. Ripple. 2004. Range contractions of North American carnivores and ungulates. *BioScience* 54 (2): 123–38.

Loarie, S. R., P. B. Duffy, H. Hamilton, G. P. Asner, C. B. Field, and D. D. Ackerly. 2009. The velocity of climate change. *Nature* 462:1052–55.

McCune, B., and D. Keon. 2002. Equations for potential annual direct incident radiation and heat load. *Journal for Vegetation Science* 13 (4): 603–6.

National Park Service (NPS). 2011. *A Call to Action: Preparing for a Second Century of Stewardship and Engagement.* Report (August).

North American Land Change Monitoring System. 2013. 2010 land cover of North America at 250 meters. https://www.sciencebase.gov/catalog/item/54a19635e 4b0bb7b6f9a1a55.

Noss, R. F. 1987. From plant communities to landscapes in conservation inventories: A look at The Nature Conservancy (USA). *Biological Conservation* 41:11–37.

Noss, R. F., H. B. Quigley, M. G. Hornocker, T. Merrill, and P. C. Paquet. 1996. Conservation biology and carnivore conservation in the Rocky Mountains. *Conservation Biology* 10(4): 949–63.

Oyler, J. W., A. Ballantyne, K. Jencso, M. Sweet, and S. W. Running. 2014. Creating a topoclimatic daily air temperature dataset for the conterminous United States using homogenized station data and remotely sensed land skin temperature. *International Journal of Climatology*. doi: 10.1002/joc.4127.

Rehfeldt, G. E., N. L. Crookston, C. Saenz-Romero, and E. M. Campbell. 2012. North American vegetation model for land-use planning in a changing climate: A solution to large classification problems. *Ecological Applications* 221 (1): 119–41.

Soller, D. R., M. C. Reheis, C. P. Garrity, and D. R. Van Sistine. 2009. Map database for surficial materials in the conterminous United States. *US Geological Survey Data Series* 425.

Stats Canada. 2014. Census population and dwelling counts, for Canada, provinces, and territories, 2011 and 2006 censuses. http://www12.statcan.gc.ca/census -recensement/index-eng.cfm.

Stein, B. A., P. Glick, N. Edelson, and A. Staudt, eds. 2014. *Climate-Smart Conservation: Putting Adaptation Principles into Practice*. Washington, DC: National Wildlife Federation.

Theobald, D. M. 2010. Estimating changes in natural landscapes from 1992 to 2030 for the conterminous United States. *Landscape Ecology* 25 (7): 999–1011.

Theobald, D. M. 2013. A general model to quantify ecological integrity for landscape assessments and US application. *Landscape Ecology*. doi: 10.1007/ s10980-013-9941-6.

Theobald, D. M., D. Harrison-Atlas, W. B. Monahan, and C. M. Albano. 2015. Ecologically-relevant maps of landforms and physiographic diversity for climate adaptation planning. *PLOS ONE* 10 (12): e0143619. doi: 10.1371/ journal.pone.0143619.

Tingley, M. W., E. S. Darling, and D. S. Wilcove. 2014. Fine- and coarse-filter conservation strategies in a time of climate change. *Annals of the New York Academy of Sciences* 1322:92–109.

Tomback, D. F., S. F. Arno, and R. E. Keane. 2001. *Whitebark Pine Communities: Ecology and Restoration*. Washington, DC: Island Press.

Watson, J. E. M., T. Iwamura, and N. Butt. 2013. Mapping vulnerability and conservation adaptation strategies under climate change. *Nature Climate Change*. doi: 10.1038/NCLIMATE2007.

PART 3

Ecological Consequences and Vulnerabilities

Climate and human land use have ebbed and flowed for millennia and have been agents of natural selection to which many organisms have adapted. The key question under the current rapid rate of change is what degree of *exposure* to these changes elicits strong negative responses in ecosystem processes and biodiversity. This section of the book examines past and potential future ecological responses to these exposures to identify the organisms and ecosystems that are most vulnerable to global change and that are, therefore, high priorities for management action.

As introduced in part 1, an approach for assessing vulnerability has emerged from the Intergovernmental Panel on Climate Change. With climate and land use as elements of exposure, *sensitivity* denotes the tolerance of an organism or ecosystem process to the change in exposure. Exposure and sensitivity are combined to gauge *potential impact*. The impact of a change in climate, however, may be mediated by an organism or system's *adaptive capacity*, which refers to coping mechanisms that an organism may be able to employ, or the ability of natural systems to persist in the face of change. *Vulnerability*, then, is a function of the potential impact of the climate or land use change on an organism or process and its adaptive capacity. The methods of assessing each of the elements of vulnerability differ in level of refinement, with adaptive capacity being the most embryonic. Each of the chapters in this section quantifies exposure, sensitivity, and potential impact. Only chapter 11 (and chapter 6 in part 2) includes consideration of adaptive capacity using quantitative methods.

Because ecological systems are complex and include many organisms and processes, a major challenge to assessing vulnerability is identifying which organisms and processes (i.e., conservation targets) can be realistically analyzed. We chose to analyze a range of ecological response variables that span from "coarse" to "fine" filter. Individual species are often of highest interest to resource managers and the public. These species-specific analyses are called fine-filter conservation approaches. However, limited knowledge or practical limitations typically reduces the number of species that can be considered to a dozen or less. Consequently, analyses are often done on coarser levels of biodiversity, such as communities, biomes, or even landforms (e.g., valley bottoms of mountain plateaus). These coarse-filter elements of biodiversity are sometimes of interest in their own right (e.g., valley bottoms with good soils and high agriculture potential) or because they support many species that are of high interest.

This part considers ecological processes, such as runoff and primary productivity (chap. 7); vegetation at the species, community, and biome levels (chaps. 8–11), and coldwater fish species (chap. 12). Although our core team of scientists focused on terrestrial ecosystems, our management partners were also quite interested in considering aquatic systems. Consequently, we invited fisheries experts from both the Rocky Mountains and the Appalachians to synthesize their previous work on climate and land use vulnerability of coldwater fish. The chapter on fish (chap. 12) is a good transition into the final part of the book on management because the chapter concludes with management case studies of fish in Yellowstone and Great Smoky Mountains national parks.

An important finding from this part of the book is that species and ecosystems vary dramatically in their response to climate and land use changes. Although some species are highly vulnerable, others may benefit under the projected changes. This result highlights again the need for rigorous assessment of vulnerability at multiple spatial scales, including the finer scales within which management is typically executed.

Chapter 7

Potential Impacts of Climate and Land Use Change on Ecosystem Processes in the Great Northern and Appalachian Landscape Conservation Cooperatives

Forrest Melton, Jun Xiong, Weile Wang, Cristina Milesi, Shuang Li, Ashley Quackenbush, David M. Theobald, Scott J. Goetz, Patrick Jantz, and Ramakrishna Nemani

Ecosystem processes are the physical, chemical, and biological actions or events that link organisms and their environment. These processes include water and nutrient cycling, plant growth and decomposition, and regulation of community dynamics (Millennium Ecosystem Assessment 2003). The ecological characteristics of many parks and protected areas are dependent on the ecosystem functions that result from interactions between ecosystem processes, characteristics, and structures. Ecosystem functions, such as the regulation of water flows, soil retention and formation, and the provisioning of habitat and maintenance of biological diversity, in turn, provide the foundation for the ecosystem services supported by parks and protected areas (Hansen and DeFries 2007). As such, the preservation of ecosystem processes can be an important conservation target that complements conservation goals for species and habitats. Defining these targets is the first step in the Climate-Smart Conservation framework (Glick, Stein, and Edelson 2011; Stein et al. 2014).

While some aspects of climate change are associated with a large degree of uncertainty, the Third National Climate Assessment concluded that ecosystems and the services they provide are already being affected by climate change and that it is highly likely that future climate change will accelerate the disruption of ecosystem processes with consequences for ecosystem functions and associated ecosystem services (Melillo, Richmond, and Yohe

2014; Groffman et al. 2014; Grimm et al. 2013). In protected areas such as national parks, recent climate conditions are already pushing the limits of the historical range of variability (Monahan and Fisichelli 2014), which in turn has cascading effects on park resources and values (Monahan et al. 2013; Fisichelli et al. 2015). However, untangling ecosystem response to climate change can be especially complex, and relying on first principles alone can quickly lead to divergent response scenarios. For example, in montane ecosystems, the projected increase in temperatures is highly likely to accelerate melt of the snowpack in the spring and extend the growing season, which could lead to increases in overall vegetation productivity by increasing the number of suitable days for vegetation growth. However, the earlier snowmelt could also lead to reduced soil moisture and increased vegetation water stress during what is normally the peak of the growing season, substantially limiting vegetation growth through a key portion of the growing season and reducing overall vegetation productivity. Ecosystem models provide a useful tool that can be applied to deepen our understanding of possible ecosystem responses to changes in climate conditions and can be used to calculate the net balance of different possible responses to climate drivers.

Chapters 4 through 6 explored the exposure of ecosystems to recent and projected changes in climate and land use in the Great Northern Landscape Conservation Cooperative (Great Northern LCC) and future climate conditions in the Appalachian LCC. In this chapter, we extend these analyses to examine changes in ecosystem processes through the application of ecosystem models to the Great Northern and Appalachian LCCs and to six of the protected area centered ecosystems (PACEs) that are contained within, or immediately adjacent to, these LCCs. These analyses are designed to provide insights into the potential impact of recent and projected changes in climate and land use on ecosystem processes. In these analyses, we leverage a widely used and well validated ecosystem model and apply recently developed high-resolution projections of climate and land use change for the United States to explore the impacts of these changes at a spatial resolution of 30 arc-seconds, or 0.25 square mile (800 × 800 meters). Performing the analysis at this management-relevant spatial resolution is important because it allows us to capture the effects of variations in elevation, slope, soil texture, and biome type, which play important roles in regulating the response and associated vulnerability of ecosystems to changes in climate and land use conditions.

In our analysis, we focus on indicators of water cycling and vegetation productivity that can be modeled with biogeochemical cycle models

TABLE 7-1. Ecosystem indicators modeled using the Terrestrial Observation and Prediction System.

Variable	Definition	Units
Snow water equivalent	The amount of water contained in the snowpack, commonly expressed as the depth of water that would result if the entire snowpack was instantaneously melted	Millimeters of water
Soil moisture	The amount of water contained in the soil (within the root zone in this study)	Kilograms of water per cubic meter of soil
Evapotranspiration (ET)	The amount of water transferred to the atmosphere from the land surface through the combined processes of evaporation and transpiration	Equivalent depth of water in millimeters if the total ET was spread evenly across the land surface
Runoff	The amount of water that leaves the ecosystem through surface or sub-surface flow	Depth of water across the land surface in millimeters
Gross primary production	The gross uptake of carbon by plants through photosynthesis	Kilograms (or grams) of carbon per square meter

and for which simulations with the model used in our analyses have been previously shown to accurately reproduce observed monthly and seasonal patterns in a range of ecosystem types. Indicators of the response of hydrologic processes and carbon cycling include snow water equivalent, soil moisture, evapotranspiration, runoff, and gross primary production (table 7-1). Evapotranspiration serves as an important indicator of plant response to water availability, while runoff provides an indicator that summarizes the net water balance at the ecosystem level.

We use gross primary production (GPP) as an indicator to capture the response of vegetation productivity and carbon cycling to the changes in climate and land use. GPP is closely related to net primary productivity (which accounts for plant respiration) and net ecosystem productivity (which accounts for decomposition and heterotrophic respiration). GPP is a key measure of vegetation productivity and plant growth, and it integrates the ecosystem response to temperature, precipitation, topography, soils, and water and nutrient availability, making it a highly useful indicator of ecosystem response to changes in climate and land use conditions.

Modeling Approach

Our analysis of impacts of climate and land use change on ecosystems in the United States was conducted using the Terrestrial Observation and Prediction System (TOPS; Nemani et al. 2009). TOPS is a modeling framework that integrates and preprocesses Earth Observation Satellite data fields so that land surface models can be run in near real time or used to generate short- and long-term forecasts (Nemani et al. 2007, 2009). Modeling of ecosystem response to climate and land use change utilized the TOPS-Biogeochemical Cycle (TOPS-BGC) model, which is based on the BIOME-BGC model (version 4.1.2). Brief overviews of BIOME-BGC and TOPS-BGC are included in this chapter, and additional details are provided in Nemani et al. (2007, 2009), with details on BIOME-BGC available in Thornton (Thornton, Running, and White 1997; Thornton et al. 2002; Thornton and Rosenbloom 2005).

BIOME-BGC and TOPS-BGC have been widely used to evaluate ecosystem response to climate and to monitor carbon and hydrologic cycling at regional to global scales (Nemani et al. 2009). BIOME-BGC was developed specifically to estimate fluxes of carbon, water, and nutrients from ecosystems, and the model can be driven with either gridded climate observations or climate forecasts. BIOME-BGC currently includes parameterizations for six primary plant functional types or biomes, including evergreen needleleaf forests, evergreen broadleaf forests, deciduous broadleaf forests, deciduous needleleaf forests, shrublands, and grasslands/croplands. Within BIOME-BGC, absorption of radiation by the canopy is driven by leaf area, and the model calculates photosynthesis and carbon assimilation based on the Farquhar model (Farquhar, von Caemmerer, and Berry 1980). BIOME-BGC simulates stomatal conductance as a function of incident radiation, vapor pressure deficit, leaf water potential, and minimum temperature. Stomatal conductance, however, is not modeled to respond to changes in atmospheric CO_2 concentrations because the model used in this analysis uses fixed atmospheric CO_2 concentrations.

TOPS-BGC is derived from BIOME-BGC and uses the ecophysiological variables from BIOME-BGC for each plant functional type (White et al. 2000), but it streamlines model components in BIOME-BGC that deal with dynamic carbon allocation and the nitrogen cycle (White and Nemani 2004). TOPS-BGC relies on satellite measurements of leaf area index and the fraction of photosynthetically active radiation absorbed as inputs and estimates photosynthesis using a production efficiency model, similar to the approach employed in the GPP algorithm of the National Aeronautics

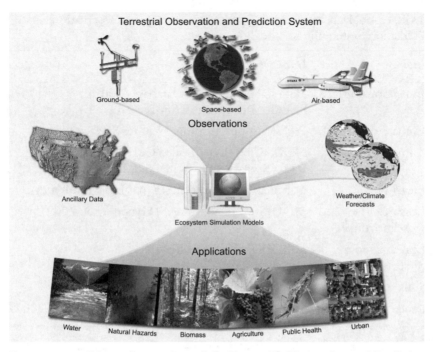

FIGURE 7-1   Schematic overview of the Terrestrial Observation and Prediction System architecture.

and Space Administration (NASA) Moderate Resolution Imaging Spectro-radiometer (MODIS) GPP data product (Zhao et al. 2005; Heinsch et al. 2006) but with improvements to refine the light use efficiency model for different land cover types (Yang et al. 2007). TOPS-BGC also includes a number of improvements to the BIOME-BGC hydrologic modules (Ichii et al. 2008) as well as modifications to improve characterization of root-ing zone depth within the model (Ichii et al. 2009). Figure 7-1 provides a schematic overview of the TOPS-BGC architecture.

Data Sets and Model Validation

As inputs, TOPS-BGC requires data sets describing maximum tempera-ture, minimum temperature, precipitation, solar radiation, vapor pressure deficit, land cover, soils, topography, and initial conditions for leaf area index. The data sets used for these variables in our analysis are described below and summarized in table 7-2.

TABLE 7-2. Data sets used as inputs to the Terrestrial Observation and Prediction System in the modeling analysis.

Input	Data Set
Impervious surface area	National Land Cover Database (2001), SERGoM Impervious Surface Area
Climate (historical)	PRISM 800m Monthly Weather Surfaces
Climate (forecast)	NEX-DCP30 Downscaled CMIP5 800m Scenarios for RCP 4.5, 8.5
Elevation	National Elevation Dataset (resampled to 800m)
Leaf area index (baseline run)	MODIS MOD15A2 LAI (Myneni et al. 2002)
Leaf area index (forecast)	MODIS LAI Climatology
Soils	U.S. STATSGO2 database
Land cover	MODIS MOD12Q1 Land cover (Friedl et al. 2002)

Climate Data

To drive the model over the analysis period, we used the Parameter-elevation Regressions on Independent Slopes Model (PRISM) gridded historical climate observation data set (Daly et al. 2000) for the climate runs from 1950 to 2005. For the period from 2006 through 2100, we used climate projections for the coterminous United States from the NASA Earth Exchange Downscaled Climate Projections (NEX-DCP30) data set (Thrasher et al. 2013). The NEX-DCP30 data set contains downscaled climate projections derived from the Coupled Model Intercomparison Project Phase 5 (CMIP5) data set (Taylor, Stouffer, and Meehl 2012). The CMIP5 data set contains the historical climate experiments (1950–2005) and climate projections (2006–2100) produced using coupled global general circulation models and four greenhouse gas emissions scenarios known as representative concentration pathways (RCPs). The CMIP5 data set was prepared for the Fifth Assessment Report of the International Panel on Climate Change (IPCC) and provided the physical basis for future climate conditions evaluated in that report. The NEX-DCP30 data set includes downscaled data from both the CMIP5 historical experiments (1950–2005) and future climate projections (2006–2100) at a monthly time step and a spatial resolution of 30 arc-seconds, which is approximately 0.5 mile (800 meters), or 0.25 square mile per pixel. We used the maximum temperature, minimum temperature, and precipitation data from NEX-DCP30 and derived data for solar radiation and vapor pressure deficit following the methods outlined in Thornton (Thornton, Running, and White 1997; Thornton and Running 1999).

In our analysis, we used the ensemble average of data from all available models for RCP 4.5 and 8.5 to evaluate the expected possible range of ecosystem response to future climate conditions. RCP 4.5 is based on a greenhouse gas emissions scenario that assumes coordinated and effective global action to slow greenhouse gases in the short term and effectively represents the "best case" achievable given current global atmospheric concentrations of greenhouse gases. RCP 8.5 represents the "business as usual" scenario, with continued rapidly increasing levels of greenhouse gases in the atmosphere and with concentrations more than doubling concentrations relative to the already high concentrations as of 2006. As of 2015, global emissions of greenhouse gases have been tracking or exceeding the trajectory described by RCP 8.5.

Land Cover Data

Land cover data for the historical modeling experiment (1950–2006) and baseline model runs (2001–2010) were derived from the National Land Cover Database 2001 (2011 edition) (Homer et al. 2004). Land cover data for the future scenarios incorporated decadal forecasts of impervious surface area from the Spatially Explicit Regional Growth Model (SERGoM) for the year 2010 to 2100 (Bierwagen et al. 2010). SERGoM (Theobald and Hobbs 1998; Theobald 2005) forecasts change as a function of historical growth rates and patterns and was originally developed specifically to model land use changes in rural landscapes and predict housing density changes at urban to rural densities. To simulate future IPCC scenarios, SERGoM was modified with a projection model that incorporated demographic parameters (birth/death, immigration/emigration) into a gravity-based model in which distance to the nearest urban core was estimated using road infrastructure capacity.

Details on linking the land use outputs from SERGoM with TOPS-BGC can be found in Goetz et al. (2009). The primary output from SERGoM used in TOPS-BGC is impervious surface area (ISA), which is closely correlated with development intensity. In summarizing results from our analysis, we also used data from the US Geological Survey's National

Gap Analysis Program (GAP) Land Cover data set (Jennings 2000) to identify the dominant ecosystem type associated with each grid cell. We then used this information to calculate summaries across each PACE to evaluate potential impacts on regions currently associated with each of the major ecosystem types in each PACE.

Soils and Topography

Data on soil texture and soil depth were derived from the US STATSGO2 database (NRCS 2015). TOPS includes complete US soil texture and rock depth grids at a spatial resolution of about 0.6 mile (1 kilometer). TOPS-BGC does not directly utilize ISA as an input parameter; however, ISA directly reduces soil water-holding capacity, and increasing ISA has been clearly shown to increase runoff (e.g., Carlson 2004; Rose and Peters 2001). To capture the influence of increasing ISA on runoff, we made the simplifying assumption that on a per pixel basis ISA is inversely related to total soil water-holding capacity. For example, a pixel with 50 percent ISA has effectively lost 50 percent of the available soil volume available to absorb and store precipitation. Elevation data is derived from the approximately 100-foot (30-meter) US Geological Survey National Elevation Dataset (NED; Gesch et al. 2002). NED was resampled to a resolution of 30 arc-seconds using a nearest neighbor spatial resampling algorithm and calculating the average elevation of all approximately 100-foot pixels within each 0.25 square mile as the pixel value.

Leaf Area Index

TOPS is capable of using simulated seasonal plant growth patterns, climatologies derived from historic satellite observations, or direct satellite observations of the land surface to estimate leaf area index and parameterize vegetation conditions within the model to track the observed seasonal patterns in vegetation growth. For this analysis, we used the MODIS 8-Day Leaf Area Index product (Myneni et al. 2002) data record from January 2001 through December 2010 to capture leaf area and phenological dynamics. A monthly climatology was calculated from the ten-year record, and data were resampled from about 0.6 mile (1 kilometer) to 30 arc-seconds. The climatology derived from satellite observations was used in the long-term ecosystem simulations from 1950 to 2100. It was also used in the model runs from 2000 to 2010 to assess model validation and accuracy.

Model Validation and Accuracy

While TOPS-BGC has been validated in a number of previous studies, prior to conducting the long-term simulations we conducted simulations for the period spanning 2000 to 2010 and compared results against observations from monitoring stations across the United States. Estimates of evapotranspiration and GPP were compared against observations collected at thirty-six different flux tower sites operated as part of the AmeriFlux observation network (http://ameriflux.lbl.gov; Law 2007). The flux tower locations used in our comparison include sites in the Great Northern and Appalachian LCCs and span the full range of ecosystem types evaluated in our analysis. The modeled estimates for evapotranspiration compared well with evapotranspiration measured at the AmeriFlux sites (R^2 = 0.73), and the root mean square error (RMSE) at most sites ranged from less than 0.3 to 0.6 millimeters a day. Modeled estimates of GPP also compared well against the observations at AmeriFlux sites (R^2 = 0.67), with RMSE values at most sites of less than 0.5 g-C/m^2 per day. Importantly, the modeled results showed minimal bias when compared against the observed data from the AmeriFlux sites.

Runoff estimates from TOPS-BGC were compared against runoff measured at US Geological Survey streamflow gauges for twelve of the largest wild and scenic rivers in the United States, and results from the comparison showed high model skill in predicting annual runoff (R^2 = 0.99), with average errors across the twelve watersheds of less than 8 m^3/s.

Projected Changes in Temperature, Precipitation, and Land Use

As described in chapters 4 through 6, both the Great Northern LCC and Appalachian LCC regions are expected to experience increases in maximum and minimum temperatures that will push temperatures above ecologically meaningful thresholds associated with plant growth, alter the amount of precipitation that falls as rain versus snow, and influence snowmelt and runoff. Even under RCP 4.5, both minimum and maximum temperatures in the Great Northern LCC are expected to increase by more than 4 degrees F (2.5 degrees C), and by more than 9 degrees F (5 degrees C) under RCP 8.5. In the Appalachian LCC, minimum and maximum temperatures are projected to increase by more than 5 degrees F (2.8 degrees C), and by more than 10 degrees F (5.6 degrees C) under RCP 8.5. Importantly, for both of these regions, there is large agreement across models that temperatures will increase; temperature increases under RCP 8.5 are especially pronounced.

There is a greater range of projections for changes in precipitation in these two regions, although the ensemble average of all thirty-four climate models included in the NEX-DCP30 data set projects modest increases in precipitation across both LCCs through the end of this century. Within the Great Northern LCC parks and PACEs, this increase ranges from 6 to 16 percent relative to current conditions (2000–2009), while in the Appalachian LCC precipitation increases by about 10 percent across the Appalachian LCC. Increases in precipitation are projected to occur under both RCP 4.5 and 8.5, with slightly larger increases occurring on average under RCP 8.5 in both LCCs.

Projected changes in impervious surface area for each LCC and PACE are summarized in table 7-3. Overall, both current and future levels of impervious surface area within the Great Northern LCC are small, amounting to less than 0.3 percent of the total area in 2010, and 0.7 percent in 2100, even under the more rapid development scenario. The only PACE where total ISA exceeds 1 percent in 2100 is the Rocky Mountain PACE, where ISA is projected to increase from 0.7 percent in 2010 to 2.5 percent in 2100 under the business as usual scenario (RCP 8.5). Current levels of ISA in the Appalachian LCC are higher and average 2 percent across the Appalachian LCC, with a maximum ISA of 3 percent of the total area for the Delaware Water Gap PACE. Across the Appalachian LCC as a whole, projected increases in ISA amount to less than 1 percent of the total area of the LCC or PACE. The Delaware Water Gap PACE is again the exception, increasing from 3 percent in 2010 to 4 percent in 2100 under the low development scenario (RCP 4.5) and increasing to more than 7 percent ISA in 2100 under the business as usual development scenario (RCP 8.5).

The impacts of these changes in temperature, precipitation, and land use on ecosystem processes are described below and summarized in figure 7-2 and tables 7-4 and 7-5. We focus on key results across the LCCs as well as results from individual PACEs that differ from the overall patterns across the LCCs. We have left the units in our tables in metric units to facilitate comparison with other studies, and in our discussion we focus on the amount of change relative to baseline conditions from 1970 to 1999. We have also included both absolute estimates and measures of relative changes in tables 7-3 and 7-4 to facilitate comparison both with other studies and across different time periods. All comparisons expressed as percent change are calculated relative to the average of conditions from 1970 to 1999, unless otherwise stated. The strength of our modeling analysis is that it

TABLE 7-3. Changes in impervious surface area from 2010 to 2100.

		Impervious Surface Area (km²)			
LCC/PACE	Total Area (km²)	2010, B2 (RCP 4.5)	2100, B2 (RCP 4.5)	2010, A2 (RCP 8.5)	2100, A2 (RCP 8.5)
GNLCC	660,404	1,731	2,052	1,703	2,167
Glacier	22,067	45	46	43	44
Yellowstone	42,646	58	66	54	58
Rocky Mountain	9,450	66	125	67	232
ALCC	592,906	12,382	13,788	12,328	15,755
DEWA	14,057	448	574	452	1,033
Shenandoah	10,752	229	273	227	319
GRSM	15,640	318	333	315	376

Note: LCC = landscape conservation cooperative; PACE = protected area centered ecosystem; RCP = representative concentration pathway; GNLCC = Great Northern LCC; ALCC = Appalachian LCC; DEWA = Delaware Water Gap National Recreation Area; GRSM = Great Smoky Mountains National Park.

captures the direction of climate and land use change at ecologically meaningful scales. As such, in our discussion we focus on the relative changes between time periods, rather than on absolute model predictions for each variable.

Ecosystem Responses in the Great Northern LCC

Regardless of the amount of precipitation, much more of the precipitation is projected to fall as rain instead of snow, with substantial impacts to the snowpack across the Great Northern LCC (figs. 7-2a and 7-2b) under both RCP 4.5 and 8.5. Even under RCP 4.5, the net balance of the projected increases in temperatures and precipitation results in reductions of almost –20 percent in the average total annual snowpack during the period from 2020 to 2049 relative to 1970 to 1999 (table 7-4a). By 2099, the annual snowpack is projected to decrease by more than –36 percent under RCP 4.5, and by –66 percent under RCP 8.5. These changes in the snowpack occur in all seasons but are especially pronounced in the spring, with reductions of –70 percent in the snowpack in 2070–2099 under RCP 4.5. Reductions exceed –90 percent under RCP 8.5, resulting in an almost complete loss of the winter snowpack in most years by the end of this century.

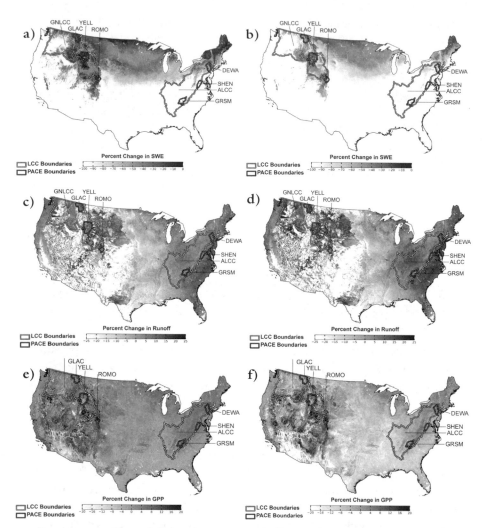

FIGURE 7-2 Percent change in ecosystem indicators from 1970–1999 to 2070–2099. The change in ecosystem indicators from 1970–1999 to 2070–2099 is shown as a percentage of the average conditions for each indicator for 1970–1999. Negative values indicate a net decrease in the indicator in 2070–2099 relative to 1970–1999. Results are included for changes in snow water equivalent (SWE) under RCP 4.5 (a) and RCP 8.5 (b), runoff under RCP 4.5 (c) and RCP 8.5 (d), and gross primary production (GPP) under RCP 4.5 (e) and RCP 8.5 (f). Key: The labels for each PACE boundary shown correspond to the National Park Service code for the park contained within each PACE: within the Great Northern LCC (GN-LCC), GLAC = Glacier National Park (NP), YELL = Yellowstone NP, ROMO = Rocky Mountain NP; within the Appalachian LCC (ALCC), GRSM = Great Smoky Mountains NP, SHEN = Shenandoah NP, and DEWA = Delaware Water Gap National Recreation Area.

Table 7-4A. Great Northern LCC ecosystem response.

Season/ Variable	Hist. 1970–1999	RCP 4.5 2020–2049	RCP 4.5 2050–2079	RCP 4.5 2070–2099	RCP 8.5 2020–2049	RCP 8.5 2050–2079	RCP 8.5 2070–2099	Percent Change RCP 4.5 2020–2049	Percent Change RCP 4.5 2050–2079	Percent Change RCP 4.5 2070–2099	Percent Change RCP 8.5 2020–2049	Percent Change RCP 8.5 2050–2079	Percent Change RCP 8.5 2070–2099
Annual													
SWE	91	73	62	58	69	45	31	–19.8	–32.0	–36.7	–24.2	–50.6	–66.1
ET	342	357	367	371	361	376	385	4.5	7.4	8.5	5.7	10.1	12.6
Soil water	311	315	316	317	316	318	320	1.1	1.6	1.7	1.4	2.1	2.7
Runoff	337	332	337	340	335	341	350	–1.4	–0.1	0.8	–0.7	1.2	3.8
GPP	631	643	641	639	642	633	621	1.8	1.5	1.2	1.6	0.2	–1.7
Winter													
SWE	233	209	182	171	200	138	99	–10.1	–21.9	–26.5	–14.1	–40.7	–57.7
ET	12	17	26	30	20	38	51	44.3	123.0	160.9	69.9	228.3	338.1
Soil water	306	321	331	335	324	343	356	4.8	8.1	9.3	5.9	12.1	16.3
Runoff	24	41	59	68	46	87	119	69.0	145.4	182.7	92.6	262.6	395.0
GPP	17	23	29	31	25	35	41	40.9	74.8	89.9	50.2	111.6	147.8
Spring													
SWE	70	34	24	21	31	14	6	–52.2	–65.6	–70.1	–56.2	–80.5	–91.2
ET	163	177	179	179	178	177	174	8.9	10.0	10.1	9.4	8.9	7.1
Soil water	370	367	360	358	366	350	338	–0.8	–2.7	–3.4	–1.3	–5.6	–8.7
Runoff	248	245	228	221	240	197	165	–1.3	–8.2	–11.0	–3.1	–20.5	–33.3
GPP	257	266	265	264	265	260	254	3.5	3.0	2.6	3.2	1.0	–1.2
Summer													
SWE	1.09	0.37	0.27	0.20	0.33	0.13	0.04	–65.6	–75.0	–81.2	–69.8	–87.9	–95.9
ET	133	123	119	117	122	115	110	–7.3	–10.1	–11.4	–8.0	–13.6	–16.9
Soil water	279	274	273	272	274	273	272	–1.8	–2.2	–2.4	–1.8	–2.3	–2.4
Runoff	50	29	26	24	28	22	21	–41.7	–48.5	–51.4	–43.9	–55.4	–58.7
GPP	317	305	297	293	303	286	272	–3.6	–6.2	–7.4	–4.4	–9.7	–14.1
Fall													
SWE	61	50	42	39	46	29	19	–18.7	–31.4	–36.6	–24.9	–53.3	–68.7
ET	35	40	43	44	41	46	49	15.9	23.6	25.8	18.9	32.2	42.2
Soil water	290	297	301	302	299	307	312	2.4	3.9	4.3	3.1	5.8	7.8
Runoff	15	18	24	27	20	34	45	16.5	59.9	76.4	33.4	125.9	195.7
GPP	41	48	50	51	49	52	53	16.7	22.5	23.7	18.9	26.7	29.8

Note: Change in indicators of ecosystem processes in the Great Northern LCC, summarized as annual and seasonal averages for four thirty-year periods, and for RCP 4.5 and 8.5. Indicators include daily average snow water equivalent (SWE) (mm), total annual/seasonal evapotranspiration (ET) (mm), daily average soil water (kg-H_2O/m^3), total annual/seasonal runoff (mm), and total annual/seasonal gross primary production (GPP) (g-C/m^2). The percent change for RCP 4.5 and 8.5 is calculated relative to the baseline period from 1970 to 1999.

TABLE 7-4B. Appalachian LCC ecosystem response.

Season/Variable	Hist. 1970–1999	RCP 4.5 2020–2049	RCP 4.5 2050–2079	RCP 4.5 2070–2099	RCP 8.5 2020–2049	RCP 8.5 2050–2079	RCP 8.5 2070–2099	Percent Change RCP 4.5 2020–2049	Percent Change RCP 4.5 2050–2079	Percent Change RCP 4.5 2070–2099	Percent Change RCP 8.5 2020–2049	Percent Change RCP 8.5 2050–2079	Percent Change RCP 8.5 2070–2099
Annual													
SWE	23	16	10	6	13	5	1	–29.3	–56.5	–73.0	–42.6	–78.8	–93.8
ET	645	684	707	720	693	721	739	6.1	9.6	11.6	7.5	11.8	14.6
Soil water	397	394	392	391	393	390	389	–0.8	–1.2	–1.6	–1.0	–1.7	–2.0
Runoff	562	556	561	559	557	561	571	–1.1	–0.2	–0.6	–0.9	–0.2	1.6
GPP	1460	1488	1478	1463	1482	1442	1395	1.9	1.2	0.2	1.5	–1.3	–4.5
Winter													
SWE	81	57	37	23	47	19	6	–28.9	–54.4	–70.9	–41.5	–76.6	–93.0
ET	69	92	105	117	99	120	134	34.2	52.5	70.1	44.8	74.8	94.5
Soil water	433	432	432	431	432	433	431	–0.3	–0.1	–0.5	–0.3	–0.1	–0.4
Runoff	153	173	186	194	180	204	218	13.3	21.8	27.2	17.6	33.8	43.0
GPP	85	102	112	120	107	121	129	20.6	32.6	42.3	26.6	42.8	52.4
Spring													
SWE	0	0	0	0	0	0	0	–	–	–	–	–	–
ET	258	259	263	261	259	259	259	0.4	1.8	1.1	0.3	0.5	0.3
Soil water	395	387	379	375	384	374	370	–2.1	–4.0	–5.0	–2.8	–5.4	–6.4
Runoff	223	202	186	174	195	168	158	–9.5	–16.8	–22.1	–12.6	–24.8	–29.1
GPP	544	542	533	523	538	516	496	–0.3	–2.1	–3.8	–1.1	–5.2	–8.9
Summer													
SWE	0	0	0	0	0	0	0	–	–	–	–	–	–
ET	219	226	230	231	227	230	233	3.1	4.9	5.4	3.7	5.3	6.3
Soil water	359	356	356	355	356	354	354	–0.7	–0.7	–0.9	–0.7	–1.2	–1.4
Runoff	89	84	88	87	85	84	86	–6.3	–1.9	–2.8	–4.6	–6.0	–4.0
GPP	629	630	617	604	624	592	560	0.1	–1.9	–4.0	–0.9	–5.8	–10.9
Fall													
SWE	10.0	6.8	2.7	1.0	4.9	0.3	0.0	–32.2	–72.9	–89.6	–51.0	–96.7	–99.8
ET	99	107	110	112	108	111	114	8.0	10.6	12.2	8.3	11.9	15.0
Soil water	401	401	401	401	400	401	401	–0.2	–0.1	–0.1	–0.2	0.0	0.0
Runoff	97	97	102	104	97	105	109	0.4	4.8	7.0	0.5	8.1	12.4
GPP	202	214	215	215	213	213	210	5.5	6.4	6.2	5.3	5.0	3.5

Note: Change in indicators of ecosystem processes in the Appalachian LCC, summarized as annual and seasonal averages for four thirty-year periods, and for RCP 4.5 and 8.5. Indicators and units are the same as those listed in table 7-4a. The percent change for RCP 4.5 and 8.5 is calculated relative to the baseline period from 1970 to 1999.

The spatial variability in ecosystem response in the Great Northern LCC that is evident in figure 7-2 follows gradients in elevation, aspect, and soil characteristics, and the patterns are also strongly influenced by current snow and runoff dynamics and the extent of future loss of the snowpack. In general, the higher elevations within the Great Northern LCC—and especially in the PACEs for Glacier and Yellowstone and the Rocky Mountain PACE—exhibit larger increases in runoff. Higher elevations with only partial loss of the snowpack are also more likely to exhibit increases in vegetation productivity through the end of the century.

Concurrent with the reductions in the winter snowpack, increasing temperatures also drive increases in evapotranspiration across the Great Northern LCC. Overall, these projected annual increases are modest, ranging from about 5 percent (under both RCP 4.5 and 8.5) over the period from 2020 to 2049, to 8 percent (RCP 4.5) to 13 percent (RCP 8.5) in 2070–2099, relative to 1970–1999 (table 7-4a). On an annual basis, the projected increases in evapotranspiration approximately offset the increases in precipitation, and as a result, annual total runoff is projected to change very slightly (–1 percent to 4 percent), if at all, across the LCC as a whole.

While the projected changes in the annual totals for evapotranspiration and runoff are relatively small, they mask large changes in the timing and seasonality of these hydrologic fluxes. Runoff, in particular, is projected to see large increases in the fall and winter seasons, with corresponding decreases in the spring and summer (table 7-4a). Even under RCP 4.5, winter runoff over the period from 2050 to 2079 is projected to increase by nearly 150 percent, and fall runoff increases 60 percent relative to 1970–1999. Under RCP 8.5, these increases are even more dramatic, with winter runoff increasing nearly 400 percent by the end of the century and with fall runoff almost doubling. Meanwhile, summer runoff is projected to decrease by more than –40 percent as early as 2020–2049 under RCP 4.5, and decline by –50 percent by the end of the century. Spring runoff is expected to change more gradually under RCP 4.5, with reductions of just over –10 percent by the end of the century. Under the more rapid warming associated with RCP 8.5, however, spring runoff also declines sharply, dropping more than –20 percent by 2050–2079, and by –33 percent by the end of the century. Under RCP 8.5, summer runoff declines as well, with reductions of –44 percent expected by 2020–2049, and declining further to –58 percent of current levels by 2070–2099.

The projected changes in vegetation productivity, as measured by GPP, follow patterns similar to the projected changes in evapotranspiration and runoff. On an annual basis, our results suggest that vegetation productivity is likely to change only slightly for the Great Northern LCC as a whole, and projected changes range from a small increase (1.3 percent) by 2070–2099 under RCP 4.5 to a small decrease (–1.7 percent) under RCP 8.5 (table 7-4a). However, the more important pattern is the pronounced shift in the seasonality of vegetation growth. At present, 50 percent of total annual GPP in the Great Northern LCC occurs in summer months, and summer GPP is projected to steadily decline in the coming decades, with reductions in 2070–2099 ranging from –7 percent (RCP 4.5) to –14 percent (RCP 8.5).

On an annual basis, these declines are projected to be offset by increases in GPP in the fall and winter, at least in terms of overall vegetation productivity across the LCC. With the increase in temperature and resultant increases in snowmelt and runoff, GPP increases substantially during these seasons relative to 1970–1999. Under both RCP 4.5 and 8.5, winter GPP is predicted to increase by 40 to 50 percent by 2020–2040, and increase by 90 percent (RCP 4.5) to nearly 150 percent (RCP 8.5) by 2070–2099. GPP in the fall months also increases from 24 percent (RCP 4.5) to 30 percent (RCP 8.5) by 2070–2099, and together these increases represent a substantial extension of the growing season, largely driven by increases in minimum temperature and the number of snow-free days. Spring GPP remains relatively consistent, with projected changes by the end of this century ranging from 2.5 percent (RCP 4.5) to –1 percent (RCP 8.5).

The relatively moderate changes in spring and summer GPP relative to the projected changes in runoff are explained in part by the buffering of these changes through water storage by soils in the root zone. Overall, changes in soil moisture in all seasons are small through the period from 2070 to 2099 under both RCPs, with the largest changes occurring in the winter (16 percent) and spring (–9 percent) under RCP 8.5 (table 7-4a). However, under RCP 8.5, the declines in soil moisture in the spring begin to accelerate toward the end of the century, suggesting that as warming continues beyond 2100, the ability of the soils to buffer changes in snowmelt runoff may be degraded.

Overall, projected land use changes and associated changes in housing density and impervious surface area have a minor role across the Great Northern LCC because of the relatively small baseline level of ISA and small, localized projected increases. However, hot spots are visible in figures 7-2c, 7-2d, 7-2e, and 7-2f; these include regions along the western edge of the Great Northern LCC boundary in Washington and located to the southeast of the Rocky Mountain PACE boundary. Projected changes in land use and increases in ISA in these localized regions drive reductions in GPP of more than –25 percent, and increases in runoff exceeding 25 percent by 2070–2099 under both RCPs. The total area represented by these hot spots of development is limited enough to have a small net influence on results for the Great Northern LCC as a whole, but the increases in ISA are very likely to have large localized impacts on ecosystem processes in the regions adjacent to these hot spots.

Results for the Yellowstone and Glacier PACE regions largely follow the same patterns described for the Great Northern LCC as a whole. Elevations and regions currently associated with mixed conifer, sagebrush steppe, and spruce-fir ecosystems exhibit similar annual and seasonal ecosystem responses to the climate and land use changes. Examples of exceptions to the general patterns in the LCC are listed in table 7-5 and include grassland ecosystems in the Glacier PACE, lodgepole and ponderosa pine ecosystems in the Rocky Mountain PACE, and alpine ecosystems in all three PACEs (representative example included in table 7-5 for the Yellowstone PACE). Grassland ecosystems exhibit increases in evapotranspiration similar to those for the Great Northern LCC but exhibit a small increase in GPP under RCP 4.5 (about 2 percent) and almost no net change under RCP 8.5 through 2070–2100. Summer GPP for grasslands also declines under RCP 8.5, but the decline is only half of the average for the Great Northern LCC as a whole.

TABLE 7-5. Ecosystem response by vegetation type.

Vegetation Type/ Variable	Hist. 1970–1999	RCP 4.5 2020–2049	RCP 4.5 2050–2079	RCP 4.5 2070–2099	RCP 8.5 2020–2049	RCP 8.5 2050–2079	RCP 8.5 2070–2099	Percent Change RCP 4.5 2020–2049	Percent Change RCP 4.5 2050–2079	Percent Change RCP 4.5 2070–2099	Percent Change RCP 8.5 2020–2049	Percent Change RCP 8.5 2050–2079	Percent Change RCP 8.5 2070–2099
Glacier, Grasslands													
SWE	60	53	42	39	50	32	20	–12	–30	–34	–16	–47	–67
ET	382	410	421	424	412	435	453	7.3	10.2	11.0	7.8	13.8	18.5
Soil water	414	425	429	428	425	428	430	2.5	3.6	3.4	2.7	3.5	3.7
Runoff	231	235	234	232	235	230	230	1.8	1.4	0.3	1.7	–0.3	–0.4
GPP	590	605	603	602	602	597	586	2.4	2.2	2.0	2.0	1.1	–0.7
Yellowstone, Alpine													
SWE	266	216	209	204	215	197	190	–19	–212	–23	–19	–26	–29
ET	330	350	363	365	355	374	389	6.0	10.0	10.7	7.6	13.5	18.0
Soil water	245	249	252	251	250	252	255	1.9	2.9	2.8	2.4	3.2	4.1
Runoff	735	733	759	760	745	777	805	–0.2	3.2	3.5	1.3	5.7	9.5
GPP	274	276	277	278	277	279	281	0.8	1.4	1.6	1.2	1.9	2.8

TABLE 7-5. (*Continued*)

Vegetation Type/ Variable	Hist. 1970–1999	RCP 4.5 2020–2049	RCP 4.5 2050–2079	RCP 4.5 2070–2099	RCP 8.5 2020–2049	RCP 8.5 2050–2079	RCP 8.5 2070–2099	Percent Change RCP 4.5 2020–2049	Percent Change RCP 4.5 2050–2079	Percent Change RCP 4.5 2070–2099	Percent Change RCP 8.5 2020–2049	Percent Change RCP 8.5 2050–2079	Percent Change RCP 8.5 2070–2099
Rocky Mountain, Alpine													
SWE	228	183	177	178	182	167	159	−20	−23	−22	−20	−27	−31
ET	354	393	407	411	398	416	428	10.9	14.7	15.8	12.4	17.5	20.6
Soil water	279	283	284	285	284	285	286	1.3	1.8	2.1	1.7	2.2	2.7
Runoff	617	594	604	614	600	613	627	−3.7	−2.0	−0.5	−2.7	−0.5	1.7
GPP	427	450	458	461	454	460	458	5.5	7.3	8.1	6.4	7.7	7.2
Rocky Mountain, Lodgepole Pine													
SWE	75	61	58	57	60	47	35	−19	−23	−24	−19	−37	−53
ET	347	366	375	378	369	388	402	5.7	8.1	9.1	6.4	12.1	16.0
Soil water	253	256	258	259	257	262	268	1.2	2.0	2.2	1.5	3.8	5.9
Runoff	259	252	258	262	256	254	252	−2.7	−0.3	1.0	−1.3	−1.9	−2.9
GPP	701	729	746	751	735	762	775	4.0	6.5	7.1	4.9	8.8	10.6
Rocky Mountain, Ponderosa Pine													
SWE	20	11	8	8	10	5	2	−43	−57	−62	−48	−77	−89
ET	289	287	295	297	290	301	302	−0.5	2.1	3.0	0.3	4.1	4.6
Soil water	193	201	203	203	202	205	207	4.3	5.2	5.5	4.6	6.3	7.1
Runoff	193	204	207	208	204	204	204	5.9	7.4	8.1	5.9	5.8	6.0
GPP	755	816	838	845	821	859	870	8.0	11.0	11.9	8.8	13.8	15.3
Delaware Water Gap, Hemlock-Hardwood Forest													
SWE	61	56	45	41	53	27	12	−8.7	−27	−34	−13	−56	−80
ET	568	605	626	636	611	657	685	6.5	10.2	12.0	7.6	15.7	20.6
Soil water	402	402	401	400	401	398	395	−0.2	−0.4	−0.5	−0.2	−1.1	−1.9
Runoff	615	622	629	632	626	627	628	1.1	2.2	2.7	1.8	1.9	2.0
GPP	1475	1511	1504	1503	1509	1483	1453	2.5	1.9	1.9	2.3	0.6	−1.5
Shenandoah, Cove Forest													
SWE	14.6	1.7	0.6	0.4	1.3	0.1	0.0	−89	−96	−98	−91	−99	−100
ET	674	728	749	751	734	761	776	7.9	11.1	11.3	8.9	12.8	15.0
Soil water	355	348	348	348	348	347	347	−2.0	−1.9	−2.0	−2.0	−2.2	−2.2
Runoff	617	577	606	602	584	603	622	−6.4	−1.8	−2.5	−5.3	−2.2	0.9
GPP	1691	1737	1735	1730	1735	1705	1658	2.7	2.6	2.3	2.6	0.8	−1.9
Great Smoky, Northern Hardwood													
SWE	36.7	7.3	1.7	1.2	4.5	0.2	0.0	−80	−95	−97	−88	−99	−100
ET	779	870	904	909	881	921	945	11.7	16.0	16.6	13.1	18.2	21.2
Soil water	412	396	395	395	396	392	391	−3.9	−4.0	−4.1	−3.8	−4.7	−4.9
Runoff	1150	1056	1087	1087	1074	1070	1082	−8.2	−5.4	−5.4	−6.6	−6.9	−5.9
GPP	1669	1732	1745	1745	1732	1744	1738	3.8	4.6	4.6	3.8	4.5	4.1

TABLE 7-5. (*Continued*)

Vegetation Type/ Variable	Hist. 1970–1999	RCP 4.5 2020–2049	RCP 4.5 2050–2079	RCP 4.5 2070–2099	RCP 8.5 2020–2049	RCP 8.5 2050–2079	RCP 8.5 2070–2099	Percent Change RCP 4.5 2020–2049	Percent Change RCP 4.5 2050–2079	Percent Change RCP 4.5 2070–2099	Percent Change RCP 8.5 2020–2049	Percent Change RCP 8.5 2050–2079	Percent Change RCP 8.5 2070–2099
Great Smoky, Spruce-Fir													
SWE	62.8	26.6	8.4	5.6	18.5	1.0	0.0	−58	−87	−91	−71	−98	−100
ET	724	823	872	881	837	898	923	13.7	20.5	21.7	15.6	24.1	27.5
Soil water	397	379	376	374	379	371	370	−4.6	−5.4	−5.8	−4.6	−6.6	−6.9
Runoff	1298	1188	1208	1203	1203	1182	1194	−8.5	−7.0	−7.3	−7.3	−9.0	−8.1
GPP	1444	1516	1528	1530	1515	1533	1534	5.0	5.9	6.0	5.0	6.2	6.2
Great Smoky, Cove Forest													
SWE	5.7	0.1	0.0	0.0	0.0	0.0	0.0	−99	−100	−100	−99	−100	−100
ET	785	835	857	860	843	866	882	6.4	9.1	9.5	7.4	10.3	12.4
Soil water	340	335	335	335	335	334	334	−1.7	−1.4	−1.5	−1.5	−1.9	−2.0
Runoff	782	752	784	786	770	776	788	−3.9	0.3	0.5	−1.6	−0.8	0.8
GPP	1715	1765	1760	1753	1762	1721	1670	2.9	2.6	2.2	2.7	0.3	−2.7
Great Smoky, Oak Forest													
SWE	6.2	0.1	0.0	0.0	0.0	0.0	0.0	−99	−100	−100	−100	−100	−100
ET	776	825	847	850	833	856	872	6.4	9.2	9.5	7.4	10.3	12.4
Soil water	331	325	326	326	326	325	324	−1.8	−1.5	−1.5	−1.5	−1.9	−2.0
Runoff	774	744	776	777	762	768	780	−3.8	0.3	0.5	−1.5	−0.8	0.8
GPP	1723	1775	1771	1765	1772	1734	1687	3.0	2.8	2.4	2.8	0.6	−2.1

Note: Examples of change in indicators of ecosystem processes in regions associated with key vegetation types for each PACE, summarized as annual averages for four thirty-year periods, and for RCP 4.5 and 8.5. Indicators and units are the same as those listed in table 7-4a. The percent change for RCP 4.5 and 8.5 is calculated relative to the baseline period from 1970 to 1999.

High elevations associated with alpine ecosystems have reductions in snow water equivalent under RCP 8.5 that are approximately half of those of the Great Northern LCC as a whole. Like the LCC, they also have large projected increases in evapotranspiration. Increases in runoff for alpine ecosystems in the Yellowstone PACE are approximately double the average for the Great Northern LCC overall, but instead of declining GPP, alpine ecosystems exhibit projected increases in annual average GPP of 6 to 8 percent under both RCPs in all time periods, with GPP generally increasing as temperatures warm and the snowpack is reduced (table 7-5).

Regions in the Rocky Mountain PACE currently associated with lodgepole pine and ponderosa pine ecosystems follow the overall Great Northern LCC patterns for reductions in the snowpack and increasing evapotranspiration and runoff. However, warming temperatures and a positive net water balance also favor these ecosystems, and GPP steadily increases under both RCPs, with a net increase by 2070–2099 of as much as 11 percent for lodgepole pine and 15 percent for ponderosa pine ecosystems under RCP 8.5.

Ecosystem Responses in the Appalachian LCC

As in the Great Northern LCC, warming temperatures across the Appalachian LCC are projected to drive large reductions in the winter snowpack. By 2050–2079, reductions in the snowpack are projected to exceed –50 percent under both RCP 4.5 and 8.5. By 2070–2099, the snowpack is projected to decline by more than –70 percent under RCP 4.5. Under RCP 8.5, the snowpack will be almost completely gone by the end of the century, with declines of more than –90 percent, indicating that with warming temperatures the Appalachians will be largely snow-free during most years (table 7-4b).

Unlike the Great Northern LCC, however, in the Appalachian LCC the melt of the snowpack plays a smaller role in regulating the annual hydrological cycle. During the period from 1970 to 1999, winter snow accounted for nearly 90 percent of the total annual snowpack, and snow rarely persisted well into the spring season in the model simulations. As a result, the balance between changes in precipitation and evapotranspiration becomes more important in determining the resilience of these ecosystems to climate change. Like the Great Northern LCC, there is a wide range in the projections for future precipitation patterns across different models and RCPs. However, the projection for the Appalachian LCC from the ensemble average of all models is for modest increases in annual precipitation under RCP 4.5 and slightly larger increases in annual precipitation under RCP 8.5.

The effect of the warmer winter temperatures and generally increasing precipitation is observable in the projected changes in winter runoff, with increases by 2070–2099 from 27 percent (RCP 4.5) to 43 percent (RCP 8.5). Increases in runoff are also projected to be evident in the short term, with increases in the period from 2020 to 2049 averaging from 13 percent (RCP 4.5) to 18 percent (RCP 8.5). Corresponding with the increase in winter runoff and loss of the snowpack across the Appalachian LCC are large reductions in runoff during the spring, and smaller but still meaningful reductions in summer runoff. Changes in spring runoff are projected to exceed –22 percent under RCP 4.5 and –29 percent under RCP 8.5 by 2070–2099, with reductions in annual spring runoff of approximately –10 percent by 2020–2049 under both RCPs (table 7-4b). Reductions in summer runoff are smaller but still noteworthy because they represent average changes, and year-to-year variability in minimum streamflows has important ecological consequences in Appalachian LCC ecosystems. Unlike trends in other hydrologic indicators, summer streamflows exhibit the largest reductions during the period from 2020 to 2049, with average reductions ranging from –6 percent (RCP 4.5) to –5 percent (RCP 8.5), reflecting the larger projected increases in precipitation under RCP 8.5.

With warmer temperatures and increased moisture, the TOPS-BGC model also projects increases in evapotranspiration across the Appalachian LCC in all seasons. Total annual changes in evapotranspiration by 2070–2099 exceed 11 percent under RCP 4.5 and approach 15 percent under RCP 8.5 (table 7-4b). The largest increases occur during the fall and winter seasons, when evapotranspiration has been low historically and limited by minimum temperatures that strongly control the current length of the growing season. However, summer evapotranspiration is also projected to increase moderately, with increases approaching 5 to 6 percent annually by 2070–2100 under both RCP 4.5 and 8.5.

As in the Great Northern LCC, the TOPS-BGC model projects small changes in root zone soil moisture across the Appalachian LCC, with annual changes in average soil moisture of approximately –2 percent in 2070–2100 relative to 1970–1999 under both RCP 4.5 and 8.5 (table 7-4b). The only season with larger reductions in soil moisture is the spring, with declines in average spring soil moisture of –5 percent (RCP 4.5) to –6 percent (RCP 8.5) by 2070–2099. The results indicate that soils in the root zone are able to at least partially offset changes in snowmelt patterns. However, in the Appalachian LCC, the overall trend in annual soil moisture is gradually but steadily negative, while both annual average runoff and evapotranspiration increase slightly, indicating that the net positive water balance for the Appalachian LCC is being driven by the projected increases in precipitation. This highlights the importance of improving climate projections for precipitation in the region, because results for the Appalachian LCC are highly sensitive to future precipitation patterns and, specifically, the net balance between increases in runoff and evapotranspiration and changes in annual and seasonal precipitation.

Annual vegetation productivity in the Appalachian LCC is projected to increase slightly in the short term under both RCP 4.5 and 8.5, with increases of 1.5 to 2 percent over the period from 2020 to 2049, relative to 1970–1999 (table 7-4b). As with the changes in evapotranspiration, much of the increase in annual GPP is driven by increases in the winter and fall as the growing season is extended. As warming accelerates in the second half of the century, these increases are offset by growing declines in GPP in the spring and summer, resulting in no measurable change in annual GPP by 2070–2099 under RCP 4.5 and a net reduction in GPP of –4.5 percent under RCP 8.5. Similar to patterns in the Great Northern LCC, results for GPP in the Appalachian LCC also indicate a shift in the seasonal patterns in vegetation growth, with GPP in the spring and summer declining by –4 percent (RCP 4.5) to as much as approximately –10 percent (RCP 8.5) by 2070–2099.

Unlike the Great Northern LCC, spatial patterns in indicators of ecosystem processes in the Appalachian LCC are much more consistent across the LCC (fig. 7-2). In general, vegetation productivity exhibits modest increases across the higher elevations associated with the Appalachian and Great Smoky Mountain ranges in the eastern half of the LCC, with small decreases in vegetation productivity predominant in the warmer, lower elevations in the western half of the LCC. Figures 7-2c through 7-2f also indicate regions where urban expansion and increasing impervious surface area are driving land use conversion, resulting in large but localized decreases in vegetation productivity. Close inspection of figure 7-2 also reveals that these areas of increasing impervious surface area are associated with larger increases in runoff, approaching increases of 20 to 25 percent annually by 2070–2099. While these areas are not large enough to affect the results for the Appalachian LCC as a whole, they are likely to have important impacts at local scales, especially the projected increases in housing density and associated ISA in areas bordering the Shenandoah PACE to the northeast, and regions adjacent to the southeast border of the Delaware Water Gap PACE.

Ecosystem responses for the pine-oak forest, calcareous forest, and oak woodland ecosystems that are prevalent within the Delaware Water Gap and Shenandoah PACEs follow very similar patterns to those for the Appalachian LCC as a whole. Regions currently associated with cove forest and hemlock-hardwood ecosystems also follow the LCC responses for increasing evapotranspiration and small declines in soil moisture. However, regions in which these ecosystems are dominant have larger projected increases in annual GPP through 2100 under RCP 4.5 (2 to 3 percent), and projected reductions in GPP through 2070–2100 are –2 percent, less than half of the projected declines in GPP for the LCC as a whole (table 7-5).

In the Great Smoky Mountains PACE, regions associated with dry calcareous and mountain pine forests follow the overall Appalachian LCC patterns. However, areas of the Great Smoky Mountains PACE associated with Appalachian spruce-fir and northern hardwood forests exhibit a different response from the LCC as a whole (table 7-5). Projected changes in evapotranspiration are substantially larger under both RCPs, with maximum increases through 2070–2099 under RCP 8.5 reaching 17 percent for northern hardwood forests and 28 percent for Appalachian spruce-fir ecosystems. Annual runoff is projected to decline in these ecosystems by –5 to –8 percent over the period from 2020 to 2049 under both RCPs, with declines in annual runoff exceeding –8 percent by 2070–2099 under RCP 8.5. However, declines in summer runoff are projected to be less than the LCC as a whole and instead exhibit a net increase of 1 to 2 percent by 2070–2099 under both RCPs. Instead of the declines in annual GPP for the LCC as a whole, with warmer temperatures and a favorable water balance, GPP for regions associated with both spruce-fir and northern hardwood forests is projected to increase throughout the rest of this century, increasing by 4 to 6 percent by 2070–2099 under both RCPs.

Implications for Ecosystem Vulnerability

The results from the modeling analysis conducted using TOPS-BGC were driven with the recently released projections of future changes in climate and land use, and they highlight the vulnerability of ecosystem processes in the Great Northern and Appalachian LCCs to these emerging changes (table 7-6). In the Great Northern LCC, the projected increases in temperature drive the most important aspects of vulnerability via changes in timing of snowmelt and runoff. In the Appalachian LCC, vulnerability to climate changes is driven by increases in temperature with associated increases in evapotranspiration and runoff, but the net effects are partly mitigated by increases in precipitation, resulting in a net positive water balance for much of the Appalachian LCC through 2050 and possibly beyond.

In the Great Northern LCC, the projected changes indicate an important shift in the seasonality of runoff and vegetation productivity. As the winter temperatures warm, the growing season lengthens and more precipitation falls as rain than snow. As a result, evapotranspiration, runoff, and vegetation productivity all increase across the LCC in the fall and

Table 7-6. Summary of key patterns and trends.

Indicator	Key Trends and Patterns
Great Northern LCC	
Temperature	Models consistently predict increasing maximum and minimum temperatures. Increase exceeds 5 degrees C (9 degrees F) by 2099 under RCP 8.5.
Precipitation	Modest increases (6 to 16 percent) predicted by 2099 under the ensemble average, with a range of prediction across different climate models.
Snow water equivalent	Large reductions in snowpack projected, even in near term (2020–2049) under RCP 4.5. Important changes in timing of snow accumulation and snowmelt.
Soil moisture	Moderate but consistent declines in spring and summer soil moisture. Soil moisture serves as a reservoir that buffers changes in timing of snowmelt.
Evapotranspiration (ET)	Under ensemble average of climate models, projected declines in summer ET due to limited moisture availability and vegetation water stress. Increases in all other seasons.
Runoff	Spatially variable. Small overall changes in annual runoff but important changes in timing of runoff. Large increases projected in fall and winter, and important decreases in runoff during spring and summer seasons.
Vegetation productivity	Spatially variable across gradients in elevation. Small changes in annual productivity but important changes in seasonal patterns. Projected declines during summer due to vegetation water stress indicate vulnerability to disturbance by wildfire, insects, and pathogens.
Appalachian LCC	
Temperature	Models consistently predict increasing maximum and minimum temperatures. Increase exceeds 5.3 degrees C (9.5 degrees F) by 2099 under RCP 8.5.
Precipitation	Modest increases (7 to 12 percent) by 2099 predicted by ensemble average under RCP 8.5, but large range in projections across different climate models.
Snow water equivalent	Large reductions in snowpack projected, exceeding 70 percent by 2070–2099 under RCP 4.5 and 90 percent under RCP 8.5. Appalachians snow-free in most years by the end of this century.
Soil moisture	Projected changes in soil moisture through the end of this century are small, as long as precipitation meets or exceeds ensemble average projections for precipitation.
Evapotranspiration	Increases in all seasons, with larger relative increases in fall and winter.
Runoff	Small overall changes in annual runoff but important changes in seasonality of runoff. Large increases in winter runoff are offset by important decreases in runoff during spring and summer.
Vegetation productivity	Modest increases on annual basis through 2049 under both RCPs. Under RCP 8.5, declines projected by 2070–2099, with accelerating declines in summer productivity in the second half of this century.

winter. However, large decreases in runoff and vegetation growth during the spring and summer, when peak vegetation growth normally occurs, largely offset these gains. Reductions in summer vegetation growth, in particular, are driven by concurrent increases in maximum temperature and reductions in water availability, which are associated with earlier melt of the snowpack and reductions in runoff during the spring and summer.

This shift in seasonality, in turn, is likely to drive future changes in species composition within ecosystems that are not captured by the model. One limitation of the TOPS-BGC model is that it is parameterized at the biome level for major plant functional types and thus does not capture the dynamics associated with species movement and replacement within an ecosystem. TOPS-BGC also uses a fixed atmospheric carbon dioxide concentration and does not simulate stomatal response to increasing carbon dioxide levels and associated feedbacks to photosynthesis and plant water use efficiency that might favor C3 plant species in an ecosystem (Field, Jackson, and Mooney 1995). Despite the limitations of the model, sustained decreases in GPP, especially during the peak of the growing season, are a strong indicator that the dominant vegetation types are susceptible to increased mortality and replacement by other species better adapted to the new climate conditions. Evidence that this is already occurring has been reported by Van Mantgem and Stephenson (2007), who documented significant increases in mortality in forests across the western United States. Increasing outbreaks of bark beetles in forests in the Great Northern LCC (chaps. 10, 14, and 15) are also consistent with the projected declines in GPP, as the reductions in GPP are driven by increased plant water stress and reduced moisture availability in the summer. Plant water stress that is significant enough to reduce growth and vegetation productivity is also likely to increase vulnerability to attacks by bark beetles and other insects (e.g., Breshears et al. 2009).

Our results suggest that as hydrologic patterns shift in response to climate change, forests in the Great Northern LCC will be increasingly vulnerable to summer water stress and associated increases in mortality. While declines in GPP may indicate higher ecosystem vulnerability to climate change, increases in GPP can also indicate risk for ecosystem change through species turnover and replacement. Higher projected GPP indicates not only that future conditions may be more favorable for the current biome type, but also that climate conditions are increasingly favorable for vegetation growth in general. This would potentially allow lower-elevation species that may previously have been excluded by minimum temperatures or persistence of the snowpack to expand upslope, displacing species that are currently present. Chapters 9 and 10 address this potential for key species in the Great Northern LCC to shift in range toward higher elevations or northward toward higher latitudes.

Another limitation of the TOPS-BGC model is that, at present, it does not dynamically adjust for fire or other disturbance events. Westerling et al. (2006, 2011) have shown that intervals will decrease across the Greater Yellowstone Ecosystem, increasing the frequency of wildfires. Our projected declines in summer GPP are highly consistent with this finding, as our model projects increasing moisture limitations on vegetation growth and associated increases in plant water stress during the summer fire season. The persistent patterns of declining summer GPP and runoff from the TOPS-BGC analysis may offer insights into the pattern of increasing fire risk in the future across the Great Northern LCC landscape. Large increases in fire frequency would also accelerate both reductions in GPP and conversion of species and ecosystem types across the landscape.

Finally, for the Great Northern LCC as whole, while total impervious area is projected to double, changes in land use and impervious surface area affect less than 0.3 percent of the LCC. As described in chapter 6, these changes have important consequences for habitat fragmentation and may add barriers to migration of key species. The consequences for ecosystem processes, however, are localized, and the impacts associated with changes in climate conditions across the LCC are much larger.

Over the next century, the projected changes in ecosystem processes in the Appalachian LCC are more moderate, especially when compared against the projected changes across the Great Northern LCC. They are also more consistent spatially across the Appalachian LCC, although some variation across elevations is observed, with future declines in runoff and GPP more prevalent at lower elevations across the western half of the LCC. By the end of this century, snow is projected to become increasingly rare across the Appalachian LCC, and a shift in the timing of runoff and vegetation growth is also projected to occur in Appalachian LCC ecosystems as temperatures warm. However, in the Appalachian LCC, increases in evapotranspiration driven by warming temperatures appear to be largely offset by increases in precipitation, resulting in relatively small changes in annual runoff and GPP. Unfortunately, projections of future precipitation patterns are associated with higher uncertainty and have a wider range of projections across different models. Our results suggest that trends in precipitation should be closely monitored across Appalachian LCC parks and ecosystems, as without increases in precipitation, the LCC's ecosystems are much more likely to be vulnerable to accelerating reductions in soil moisture, runoff, and vegetation productivity.

This modeling analysis, done at a monthly time step, does not provide direct insights into the potential for increases in peak flows and flooding, which would require analysis of precipitation patterns at finer temporal scales. However, recent increases in the frequency of extreme weather events, and heavy precipitation events in particular, have been documented across the eastern United States (Melillo, Richmond, and Yohe 2014). Future extreme events would be likely to occur on top of overall increases in average winter runoff, and thus the trends documented in our analysis would suggest an increasing level of vulnerability across the Appalachian LCC to increases in extreme precipitation events.

As in the Great Northern LCC, our analysis does not capture changes in species composition or account for rates of invasion by nonnative species. Work described in chapters 8 and 11 explores the vulnerability of Appalachian LCC ecosystems to shifts in species in ranges. The projected changes in runoff and vegetation productivity during the spring and summer months would clearly indicate a vulnerability to replacement of species that are currently dominant in the LCC with species that are better adapted to future climate conditions. It is possible that shifts in species composition and even biome types across the LCC will in part offset the projected trends in runoff and GPP, but they would be associated with the disruption of trophic webs and a range of ecosystem services.

Conclusion

We conducted the modeling analysis described in this chapter to explore the potential changes in key ecosystem indicators to future changes in climate and land use. Our analysis applied a widely used ecosystem model and used data sets that capture spatial variations in topography, soils, and biome types. Our results provide insights into the likely types of changes in ecosystem processes that can be expected across the Great Northern and Appalachian LCCs through the end of this century. Over the coming decades, predicted changes in evapotranspiration, snow water equivalent, runoff and vegetation productivity for many regions in both LCCs exceed 10% in one or more seasons, even under RCP 4.5, suggesting important and observable changes in the seasonality of these processes. Sustained changes of this magnitude are consistent with increasing ecosystem vulnerability to insect outbreaks, wildfire, and invasive species, with corresponding changes in species composition and a range of ecosystem services. As a result, it will be important to consider these potential changes in ecosystem processes in establishing conservation targets and developing monitoring plans. It is our hope that the results described in this chapter will provide information that is useful for these planning efforts.

References

Bierwagen, B. G., D. M. Theobald, C. R. Pyke, A. Choate, P. Groth, J. V. Thomas, and P. Morefield. 2010. National housing and impervious surface scenarios for integrated climate impact assessments. *Proceedings of the National Academy of Sciences* 107 (49): 20887–92.

Breshears, D. D., O. B. Myers, C. W. Meyer, F. J. Barnes, C. B. Zou, C. D. Allen, N. G. McDowell, and W. T. Pockman. 2009. Tree die-off in response to global change-type drought: Mortality insights from a decade of plant water potential measurements. *Frontiers in Ecology and the Environment* 7:185–89. http://dx.doi.org/10.1890/080016.

Carlson, T. N. 2004. Analysis and prediction of surface runoff in an urbanizing watershed using satellite imagery. *Journal of the American Water Resources Association* 40 (4): 1087.

Christensen, N. S., A. W. Wood, N. Voisin, D. P. Lettenmaier, and R. N. Palmer. 2004. The effects of climate change on the hydrology and water resources of the Colorado River basin. *Climatic Change* 62 (1–3): 337–63.

Dale, V. H., S. Brown, R. A. Haeuber, N. T. Hobbs, N. Huntly, R. J. Naiman, W. E. Riebsame, M. G. Turner, and T. J. Valone. 2000. Ecological principles and guidelines for managing the use of land. *Ecological Applications* 10 (3): 639–70.

Daly, C., G. H. Taylor, W. P. Gibson, T. W. Parzybok, G. L. Johnson, and P. A. Pasteris. 2000. High-quality spatial climate data sets for the United States and beyond. *Transactions of the ASAE—American Society of Agricultural Engineers* 43 (6): 1957–62.

Farquhar, G. D., S. von Caemmerer, and J. A. Berry. 1980. A biochemical model of photosynthetic CO_2 assimilation in leaves of C3 species. *Planta* 149 (1): 78–90.

Field, C. B., R. B. Jackson, and H. A. Mooney. 1995. Stomatal responses to increased CO2: Implications from the plant to the global scale. *Plant, Cell & Environment* 18 (10): 1214–25.

Fisichelli, N. A., G. W. Schuurman, W. B. Monahan, and P. S. Ziesler. 2015. Protected area tourism in a changing climate: Will visitation at US national parks warm up or overheat? *PLOS ONE* 10 (6): e0128226. doi: 10.1371/journal.pone.0128226.

Gesch, D., M. Oimoen, S. Greenlee, C. Nelson, M. Steuck, and D. Tyler. 2002. The National Elevation Dataset. *Photogrammetric Engineering and Remote Sensing* 68 (1): 5–32.

Glick, P., B. A. Stein, and N. Edelson, eds. 2011. *Scanning the Conservation Horizon: A Guide to Climate Change Vulnerability Assessment*. Washington, DC: National Wildlife Federation.

Goetz, S., F. Melton, W. Wang, C. Milesi, and D. Theobald. 2009. Modeling strategies for adaptation to coupled climate and land use change in the United States. In *Proceedings of the World Bank 2009 Marseille Cities and Climate Change Urban Research Symposium*. World Bank. http://www.urs2009.net.

Grimm, N. B., F. S. Chapin III, B. Bierwagen, P. Gonzalez, P. M. Groffman, Y. Luo, F. Melton, K. Nadelhoffer, A. Pairis, P. A. Raymond, J. Schimel, and C. E. Williamson. 2013. The impacts of climate change on ecosystem structure and function. *Frontiers in Ecology and the Environment* 11 (9): 474–82.

Groffman, P. M., P. Kareiva, S. Carter, N. B. Grimm, J. Lawler, M. Mack, V. Matzek, and H. Tallis. 2014. Ecosystems, biodiversity, and ecosystem services. Chap. 8 in *Climate Change Impacts in the United States: The Third National Climate Assessment*, edited by J. M. Melillo, T. C. Richmond, and G. W. Yohe, 195–219. US Global Change Research Program. doi: 10.7930/J0TD9V7H.

Hansen, A. J., and R. DeFries. 2007. Ecological mechanisms linking protected areas to surrounding lands. *Ecological Applications* 17 (4): 974–88.

Hansen, A. J., J. J. Rotella, M. L. Kraska, and D. Brown. 2000. Spatial patterns of primary productivity in the Greater Yellowstone Ecosystem. *Landscape Ecology* 15:505–22.

Heinsch, F. A., Z. Maosheng, S. W. Running, J. S. Kimball, R. R. Nemani, K. J. Davis, P. V. Bolstad, et al. 2006. Evaluation of remote sensing based terrestrial productivity from MODIS using regional tower eddy flux network observations. *IEEE Transactions on Geoscience and Remote Sensing* 44 (7): 1908–25.

Homer, C., C. Huang, L. Yang, B. Wylie, and M. Coan. 2004. Development of a 2001 national land-cover database for the United States. *Photogrammetric Engineering & Remote Sensing* 70 (7): 829–40.

Ichii, K., W. Wang, H. Hashimoto, F. Yang, P. Votava, A. R. Michaelis, and R. R. Nemani. 2009. Refinement of rooting depths using satellite-based evapotranspiration seasonality for ecosystem modeling in California. *Agricultural and Forest Meteorology* 149 (11): 1907–18.

Ichii, K., M. A. White, P. Votava, A. Michaelis, and R. R. Nemani. 2008. Evaluation of snow models in terrestrial biosphere models using ground observation and satellite data: Impact on terrestrial ecosystem processes. *Hydrological Processes* 22 (3): 347–55.

Jennings, M. D. 2000. Gap analysis: Concepts, methods, and recent results. *Landscape Ecology* 15 (1): 5–20.

Law, B. 2007. AmeriFlux network aids global synthesis. *Eos, Transactions American Geophysical Union*, 88 (28): 286–86.

Melillo, J. M., T. C. Richmond, and G. W. Yohe, eds. 2014. The Third National Climate Assessment. US Global Change Research Program. doi: 10.7930/J0TD9V7H.

Millennium Ecosystem Assessment. 2005. *Ecosystems and Human Well-being: Synthesis*. Washington, DC: Island Press.

Monahan, W. B., T. Cook, F. Melton, J. Connor, and B. Bobowski. 2013. Forecasting distributional responses of limber pine to climate change at management-relevant scales in Rocky Mountain National Park. *PLOS ONE* 8 (12): e83163. doi: 10.1371/journal.pone.0083163.

Monahan, W. B., and N. A. Fisichelli. 2014. Climate exposure of US national parks in a new era of change. *PLOS ONE* 9 (7): e101302. doi: 10.1371/journal.pone.0101302.

Myneni, R. B., S. Hoffman, Y. Knyazikhin, J. L. Privette, J. Glassy, Y. Tian, Y. Wang, X. Song, Y. Zhang, G. R. Smith, et al. 2002. Global products of vegetation leaf area and fraction absorbed PAR from year one of MODIS data. *Remote Sensing of Environment* 83 (1): 214–31.

Nemani, R., H. Hashimoto, P. Votava, F. Melton, W. L. Wang, A. Michaelis, L. Mutch, C. Milesi, S. Hiatt, and M. White. 2009. Monitoring and forecasting protected area ecosystem dynamics using the Terrestrial Observation and Prediction System (TOPS). *Remote Sensing of Environment* 113 (7): 1497–1509.

Nemani, R. R., C. D. Keeling, H. Hashimoto, W. M. Jolly, S. C. Piper, C. J. Tucker, R. B. Myneni, and S. W. Running. 2003. Climate-driven increases in global terrestrial net primary production from 1982 to 1999. *Science* 300 (5625): 1560–63.

Nemani, R., P. Votava, A. Michaelis, M. White, F. Melton, J. Coughlan, K. Golden, H. Hashimoto, K. Ichii, L. Johnson, et al. 2007. Terrestrial Observation and Prediction System (TOPS): Developing ecological nowcasts and forecasts by integrating surface, satellite and climate data with simulation models. In *Research and Economic Applications of Remote Sensing Data Products*, edited by U. Aswathanarayana and R. Balaii. London: Taylor and Francis.

NRCS (Natural Resources Conservation Service), US Department of Agriculture, Soil Survey Staff. 2015. Web soil survey. http://websoilsurvey.nrcs.usda.gov.

Rose, S., and N. E. Peters 2001. Effects of urbanization on streamflow in the Atlanta area (Georgia, USA): A comparative hydrological approach. *Hydrological Processes* 15 (8): 1441–57.

Stein, B. A., P. Glick, N. Edelson, and A. Staudt, eds. 2014. *Climate-Smart Conservation: Putting Adaptation Principles into Practice*. Washington, DC: National Wildlife Federation.

Taylor, K. E., R. J. Stouffer, and G. A. Meehl. 2012. An overview of CMIP5 and the experiment design. *Bulletin of the American Meteorological Society* 93 (4): 485–98.

Theobald, D. M. 2005. Landscape patterns of exurban growth in the USA from 1980 to 2020. *Ecology and Society* 10:32. http://www.ecologyandsociety.org/vol10/iss1/art32/.

Theobald, D. M., and N. T. Hobbs. 1998. Forecasting rural land-use change: A comparison of regression- and spatial transition-based models. *Geographical and Environmental Modelling* 2:65–82.

Thornton, P. E., B. E. Law, H. L. Gholz, K. L. Clark, E. Falge, D. S. Ellsworth, A. H. Goldstein, R. K. Monson, D. Hollinger, M. Falk, J. Chen, and J. P. Sparks. 2002. Modeling and measuring the effects of disturbance history and climate on carbon and water budgets in evergreen needleleaf forests. *Agricultural and Forest Meteorology* 113:185–222.

Thornton, P. E., and N. A. Rosenbloom. 2005. Ecosystem model spin-up: Estimating steady state conditions in a coupled terrestrial carbon and nitrogen cycle model. *Ecological Modelling* 189:25–48.

Thornton, P. E., and S. W. Running. 1999. An improved algorithm for estimating incident daily solar radiation from measurements of temperature, humidity, and precipitation. *Agricultural and Forest Meteorology* 93:211–28.

Thornton, P. E., S. W. Running, and M. A. White. 1997. Generating surfaces of daily meteorological variables over large regions of complex terrain. *Journal of Hydrology* 190:214–51.

Thrasher, B., J. Xiong, W. Wang, F. Melton, A. Michaelis, and R. Nemani. 2013. Downscaled climate projections suitable for resource management. *Eos, Transactions American Geophysical Union* 94 (37): 321–23.

Van Mantgem, P. J., and N. L. Stephenson. 2007. Apparent climatically induced increase of tree mortality rates in a temperate forest. *Ecology Letters* 10 (10): 909–16.

Wang, W., J. Dungan, H. Hashimoto, A. R. Michaelis, C. Milesi, K. Ichii, and R. R. Nemani. 2011. Diagnosing and assessing uncertainties of terrestrial ecosystem models in a multimodel ensemble experiment: 1. Primary production. *Global Change Biology* 17 (3): 1350–66.

Westerling, A. L., H. G. Hidalgo, D. R. Cayan, and T. W. Swetnam. 2006. Warming and earlier spring increase western US forest wildfire activity. *Science* 313 (5789): 940–43.

Westerling, A. L., M. G. Turner, E. A. Smithwick, W. H. Romme, and M. G. Ryan. 2011. Continued warming could transform Greater Yellowstone fire regimes by mid-21st century. *Proceedings of the National Academy of Sciences of the United States of America* 108 (32): 13165–70.

White, M. A., and R. R. Nemani. 2004. Soil water forecasting in the continental United States: Relative forcing by meteorology versus leaf area index and the effects of meteorological forecast errors. *Canadian Journal of Remote Sensing* 30:717–30.

White, M. A., and Nemani, R. R. 2006. Real-time monitoring and short-term forecasting of land surface phenology. *Remote Sensing of Environment* 104 (1): 43–49.

White, M. A., P. E. Thornton, S. W. Running, and R. R. Nemani. 2000. Parameterization and sensitivity analysis of the BIOME-BGC terrestrial ecosystem model: Net primary production controls. *Earth Interactions* 4:1–85.

Yang, F., K. Ichii, M. A. White, H. Hashimoto, A. R. Michaelis, P. Votava, A. Zhu, A. Huete, S. W. Running, and R. R. Nemani. 2007. Developing a continental-scale measure of gross primary production by combining MODIS and Ameri-Flux data through Support Vector Machine approach. *Remote Sensing of Environment* 110 (1): 109–22.

Zhao, M., F. A. Heinsch, R. R. Nemani, and S. W. Running. 2005. Improvements of the MODIS terrestrial gross and net primary production global data set. *Remote Sensing of Environment* 95 (2): 164–76.

Chapter 8

Potential Impacts of Climate Change on Vegetation for National Parks in the Eastern United States

Patrick Jantz, William B. Monahan,
Andrew J. Hansen, Brendan M. Rogers,
Scott Zolkos, Tina Cormier, and Scott J. Goetz

Forests in the eastern United States have a long history of change related to climate and land use. Eighteen thousand years ago, temperatures were considerably lower and glaciers covered much of the area where deciduous forests currently grow. As glaciers retreated and temperatures rose, tree species advanced from southern areas (Delcourt and Delcourt 1988) and may also have dispersed from low-density populations near the edge of the Laurentide ice sheet (McLachlan, Clark, and Manos 2005). A variety of other processes have also influenced the distribution of tree species. Derechos, tornadoes, and fires cause frequent, small- to intermediate-scale disturbances that are important influences on canopy structure and species composition, while larger disturbances, such as hurricanes, cause less frequent but more extensive changes (Dale et al. 2001).

Regional- to continental-scale disturbances during the past several centuries include extensive deforestation by European settlers (Cronon 1983; Houghton and Hackler 2001; Thompson et al. 2013) and invasive pests and pathogens, such as chestnut blight (*Cryphonectria parasitica*) and hemlock woolly adelgid (*Adelges tsugae*), that cause severe declines in populations of individual species. Many forests in the eastern United States are still recovering from postsettlement deforestation, often with altered species composition (Bürgi, Russell, and Motzkin 2000). Fire suppression has changed the abundance and distribution of tree species across large areas (Nowacki and Abrams 2008). Drought has historically played a role in eastern US forests (e.g., a drought five thousand years ago affecting hemlock distribution; Haas and McAndrews 2000) and may increasingly do so in the future—for example, it has been suggested that recent drought induced die-offs in the Southeast are indicative of future climate risks to forests as temperatures rise (Allen et al. 2010).

As a whole, these events have significantly altered the relative abundance and geographic distribution of eastern tree species. The landscape we see today was formed by a combination of these processes, and the replacement of forest with agricultural and residential land uses has resulted in widespread forest fragmentation (Riitters, Coulston, and Wickham 2012). Assessing how these forested landscapes might respond to climate change across the Appalachian Landscape Conservation Cooperative (Appalachian LCC) region, and across the large protected area centered ecosystems (PACEs) within it, is the primary goal of this chapter. Climate and land use change are projected to occur at unprecedented rates in the Appalachian LCC (chap. 5; US Environmental Protection Agency 2009). Land managers stand to benefit from plans that anticipate future conditions.

There are no recent analogs for the degree of climate warming expected for the twenty-first century. Therefore, much of our understanding of potential climate change impacts on North American forests is derived from environmental models (table 8-1). Dynamic global vegetation models have been used for assessing broad vegetation response to climate change but are inadequate for assessing responses of individual species. Species distribution models were developed specifically for understanding species-environment relationships and are now widely used for assessing potential effects of climate change on species distributions (table 8-1; Prasad, Iverson, and Liaw 2006; Iverson et al. 2008; McKenney et al. 2011; Potter and Hargrove 2013; Pederson et al. 2014).

Species distribution models are statistical models of the relationship between species distributions and descriptors of environmental conditions, such as temperature, precipitation, light regime, and soil properties (Franklin 2009). When used with gridded environmental data, predictive maps of species distributions can be generated. Species distribution models rely on the idea of the fundamental niche (where a species could exist), which hypothesizes that a species' physiological requirements can only be met by particular combinations of environmental conditions. Additional limiting factors, such as competition, disturbance, and legacy populations, are known to influence the realized niche (where a species actually exists) (Hutchinson 1957). Although the limitations of species distribution models are well known—for example, they generally do not include biotic interactions, population dynamics, dispersal, or evolutionary adaptation (Pear-

TABLE 8-1. Summary of studies assessing potential tree species or biome response to climate change in the eastern United States.

Study	Method	Response Variable	Extent	Spatial Resolution	Primary Findings
Iverson et al. (2008)	Random forests	Species importance values	Eastern North America	20 km	Suitability loss for species in spruce-fir and northern hardwood forest types
McKenney et al. (2011)	Bioclimatic	Species presence–absence	North America	10 km	Suitability loss for most species in eastern United States
Potter and Hargrove (2013)	Ecoregion matching	Species presence–absence	Global	2 km	Suitability loss for most species in eastern United States
Gonzalez et al. (2010)	Dynamic global vegetation model	Biome	Global	50 km	Relatively high confidence for northeastern biome shift
Rehfeldt et al. (2012)	Random forests	Biome	North America	1 km (at equator)	Loss of eastern subalpine forest and tundra, minimal change in temperate deciduous forest, stable boundary between eastern temperate deciduous forest and deciduous and evergreen forest

son and Dawson 2003)—they remain a valuable tool for understanding first-order effects of climate change on tree species by identifying broad bioclimatic areas of the landscape within which finer-scale distributional changes are anticipated to occur (Jackson et al. 2009).

Species distribution models can inform step two of the Climate-Smart Conservation framework, in which climate impacts and vulnerabilities are evaluated (chap. 2; Stein and Glick 2011). By estimating changes in climate suitability, species distribution models effectively combine climate exposure and species' inherent sensitivity to climate change. The combination of these properties represents the potential impact of climate change. The adaptive capacity of a species can be incorporated to further evaluate vulnerability to climate change (chap. 11).

Three groups have used species distribution models to model changing climate suitability for eastern US tree species: Iverson et al. (2008), McKenney et al. (2011), and Potter and Hargrove (2013). The work by Iverson et al. (2008) is an important component of US Forest Service climate adaptation evaluation and reporting (Butler et al. 2015). In a comparison of results for thirty-five species common to the work of McKenney et al. (2011) and Iverson et al. (2008), Zolkos et al. (2015) found that more than 40 percent of species were projected to lose suitable habitat space in the Appalachian LCC (fig. 8-1). The studies agreed on eight of ten species projected to have the largest magnitude of loss of suitable habitat space (table 8-2) under a high greenhouse gas emissions scenario, although they showed less agreement on the specific magnitude of change. Results from both groups were in agreement that suitable habitat would increase for

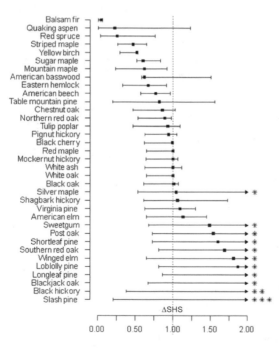

FIGURE 8-1   Change in suitable habitat space (ΔSHS) from species distribution models aggregated from two studies (Iverson et al. 2008 and McKenney et al. 2011) and driven by six general circulation models assuming a high greenhouse gas emissions scenario. Values on the x-axis show ΔSHS by 2100 across the Appalachian LCC. ΔSHS was calculated by dividing SHS in 2100 by historical SHS. Squares represent median ΔSHS. Values above one represent an increase in SHS, while values below one represent a decrease. Arrows indicate ΔSHS > 2, where * = max ΔSHS > 2; ** = max ΔSHS > 10; and *** = max and median ΔSHS > 10. (Adapted from Zolkos et al. 2015.)

TABLE 8-2. Ten tree species with highest projected loss of suitable habitat space in the Appalachian LCC ranked by Iverson et al. (2008) and McKenney et al. (2011).

Common Name	[i]ΔSHS (A1FI)	Common Name	[m]ΔSHS (A2)
Mountain maple[t]	0.08	Balsam fir[t]	0.07
Balsam fir[t]	0.56	Red spruce[t]	0.18
Yellow birch[t]	0.58	Mountain maple[t]	0.27
Striped maple[t]	0.64	Table mountain pine	0.29
Eastern hemlock[t]	0.79	Quaking aspen[t]	0.29
Sugar maple	0.8	Striped maple[t]	0.31
Quaking aspen[t]	0.82	Yellow birch[t]	0.34
Red spruce[t]	0.87	Slash pine	0.34
American beech	0.93	Eastern hemlock[t]	0.36
Longleaf pine[t]	1	Longleaf pine[t]	0.42

Note: Tree names followed by s [t] are common to both rankings. Column values show average change in suitable habitat space (ΔSHS) under high greenhouse gas emissions scenarios (A1FI for Iverson et al. 2008; A2 for McKenney et al. 2011). ΔSHS is calculated by dividing SHS in 2100 by current SHS. Values above one represent an increase in SHS; values below one represent a decrease. Values equal to one indicate no change.

hickories, oaks, and pines in the PACEs of the Appalachian LCC (Hansen et al. 2011).

Both groups also agreed on the loss of suitable habitat in PACEs for many constituents of spruce-fir and northern hardwood forests. Potter and Hargrove (2013) projected rangewide loss of suitable habitat for many of those same species. The largest declines they reported were for table mountain pine (*Pinus pungens*), silver maple (*Acer saccharinum*), American basswood (*Tilia americana*), blackjack oak (*Quercus marilandica*), and Virginia pine (*Pinus virginiana*), with more than 50 percent suitable habitat space lost by 2050.

Previous work has provided an important understanding of potential impacts of climate change at regional scales. However, the Appalachian LCC contains some of the most topographically complex terrain in the conterminous United States (Theobald et al. 2015), and much of the associated climate variability is averaged out at the scale of tens of miles (the resolution used in the above studies). Protected area managers need climate vulnerability assessments at spatial scales fine enough to resolve climate gradients within protected areas (or PACEs) but over geographic domains large enough to provide a regional ecological and management context (Thrasher et al. 2013). Newly developed data sets make these types of assessments possible, such as those featured in chapters 4, 5, and 7. Our analysis focuses on three PACEs containing four National Park Service units, thereby sampling a broad latitudinal gradient within the Appalachian LCC (chap. 1).

After consultation with National Park Service managers and reviewing management documents and scientific literature, we identified forty tree species of potential management concern. We selected trees that are canopy dominants in many of the ecological systems within focal park units, trees that provide critical ecosystem functions (e.g., mast or deep shade), and trees that are relatively rare in the Appalachian LCC but which may become more common in the future. Here we present results of distribution models for these species at management-relevant scales.

Study Areas

Although the tree species distribution models include all of the eastern United States east of the 100th meridian, for summarizing we focus on two geographic extents: (1) the entire Appalachian LCC, and (2) protected area centered ecosystems. We focus on PACEs centered on four park units: Great Smoky Mountains National Park, Shenandoah National Park, Delaware Water Gap National Recreation Area (NRA), and Upper Delaware Scenic and Recreational River. The last two National Park Service units are both fully contained within the Delaware Water Gap PACE, thus resulting in a total of three PACEs considered in our analysis.

Forests in the eastern United States transition from longleaf-slash pine and loblolly-shortleaf pine in the south to oak-pine and oak-hickory in the central Appalachians, maple-beech-birch in the northern Appalachians, and spruce-fir forests in Maine. Because of steep elevation gradients and a broad latitudinal range, the Appalachian LCC contains a large number of the species that make up these forest types. Most of the land in the Appalachian LCC (87 percent) is private with no use restrictions (US Geological Survey 2012). Four percent of the Appalachian LCC is under federal or state management for maintenance of natural processes and biodiversity, while an additional 8 percent is under federal or state management that allows multiple uses.

Tree Species

We modeled the presence and absence of forty eastern tree species (table 8-3). We selected a mix of common species, such as red maple (*Acer rubrum*), American beech (*Fagus grandifolia*), yellow birch (*Betula alleghaniensis*), white oak (*Quercus alba*), and eastern white pine (*Pinus strobus*), as well as other species that provide food resources or are associated with specific habitats of importance, such as black cherry (*Prunus serotina*), red spruce (*Picea rubens*), balsam fir (*Abies balsamea*), and eastern hemlock (*Tsuga canadensis*). Presence-absence observations were taken from over one hundred thousand US Forest Service Forest Inventory and Analysis plots surveyed in the eastern United States from 2000 to 2010 (fig. 8-2). Although our focus is on the Appalachian LCC and PACEs, we fit models using plots from across the eastern United States in order to capture a wider range of climate conditions for each tree species. For reporting,

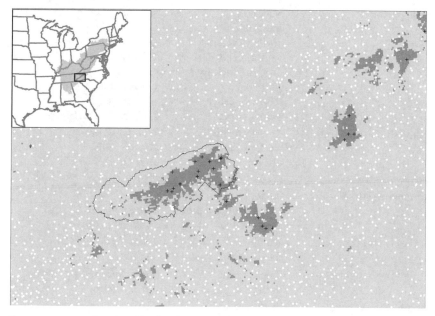

FIGURE 8-2   Species distribution model output for red spruce (*Picea rubens*) centered on Great Smoky Mountains National Park, shown in black outline. Darker gray pixels correspond to areas classified as suitable. Unoccupied Forest Inventory and Analysis (FIA) points are shown as white circles. Occupied FIA points are represented by black crosses. The outline of the Appalachian LCC is shown in transparent gray in the inset map.

TABLE 8-3. Tree species modeled in this chapter and associated forest groups.

Forest Group	Tree Species	Scientific Name
Spruce-fir	balsam fir	*Abies balsamea*
	red spruce	*Picea rubens*
Northern pines	eastern hemlock	*Tsuga canadensis*
	eastern white pine	*Pinus strobus*
Maple-beech-birch	American basswood	*Tilia americana*
	American beech	*Fagus grandifolia*
	black cherry	*Prunus serotina*
	red maple	*Acer rubrum*
	sugar maple	*Acer saccharum*
	white ash	*Fraxinus americana*
Northern hardwoods	mountain maple	*Acer spicatum*
	quaking aspen	*Populus tremuloides*
	striped maple	*Acer pensylvanicum*
	yellow birch	*Betula alleghaniensis*
Oak-hickory	black oak	*Quercus velutina*
	chestnut oak	*Quercus prinus*
	northern red oak	*Quercus rubra*
	pignut hickory	*Carya glabra*
	shagbark hickory	*Carya ovata*
	white oak	*Quercus alba*
	yellow buckeye	*Aesculus flava*
	yellow poplar	*Liriodendron tulipifera*
Elm-ash-cottonwood	American elm	*Ulmus americana*
	black walnut	*Juglans nigra*
	silver maple	*Acer saccharinum*
Central and mountain pines	Fraser fir	*Abies fraseri*
	pitch pine	*Pinus rigida*
	table mountain pine	*Pinus pungens*
	Virginia pine	*Pinus virginiana*
Loblolly-shortleaf pine	loblolly pine	*Pinus taeda*
	shortleaf pine	*Pinus echinata*
Longleaf-slash pine	longleaf pine	*Pinus palustris*
	slash pine	*Pinus elliottii*
Southern oak-hickory	black hickory	*Carya texana*
	blackjack oak	*Quercus marilandica*
	mockernut hickory	*Carya alba*
	post oak	*Quercus stellata*
	red hickory	*Carya ovalis*
	sweetgum	*Liquidambar styraciflua*
	winged elm	*Ulmus alata*

we grouped species by US Forest Service forest types (Ruefenacht et al. 2008) with minor modifications based on conditions in the Appalachian LCC (table 8-3).

Environmental Predictors of Tree Species Distributions

Climate

We used publicly available Parameter-elevation Regressions on Independent Slopes Model (PRISM) monthly climate normals (thirty-year averages of climate variables) for the 1981–2010 period (Daly et al. 2008) to represent contemporary climate conditions. We derived a set of nineteen bioclimatic variables (Hijmans et al. 2005) from the PRISM precipitation, minimum temperature, and maximum temperature data, at 0.5-mile (800-meter) resolution. To represent heat accumulation throughout the year, we created an additional variable—growing degree days—derived using an algorithm for monthly data (Sork et al. 2010).

To represent potential future climate conditions, we used ensemble means of thirty-three Coupled Model Intercomparison Project Phase 5 (CMIP5) general circulation models for two representative concentration pathways (RCPs): 4.5 and 8.5 from the National Aeronautics and Space Administration (NASA) Earth Exchange Downscaled Climate Projections (NEX-DCP30) data set (Thrasher et al. 2013). RCP 4.5 is considered a relatively low greenhouse gas emissions scenario, resulting in an atmospheric carbon dioxide concentration of 650 parts per million (PPM) by the year 2100. RCP 8.5 is considered a high-emissions scenario and corresponds more closely to our current emissions trend, rising from the current 400 PPM to 1,370 PPM by 2100.

From these gridded projections, we calculated thirty-year averages of precipitation, minimum temperature, and maximum temperature. We bias-adjusted the resulting time series using the delta, or perturbation method, by adding global climate model–simulated changes from historical climate for every future month to the PRISM normals. We generated the same set of nineteen bioclimatic variables plus growing degree days from the bias-adjusted projections. We report results for the thirty-year period centered on 2055 because it is a management-relevant planning horizon and our results are less likely to be affected by no-analog climate conditions.

Soils

Databases of soil chemical and structural properties can be used to derive indicator variables that relate to soil properties, such as nutrient and water availability, that are important determinants of plant species distributions. Information on soil properties was derived from the State Soil Geographic Dataset (STATSGO) soils database and consisted of pH, percent sand, percent silt, percent clay, bulk density, depth to bedrock, and water-holding capacity. Soil layers were extracted from a vector database of soil characteristics created by Miller and White (1998) and gridded to 886-foot (270-meter) resolution before interpolating to 0.5-mile resolution to match the bioclimatic data.

Topographic Metrics

Topographic variability greatly influences the amount of light, heat, and water available for plant growth. We calculated two topographic metrics at ~100-foot (30-meter) resolution from the US Geological Survey National Elevation Dataset: topographically distributed solar radiation and topographic wetness. Solar radiation was calculated using the "insol" package (Corripio 2003) in "R" using the approach outlined in Pierce, Lookingbill, and Urban (2005), which uses hourly calculations of hillshading for a representative day in each month as a proxy for potential clear sky direct radiation received over the course of a year. A topographic wetness index was calculated as a function of the upstream contributing area of a pixel and the slope of that pixel. Higher index values correspond to areas likely to experience hydrologic flow accumulation, such as the lower portions of stream courses. We interpolated and aggregated the ~100-foot layers to 0.5-mile resolution to match the climate data layers.

Tree Species Distribution Models

We used the "randomForest" R package (Liaw and Wiener 2002) to estimate the probability of presence for each tree species individually as a function of the climate, soil, and topographic variables. For model evaluation and to identify probability thresholds corresponding to species presences, we fit models using a random sample of 80 percent of Forest Inventory and Analysis plots, reserving the remaining 20 percent for testing and threshold calculation. To avoid collinearity issues, we identified each temperature, precipitation, and soil variable with the highest spearman rank correlation with presence-absence observations. We then added the remaining variables to the model if they were uncorrelated (spearman rho < 0.7) with one another or with the variables with the highest spearman coefficient. Potential relative radiation and topographic wetness were allowed to be in all models because they are expected to provide additional information on fine-scale topographic variability.

This set of variables was used to fit one model for each species. We then discarded variables that contributed relatively little to model performance and used the remaining subset to fit another model for each species. Model fit and performance were evaluated using sensitivity (the fraction of true presences classified as such), specificity (the fraction of true absences classified as such), the area under the receiver operating characteristic curve (AUC), an indicator of how well models discriminate between presences and absences, and slope and intercept estimates derived from calibration plots, which show how well modeled prevalence corresponds to observed prevalence. Generally, AUC values > 0.7 are considered "good." Calibration plot slope coefficient values close to one indicate good correspondence between modeled probabilities and observed prevalence.

Our methods are most similar to those of Iverson et al. (2008) but differ in our use of tree species presence-absence as the response variable as well as in variable selection methods. The approach used by Potter and Hargrove (2013) consists of a priori delineation of ecoregions using soil, climate, and solar insolation variables. Tree species presences-absences are then matched to particular ecoregions. The approach used by McKenney et al. (2011) uses bioclimatic envelopes to describe the suitable climate space associated with particular species' distributions. In addition, McKenney et al. (2011) included data for both the United States and Canada in their efforts.

For mapping and trend analysis, we fit models using all of the Forest Inventory and Analysis points to avoid potential bias related to using a single subsample of the data. Behavior was similar between models using the 80 percent subsample and the full set of plots, and the variables selected for each were identical, indicating that model evaluations using the 20 percent testing data set are reasonable indicators of the performance of the models using the full set of plots. NASA Earth Exchange computing resources were used to create potential relative radiation and bioclimatic variables as well as random forest models.

Impact of Climate Change on Eastern Tree Species

Southern oak, hickory, and pine species, along with those in the elm-ash-cottonwood forest type, were projected to have the largest net increase in suitability in the Appalachian LCC as well as in individual PACEs (fig. 8-3). Northern hardwoods, northern pines, and members of the maple-beech-birch forest type were projected to have the largest decreases across the Appalachian LCC. High-elevation species with smaller areas of suitable habitat within PACEs, such as red spruce and balsam fir, were projected to have small absolute but large relative changes. Red spruce in Great Smoky Mountains National Park, for example, was projected to lose more than 45 percent of its currently suitable habitat by 2055 under both emissions scenarios.

In Delaware Water Gap NRA, yellow birch and silver maple, both in the northern hardwood group, were projected to lose the most suitable habitat, and gains were expected to be largest for elm-ash-cottonwood and oak-hickory species. This is in contrast to Great Smoky Mountains and Shenandoah national parks, where suitability gains were expected to be largest for southern oak-hickory species. Sugar maple was projected to have the greatest net loss of suitable habitat across the Appalachian LCC. Eastern hemlock, eastern white pine, and yellow birch were projected to have the highest local losses in individual PACEs. Eastern white pine and red maple were frequently in the top three species expected to lose suitable climate space within PACEs.

Changes in suitability were predictably less extreme under the low greenhouse gas emissions scenario (RCP 4.5). Both net gains and net losses in suitability tended to be smaller than for the high-emissions scenario. This was true across all PACEs. In most cases, the direction of change across scenarios was the same, except for yellow poplar across the Appalachian LCC and black cherry in the Shenandoah PACE. The maximum temperature of the warmest month was most frequently selected as the most important variable. An index of topographic wetness, mean temperature of the wettest quarter, mean temperature of the driest quarter, and potential relative radiation were also selected multiple times. Temperature, either its magnitude or variability, was disproportionately represented among the most important variables selected, making up ten of the thirteen climate variables chosen as most important for at least one species. Precipitation was infrequently chosen as the most important variable.

Based on AUC scores, most of the models were able to reliably differentiate between species presences and absences (table 8-4; fig. 8-2). How-

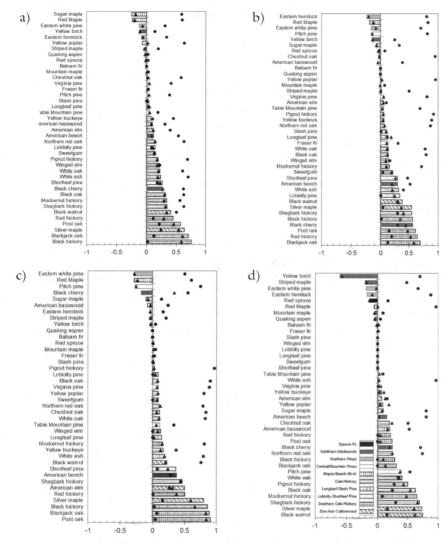

FIGURE 8-3   Net change in suitable climate space for forty eastern tree species across the Appalachian LCC grouped into ten forest types. Changes were calculated using a high representative concentration pathway (RCP 8.5) greenhouse gas emissions scenario. Values on the x-axis show net change in suitable area for each species by the year 2055 normalized by area. Positive values indicate increases in suitable area. Negative values indicate decreases. Triangle markers show expected net change under low RCP emissions (RCP 4.5). Circle markers show the fraction of suitable habitat space for the historical period. Bar patterns indicate the Forest Inventory and Analysis forest type to which each species belongs: (a) Appalachian LCC; (b) Great Smoky Mountains National Park; (c) Shenandoah National Park; (d) Delaware Water Gap National Recreation Area.

TABLE 8-4. Summaries of model performance.

Species	AUC	Threshold	Training Presence	Testing Presence	Intercept	Slope	First Variable*
Balsam fir	0.97	0.224	8579	2190	0.031	0.929	gdd (−)
Fraser fir[t]	0.67	0.002	12	5	0.003	-0.005	bio2 (−)
Red spruce	0.99	0.076	2543	660	0.043	0.911	bio5 (−)
Shortleaf pine	0.92	0.1	5439	1352	0.012	0.898	bio9 (+)
Slash pine	0.98	0.078	3003	808	0.028	0.888	bio8 (+)
Longleaf pine[t]	0.95	0.028	1741	477	-0.014	0.891	bio8 (+)
Table mountain pine[t]	0.89	0.002	108	27	0.005	-0.009	twi (−)
Pitch pine	0.95	0.016	768	206	-0.027	1.01	bio5 (−)
Eastern white pine	0.9	0.102	6623	1651	0.03	0.88	bio10 (−)
Loblolly pine	0.96	0.36	15391	3869	0.026	0.956	bio6 (+)
Virginia pine	0.95	0.066	3188	804	0.074	0.659	bio15 (−)
Eastern hemlock	0.93	0.102	4857	1195	0.027	0.829	bio5 (−)
Striped maple[t]	0.96	0.044	1834	524	0.017	0.818	bio5 (−)
Red maple	0.79	0.452	33042	8352	0.066	0.849	bio5 (−)
Silver maple	0.9	0.012	1184	308	-0.025	0.878	twi (+)
Sugar maple	0.89	0.25	14560	3613	0.029	0.898	bio5 (−)
Mountain maple[t]	0.94	0.01	796	208	0.063	0.475	bio1 (−)
Yellow buckeye[t]	0.95	0.01	522	139	0.031	0.276	twi (−)
Yellow birch	0.94	0.1	5685	1417	0.025	0.905	bio5 (−)
Pignut hickory[t]	0.86	0.132	7683	1882	0.024	0.781	twi (−)
Shagbark hickory[t]	0.89	0.086	4822	1166	0.023	0.698	bio10 (−)
Black hickory	0.97	0.054	2674	670	0.084	0.651	bio18 (−)
Mockernut hickory	0.84	0.13	8231	2050	0.219	0.001	bio9 (+)
Red hickory[t]	0.92	0.002	60	9	0.004	-0.007	bio8 (−)
American beech	0.88	0.138	8435	2057	0	0.933	bio5 (−)
White ash	0.83	0.154	10183	2489	0.022	0.808	srad (−)
Black walnut	0.89	0.064	3922	966	0.116	0.297	bio4 (+)
Sweetgum	0.95	0.32	15680	3899	0.047	0.898	srad (+)
Yellow poplar	0.91	0.196	10601	2597	0.034	0.858	bio7 (−)
Quaking aspen	0.93	0.176	9348	2334	0.055	0.817	bio11 (−)
Black cherry	0.75	0.224	16388	4130	0.036	0.791	bio12 (+)
White oak	0.85	0.256	16362	4012	0.019	0.9	twi (−)
Blackjack oak	0.89	0.018	1800	450	-0.019	0.967	bio5 (+)
Chestnut oak	0.95	0.104	4760	1211	0.023	0.886	twi (−)
Northern red oak	0.82	0.21	13148	3193	0.044	0.784	bio5 (−)
Post oak	0.9	0.124	7491	1886	0.022	0.856	bio5 (+)

TABLE 8-4. (*Continued*)

Species	AUC	Threshold	Training Presence	Testing Presence	Intercept	Slope	First Variable*
Black oak	0.86	0.16	9654	2392	0.024	0.824	srad (−)
American basswood	0.85	0.064	5054	1239	0.018	0.803	srad (−)
Winged elm	0.93	0.144	6859	1746	0.073	0.699	bio9 (+)
American elm	0.82	0.13	11798	2974	0.054	0.755	bio8 (+)

Note: "AUC" gives the area under the receiver operating characteristic curve (models with values > 0.7 are considered "good"). "Threshold" is the probability of presence value used to distinguish presences and absences and was chosen to equalize sensitivity and specificity values. "Training Presence" is the number of plots in the training data set in which a species is present. "Testing Presence" is the number of plots in the testing data set in which a species is present. "Intercept" and "Slope" refer to regression coefficients derived from the calibration plot. "First Variable" gives the most important variable and direction of its influence determined by spearman correlation. Species names followed by a [t] are those for which the "all variables" model was used for reporting; otherwise, the most parsimonious set of variables was used.

* bio1 = annual mean temperature, bio2 = mean diurnal range, bio4 = temperature standard deviation, bio5 = maximum temperature of warmest month, bio6 = minimum temperature of coldest month, bio7 = temperature annual range, bio8 = mean temperature of wettest quarter, bio9 = mean temperature of driest quarter, bio10 = mean temperature of warmest quarter, bio11 = mean temperature of coldest quarter, bio12 = annual precipitation, bio15 = precipitation coefficient of variation, bio18 = precipitation of warmest quarter, twi = topographic wetness index, srad = potential relative radiation.

ever, species with small ranges, such as Fraser fir (*Abies fraseri*) and table mountain pine did not have sufficient presence observations to yield well-calibrated models, indicated by calibration plot slope coefficients close to zero.

Global climate models project between 2.7 and 5.2 degrees F (1.5 and 2.9 degrees C) of warming across the PACEs of the Appalachian LCC by 2055 and between 1.4 and 2.3 inches (36 and 58 millimeters) of precipitation increase, depending on emissions scenario and PACE (chap. 5). Species distribution models for eastern tree species depicted a variety of responses (fig. 8-3) corresponding to a range of potential impacts from projected climate change. The low- and high-emissions scenarios start to diverge mid-century, which may explain the relatively small differences in potential impacts between scenarios assessed at 2055. After midcentury, temperatures in the high-emissions scenario are expected to increase sharply, which would likely lead to more extreme projections of potential impact.

Our results suggest that, by 2055 under a high-emissions scenario, climate conditions currently associated with southern oak-hickory forests will expand considerably across the Appalachian LCC, while those associated with species in northern hardwoods, maple-beech-birch, and high-elevation spruce and pine forests will contract. This is in agreement with findings from Iverson et al. (2008) that conditions supporting oak-hickory forests could spread far north at the expense of maple-beech-birch forests. We project the largest relative increases for slash pine (*Pinus elliottii*) and black hickory (*Carya tomentosa*) and the largest relative decreases for balsam fir, quaking aspen (*Populus tremuloides*), yellow birch, and red spruce, which again are similar to findings of Iverson et al. (2008). Trends for Great Smoky Mountains and Shenandoah national parks were more similar to those of the Appalachian LCC than trends in Delaware Water Gap NRA. This is likely due to larger projected temperature changes for more northerly latitudes. In addition, Delaware Water Gap NRA lies at the transition between oak-hickory and maple-beech-birch forest types, whereas Great Smoky Mountains and Shenandoah national parks are more central in the oak-hickory range.

Temperature drove many of the largest changes in suitability. Decreases in suitability, in particular, were projected for species with negative relationships with maximum temperature of the warmest month. A larger variety of variables were associated with species projected to see increasing suitability. However, temperature was frequently in the set of the top five most important variables for the oak and hickory species that comprise most of these cases. Precipitation was relatively less important overall, showing up as the most important variable for only three species. These patterns give us additional confidence in our tree species distribution projections because general circulation models agree much more on magnitude and trend of temperature change than they do for precipitation. This is in contrast to the results of Iverson et al. (2008), who found growing season precipitation was the most important driving variable across 134 species. We did not conduct a comprehensive variable analysis as in Iverson et al. (2008), and there are important differences between our approaches, but the relative importance of temperature and precipitation on the distribution of eastern tree species is an important area for further investigation given these results and the drastic temperature increases projected by global circulation models.

Because of the longevity of overstory trees (Runkle 2000; Busing 2005), the effects of climate change on forest composition may not be apparent until disturbance events provide opportunities for establishment of new species and release of advance regeneration. Drought-induced mortality is an important driver of change in stand composition and may increase in the future (Allen et al. 2010). As temperatures increase, trees will be subject to greater evaporative demand from a warmer atmosphere. When combined with normal drought periods, this could increase mortality of sensitive species, allowing understory release in canopy gaps. Mortality events may be even larger if drought periods become more frequent or more severe, as has been both projected (Gao et al. 2012) and observed over the past few decades (Li et al. 2011).

Fire and wind are additional abiotic disturbance agents that may play an increasingly important role in eastern forests if storm intensity increases or if droughts promote more severe fire activity. In areas with high potential impact from climate change, National Park Service managers may be faced with situations where community composition could change rapidly as canopy trees die, creating gaps exposed to prevailing climate conditions that may favor establishment or growth of different plant communities (Dale et al. 2001). While disturbance events in a changing climate may make it more difficult to maintain existing communities, they provide opportunities for managers to affect patterns of succession in ways that may help maintain ecosystem function.

Promising management applications of these models include monitoring areas of recent disturbance for early indicators of forest change associated with changing environmental conditions, especially in Delaware Water Gap NRA, where temperature increases in recent decades have been most rapid (chap. 5). In Great Smoky Mountains and Shenandoah national parks, where temperatures have not yet shown marked increases, these results may be most useful for generating management scenarios or identifying areas that may be most resistant to change to help plan for projected temperature increases.

For some species, these results may have direct management implications. For example, Eschtruth, Evans, and Battles (2013) showed a climate effect in which time to death for hemlock trees infested with hemlock woolly adelgid was considerably shorter when winter temperature in the previous year and summer drought in the current year were higher. Projected changes in hemlock suitability may be useful for identifying areas where hemlock occupies more climatically favorable positions and where it may be more likely to persist in the future. Targeting such areas for treatment may be more effective in the long run than targeting hemlock in areas where they may be most impacted by climate change. Likewise, many eastern tree species are susceptible to forest pests, and a similar approach could be used to assess management or monitoring priorities based on projected changes in suitability that may make tree species more or less vulnerable to depredation (Fisichelli et al. 2014).

These results may be improved by, and may help to improve, National Park Service science programs. Extensive inventory and monitoring data are available to help assess the performance of these models across the National Park Service system in the East. Furthermore, inventory and monitoring efforts may be combined with projections of habitat suitability change to identify sites where climate change impacts are most likely to be observed, helping such efforts to stay abreast of environmental change. Collaboration with local management partners can significantly increase the value of modeling efforts and, when combined with detailed local knowledge, may lead to greater insights into the potential impacts of climate change on eastern US forests.

Climate has and will likely continue to exert strong influences on the composition and function of eastern forests. However, because most of the land in the Appalachian LCC is privately owned, land use change may have even larger effects on forests than climate change in the short term. Eastern US forests are currently fragmented, and the least fragmented forests are primarily on public lands (Jantz and Goetz 2008), which are nonetheless threatened with isolation by development in surrounding private lands (Goetz, Jantz, and Jantz 2009). In addition to fragmentation, residential development increases the prevalence of invasive species (Gavier-Pizarro et al. 2010) and can transport some species outside their natural ranges (e.g., via landscaping; Hanberry and Hansen 2015).

For species for which suitable climate space is expected to shift to higher elevations, National Park Service and US Forest Service lands in the southeast and central Appalachians, as well as state game lands and state parks in Pennsylvania and New York, could become more important as reservoirs of species diversity and ecosystem function. For species losing low-elevation suitable climate space, fragmentation on private lands could be a significant barrier to dispersal (chap. 11).

Although most of our models were reasonably well calibrated and showed good discriminatory ability, a few were poorly calibrated and were therefore less reliable for assessing continuous changes in suitability. These species generally had low prevalence in the Forest Inventory and Analysis database. Scale mismatches between the Forest Inventory and Analysis plot observations and gridded climate data introduce additional uncertainty into the models. Trees in Forest Inventory and Analysis plots respond to microclimates that are not well captured by interpolated meteorological data. This may have caused underestimation of climatically suitable areas associated with topographic or hydrologic features such as mountain streams (Fridley 2009), especially for species with more restricted ranges (Franklin et al. 2013). In addition, the precision of Forest Inventory and Analysis plot coordinates is degraded to protect landowner identity, and in some cases plot coordinates are swapped. The effect of plot coordinate precision on species distribution model outputs was relatively minor in an analysis of western tree species (Gibson et al. 2013), but the potential effects of imprecise coordinates bear further investigation.

We did not include the effects of errors in weather station interpolations in our analysis. This is another area where additional work could reduce uncertainty in tree-climate associations. Although artifacts have been observed in PRISM interpolated products (Beier et al. 2012), they do not appear to vary systematically with geographic factors (e.g., elevation, distance to coasts) in the Northeast (Bishop and Beier 2013).

Conclusion

Our analysis assessed potential impacts of climate change on eastern tree species, generating results that we believe are reasonable and relevant for management because of a combination of previously unavailable high-resolution downscaled climate data sets from NEX-DCP30, a densely sampled Forest Inventory and Analysis database of field presence-absence observations, and machine learning algorithms of species range distributions that are robust under a range of conditions. We recognize that the actual influence of climate change on tree species is considerably more difficult to assess for any given location because it depends on a variety of interacting processes, including dispersal, disturbance, fragmentation, competition, and pest-host-climate relationships, among others. We include basic information on dispersal probability, fragmentation, and forest pests in chapter 11 in order to provide informative rankings of relative vulnerability to climate change across species that can further inform management priorities and decisions.

Acknowledgments

Work for this chapter was funded by NASA Applied Sciences Program award number 10-BIOCLIM10-0034. We would like to thank Kevin Guay for assistance with data processing, Nick Fisichelli and Nathan Piekielik for insightful reviews, National Park Service staff for input that helped guide our analysis, and Forrest Melton for assistance with data access and data processing.

References

Allen, C. D., A. K. Macalady, H. Chenchouni, D. Bachelet, N. McDowell, M. Vennetier, T. Kitzberger, A. Rigling, D. D. Breshears, E. H. (Ted) Hogg, et al. 2010. A global overview of drought and heat-induced tree mortality reveals emerging climate change risks for forests. *Forest Ecology and Management* 259:660–84.

Beier, C. M., S. A. Signell, A. Luttman, and A. T. DeGaetano. 2012. High-resolution climate change mapping with gridded historical climate products. *Landscape Ecology* 27:327–42.

Bishop, D. A., and C. M. Beier. 2013. Assessing uncertainty in high-resolution spatial climate data across the US Northeast. *PLOS ONE* 8.

Bürgi, M., E. Russell, and G. Motzkin. 2000. Effects of postsettlement human activities on forest composition in the north-eastern United States: A comparative approach. *Journal of Biogeography* 27:1123–38.

Busing, R. T. 2005. Tree mortality, canopy turnover, and woody detritus in old cove forests of the southern Appalachians. *Ecology* 86:73–84.

Butler, P. R., L. Iverson, F. R. Thompson III, L. Brandt, S. Handler, M. Janowiak, P. D. Shannon, C. Swanston, K. Karriker, J. Bartig, et al. 2015. *Central Appalachians Forest Ecosystem Vulnerability Assessment and Synthesis: A Report from the Central Appalachians Climate Change Response Framework Project*. Gen. Tech. Rep. NRS-146. Newtown Square, PA: US Department of Agriculture, Forest Service, Northern Research Station.

Corripio, J. G. 2003. Vectorial algebra algorithms for calculating terrain parameters from DEMs and solar radiation modelling in mountainous terrain. *International Journal of Geographical Information Science* 17:1–23.

Cronon, W. 1983. *Changes in the Land: Indians, Colonists, and the Ecology of New England*. New York: Hill and Wang.

Dale, V. H., L. A. Joyce, S. Mcnulty, R. P. Neilson, M. P. Ayres, M. D. Flannigan, P. J. Hanson, L. C. Irland, A. E. Lugo, C. J. Peterson, et al. 2001. Climate change and forest disturbances. *BioScience* 51:723–34.

Daly, C., M. Halbleib, J. I. Smith, W. P. Gibson, M. K. Doggett, G. H. Taylor, J. Curtis, and P. P. Pasteris. 2008. Physiographically sensitive mapping of climatological temperature and precipitation across the conterminous United States. *International Journal of Climatology* 28:2031–64.

Delcourt, H. R., and P. A. Delcourt. 1988. Quaternary landscape ecology: Relevant scales in space and time. *Landscape Ecology* 2:23–44.

Eschtruth, A. K., R. A. Evans, and J. J. Battles. 2013. Patterns and predictors of survival in *Tsuga canadensis* populations infested by the exotic pest *Adelges tsugae*: 20 years of monitoring. *Forest Ecology and Management* 305:195–203.

Fisichelli, N. A., S. R. Abella, M. Peters, and F. J. Krist. 2014. Climate, trees, pests, and weeds: Change, uncertainty, and biotic stressors in eastern U.S. national park forests. *Forest Ecology and Management* 327:31–39.

Franklin, J. 2009. *Mapping Species Distributions: Spatial Inference and Prediction*. Cambridge: Cambridge University Press.

Franklin, J., F. W. Davis, M. Ikegami, A. D. Syphard, L. E. Flint, A. L. Flint, and L. Hannah. 2013. Modeling plant species distributions under future climates: How fine scale do climate projections need to be? *Global Change Biology* 19:473–83.

Fridley, J. D. 2009. Downscaling climate over complex terrain: High finescale (< 1000 m) spatial variation of near-ground temperatures in a montane forested landscape (Great Smoky Mountains). *Journal of Applied Meteorology and Climatology* 48:1033–49.

Gao, Y., J. S. Fu, J. B. Drake, Y. Liu, and J.-F. Lamarque. 2012. Projected changes of extreme weather events in the eastern United States based on a high resolution climate modeling system. *Environmental Research Letters* 7: 044025.

Gavier-Pizarro, G. I., V. C. Radeloff, S. I. Stewart, C. D. Huebner, and N. S. Keuler. 2010. Housing is positively associated with invasive exotic plant species richness in New England, USA. *Ecological Applications* 20:1913–25.

Gibson, J., G. Moisen, T. Frescino, and T. C. Edwards. 2013. Using publicly available forest inventory data in climate-based models of tree species distribution: Examining effects of true versus altered location coordinates. *Ecosystems* 17: 43–53.

Goetz, S. J., P. Jantz, and C. A. Jantz. 2009. Connectivity of core habitat in the northeastern United States: Parks and protected areas in a landscape context. *Remote Sensing of Environment* 113:1421–29.

Gonzalez, P., R. P. Neilson, J. M. Lenihan, and R. J. Drapek. 2010. Global patterns in the vulnerability of ecosystems to vegetation shifts due to climate change. *Global Ecology and Biogeography* 19:755–68.

Haas, J., and J. McAndrews. 2000. The summer drought related hemlock (*Tsuga canadensis*) decline in eastern North America 5700 to 5100 years ago. In *Proceedings: Symposium on Sustainable Management of Hemlock Ecosystems in Eastern North America*, 81–88.

Hanberry, B. B., and M. H. Hansen. 2015. Latitudinal range shifts of tree species in the United States across multi-decadal time scales. *Basic and Applied Ecology* 16:231–38.

Hansen, A. J., C. R. Davis, N. Piekielek, J. Gross, D. M. Theobald, S. Goetz, F. Melton, and R. DeFries. 2011. Delineating the ecosystems containing protected areas for monitoring and management. *BioScience* 61:363–73.

Hijmans, R. J., S. E. Cameron, J. L. Parra, P. G. Jones, and A. Jarvis. 2005. Very high resolution interpolated climate surfaces for global land areas. *International Journal of Climatology* 25:1965–78.

Houghton, R., and J. Hackler. 2001. Changes in terrestrial carbon storage in the United States. 1: The roles of agriculture and forestry. *Global Ecology and Biogeography* 9:125–44.

Hutchinson, G. E. 1957. Concluding remarks. *Cold Spring Harbor Symposium on Quantitative Biology* 22:415–27.

Iverson, L. R., A. M. Prasad, S. N. Matthews, and M. Peters. 2008. Estimating potential habitat for 134 eastern US tree species under six climate scenarios. *Forest Ecology and Management* 254:390–406.

Jackson, S. T., J. L. Betancourt, R. K. Booth, and S. T. Gray. 2009. Ecology and the ratchet of events: Climate variability, niche dimensions, and species distributions. *Proceedings of the National Academy of Sciences of the United States of America* 106 (Supp): 19685–92.

Jantz, P., and S. Goetz. 2008. Using widely available geospatial data sets to assess the influence of roads and buffers on habitat core areas and connectivity. *Natural Areas Journal* 28.

Li, W., L. Li, R. Fu, Y. Deng, and H. Wang. 2011. Changes to the North Atlantic Subtropical High and its role in the intensification of summer rainfall variability in the southeastern United States. *Journal of Climate* 24:1499–1506.

Liaw, A., and M. Wiener. 2002. Classification and regression by randomForest. *R News: The Newsletter of the R Project* 2:18–22.

McKenney, D. W., J. H. Pedlar, R. B. Rood, and D. Price. 2011. Revisiting projected shifts in the climate envelopes of North American trees using updated general circulation models. *Global Change Biology* 17:2720–30.

McLachlan, J., J. Clark, and P. Manos. 2005. Molecular indicators of tree migration capacity under rapid climate change. *Ecology* 86:2088–98.

Miller, D. A., and R. A. White. 1998. A conterminous United States multilayer soil characteristics dataset for regional climate and hydrology modeling. *Earth Interactions* 2:1–26.

Nowacki, G. J., and M. D. Abrams. 2008. The demise of fire and "mesophication" of forests in the eastern United States. *BioScience* 58:123.

Pearson, R. G., and T. P. Dawson. 2003. Predicting the impacts of climate change on the distribution of species: Are bioclimate envelope models useful? *Global Ecology and Biogeography* 12:361–71.

Pederson, N., A. W. D'Amato, J. M. Dyer, D. R. Foster, D. Goldblum, J. L. Hart, A. E. Hessl, L. R. Iverson, S. T. Jackson, D. Martin-Benito, et al. 2014. Climate remains an important driver of post-European vegetation change in the eastern United States. *Global Change Biology*, 1–6. doi: 10.1111/gcb.12663.

Pierce, K. B., T. Lookingbill, and D. Urban. 2005. A simple method for estimating potential relative radiation (PRR) for landscape-scale vegetation analysis. *Landscape Ecology* 20:137–47.

Potter, K. M., and W. W. Hargrove. 2013. Quantitative metrics for assessing predicted climate change pressure on North American tree species. *Mathematical and Computational Forestry and Natural-Resource Sciences* 5:151–69.

Prasad, A. M., L. R. Iverson, and A. Liaw. 2006. Newer classification and regression tree techniques: Bagging and random forests for ecological prediction. *Ecosystems* 9:181–99.

Rehfeldt, G. E., N. L. Crookston, C. Saenz-Romero, and E. M. Cambell. 2012. North American vegetation model for land-use planning in a changing climate: A solution to large classification problems. *Ecological Applications* 22:119–41.

Riitters, K. H., J. W. Coulston, and J. D. Wickham. 2012. Fragmentation of forest communities in the eastern United States. *Forest Ecology and Management* 263:85–93.

Ruefenacht, B., M. V. Finco, M. D. Nelson, R. Czaplewski, E. H. Helmer, J. A. Blackard, G. R. Holden, A. J. Lister, D. Salajanu, D. Weyermann, et al. 2008. Conterminous US and Alaska forest type mapping using forest inventory and analysis data. *Photogrammetric Engineering and Remote Sensing* 74:1379–88.

Runkle, J. R. 2000. Canopy tree turnover in old-growth mesic forests of eastern North America. *Ecology* 81:554–67.

Sork, V. L., F. W. Davis, R. Westfall, A. Flint, M. Ikegami, H. Wang, and D. Grivet. 2010. Gene movement and genetic association with regional climate gradients in California valley oak (*Quercus lobata*) in the face of climate change. *Molecular Ecology* 19:3806–23.

Stein, B. A., and P. Glick. 2011. Introduction to *Scanning the Conservation Horizon: A Guide to Climate Change Vulnerability Assessment*, edited by P. Glick, B. A. Stein, and N. Edelson, 6–18. Washington, DC: National Wildlife Federation.

Theobald, D. M., D. Harrison-Atlas, W. B. Monahan, and C. M. Albano. 2015. Ecologically-relevant maps of landforms and physiographic diversity for climate adaptation planning. *PLOS ONE* 10 (12): e0143619.

Thompson, J. R., D. N. Carpenter, C. V. Cogbill, and D. R. Foster. 2013. Four centuries of change in northeastern United States forests. *PLOS ONE* 8: e72540.

Thrasher, B., J. Xiong, W. Wang, F. Melton, A. Michaelis, and R. Nemani. 2013. Downscaled climate projections suitable for resource management. *Eos, Transactions American Geophysical Union* 94:321–23.

US Environmental Protection Agency (EPA). 2009. Land-Use Scenarios: National-Scale Housing-Density Scenarios Consistent with Climate Change Storylines. EPA/600/R-08/076F. Washington, DC.

US Geological Survey, GAP Analysis Program (GAP). 2012. Protected Areas Database of the United States (PADUS), version 1.3.

Zolkos, S. G., P. Jantz, T. Cormier, L. R. Iverson, D. W. McKenney, and S. J. Goetz. 2015. Projected tree species redistribution under climate change: Implications for ecosystem vulnerability across protected areas in the eastern United States. *Ecosystems* 18 (2): 202–20.

Chapter 9

Potential Impacts of Climate Change on Tree Species and Biome Types in the Northern Rocky Mountains

Andrew J. Hansen and Linda B. Phillips

If one stands on a peak on the eastern side of the Northern Rocky Mountains on a clear day and gazes across the surrounding landscape, striking patterns of vegetation are apparent. From valley bottoms to ridgetops, vegetation grades from grassland and shrublands to open savannas, from dense tall forest to scattered clumps of krumholtz trees in the alpine above the pronounced treeline (fig. 9-1). These recurrent patterns of climatically zoned vegetation suggest that plants are a logical starting point for understanding biodiversity response to climate change. Plants, once established, are sessile and unable to move to more favorable locations and thus are strongly limited by the local climate. The predictable variation in climate with elevation explains this striking pattern of vegetation in the Rockies. To the extent that climate changes in the future, vegetation is expected to change in establishment, growth, and death rates, in canopy structure, and in the distributions of species and thus to show major shifts upward in elevational distribution.

Vegetation patterns also differ predictably west and east of the Continental Divide. While most of the Great Northern Landscape Conservation Cooperative (Great Northern LCC) region (chap. 1) has a cold, continental climate, portions of the west slope of the Rockies have a warmer, moister climate due to the Pacific maritime influence. Tree species that dominate the Coast and Cascade ranges, such as western hemlock (*Tsuga heterophylla*), mountain hemlock (*Tsuga mertensiana*), and western redcedar (*Thuja plicata*), are abundant in the maritime-influenced zones in north-

FIGURE 9-1 The Beartooth Front in Montana illustrates the typical vegetation patterns from valley bottoms to mountaintops in the eastern Rocky Mountains. (Photo by Andrew J. Hansen.)

western Montana. Under climate warming, areas east of the Continental Divide may become suitable for these maritime species if moisture levels become high enough.

Such changes in vegetation particularly interest managers both because individual plant species are valued in their own right and because vegetation provides critical habitat to other species and influences ecosystem properties, such as snow retention, runoff, soil fertility, and fire regimes. This is especially true in the Rocky Mountains because the relatively few tree species tend to dominate the major habitat types. Thus, studies of vegetation response to climate change represent both fine-filter analyses of individual plant species and coarse-filter analyses relevant to other species and ecosystem processes (Hunter, Jacobson, and Webb 1988).

Climate in the northern Rockies has varied substantially with decades of cool and wet followed by periods of warm and dry (chap. 4). Natural disturbances, such as fire and tree demography, have responded to the climate variability with, for example, local shifts in the locations of lower and upper treeline. A signal of human-induced warming may have emerged in the climate record in the 1980s, and the projected climate trajectory is toward substantial warming. Although precipitation is projected to also increase, the rise in evapotranspiration associated with the warming is projected to lead to less moisture available to plants (chaps. 4 and 7). These changes in climate are projected to result in reduced snowpack and earlier snowmelt (chap. 7). Consequently, growing conditions for plant species will likely improve in spring as a result of earlier release from snow and increased soil moisture from snowmelt, but will decline in late summer due to dry soils.

Understanding forest response to climate change within local and regional management jurisdictions is vital to designing locally relevant strategies to cope with pending changes. Resource managers can best plan, orient research, and manage if they are able to anticipate which species and ecosystems are most vulnerable to possible future change (Colwell et al. 2012; Stein et al. 2014). Patterns of vulnerability likely vary geographically across the topographically complex northern Rockies. Thus, there is a need to conduct vulnerability assessments both within local management jurisdictions and across the region as a whole.

Fortunately, several studies of vegetation response to climate change have been completed across western North America and provide a basis for vulnerability assessment in the Great Northern LCC based on published studies. Most of these studies use an approach called bioclimate envelope modeling. This approach quantifies the climate conditions where a species is currently present and projects the locations of these climate conditions under future scenarios (Huntley et al. 1995; Pearson and Dawson 2003; Guisan and Thuiller 2005) (see also chaps. 8 and 10). This approach describes the conditions under which populations of a species persist in the presence of other biota as well as climatic constraints. Possible future distributions are projected on the assumption that current envelopes reflect species' environmental preferences, which will be retained under climate change. While this approach does not necessarily predict where a species will occur in the future (Pearson and Dawson 2003), it does project one foundational filter of where a species could exist in the future: climate suitability (Serra-Diaz et al. 2014).

The results of bioclimate envelope studies are very useful to resource managers for identifying which species may be most vulnerable to climate change and for developing management strategies for these species (Hansen and Phillips 2015). Whereas managers cannot manipulate climate over large landscapes, they can manipulate many of the other factors that influence plant population viability, including establishment, genetic composition, interactions with other species, and disturbances. Thus, knowledge of climate suitability is a critical first filter for deciding where to use management actions to protect, restore, or establish tree populations under climate change. Species identified as vulnerable based on climate suitability are candidates for the additional research used in vulnerability assessment (Dawson et al. 2011), which typically are more expensive and have higher uncertainty than climate suitability analyses.

In this chapter, we summarize the results of our vulnerability assessment of vegetation response to climate change in the northern Rockies (Hansen and Phillips 2015). We drew on five published studies of tree species and biome response to climate change. Biomes are broad-scale patterns of vegetation reflecting climate, soils, and disturbance regimes. They are defined by plant life form and are relevant to management in providing guidance on changes in the distribution of grasslands, shrublands, and forest. We first briefly describe the methods used in the analysis, highlight key results, and draw implications for management in the Great Northern LCC and the two major national parks and surrounding ecosystems: Yellowstone and Glacier.

Climate Suitability Models of Tree Species and Biomes

The studies used in this synthesis were Crookston et al. (2010), Coops and Waring (2011), Gray and Hamann (2013), and Bell et al. (2014), who modeled climate suitability for tree species, and Rehfeldt et al. (2012), who modeled biome climate suitability. These studies differed to some extent in the climate scenarios projected, the global climate models used as a basis for the projections, and the statistical methods linking vegetation to climate. The influence of these differences among studies on their vegetation projections are fully evaluated in Hansen and Phillips (2015). In this chapter, we focus on the consensus results of these studies to present the major findings. We present the results for the Intergovernmental Panel on Climate Change (IPCC) Special Report on Emissions Scenarios (SRES) A2 scenario (higher warming) and point out where the results from the B1 scenario (lower warming) differ. These IPCC SRES scenarios are the IPCC 2000 and 2007 precursors to the current IPCC Coupled Model Intercomparison Project Phase 5 representative concentration pathway (RCP) 8.5 and 4.5 scenarios.

Using data from each study, we mapped locations projected to have suitable climate for a species in a historic reference period (within 1961–2010) and in the 2070–2100 period and identified locations where climate suitability during the reference period was retained versus lost in the future period. We then summarized the aerial extent of suitable climate during the reference period, loss of reference-period suitable climate in the future period, and gain in suitable climate in the future period. For newly suitable habitats, we distinguished between those near enough to currently suitable habitats to have some probability of colonization by 2100 (within 18.6 miles, or 30 kilometers) from those more distant from potential source areas. These summaries were done separately for each tree species or biome type, geographic unit, climate scenario, and study.

We assessed components of vulnerability based on climate suitability. Vulnerability has been defined as a function of exposure (the magnitude of change experienced), sensitivity (the degree to which that change impacts the system), and adaptive capacity (the ability to respond or cope with the change) (chap. 2). Bioclimate envelope approaches consider exposure (climate change) and sensitivity (species' tolerances to climate in the presence of other species) and project the potential impact of climate change on the locations of suitable climate for a species. Thus, we evaluate vulnerability based on potential impact, which is an objective output of the statistical models. However, we do not consider adaptive capacity, which typically is derived based on expert opinion. Potential impact was assessed based on the criteria and cardinal rank scoring. This resulted in summary scores for the Great Northern LCC that ranged from a minimum of −1.5 (low impact) to a high of 6.5 (high impact); scores between 1.5 and 4.5 are considered moderate impact.

Projected Shifts in Vegetation under Climate Change

Climate suitability for biome types shifted substantially between the projected reference and late-century periods (fig. 9-2). Across the Great Northern LCC, the eastern alpine tundra, Rocky Mountain subalpine conifer, Rocky Mountain montane conifer, and Great Basin shrub-grassland types that dominate during the reference period declined in suitable area throughout the century. The subalpine and montane conifer types were projected to be largely replaced in climate suitability by the nonforest types of Great Basin montane scrub, Great Basin desert scrub, Plains grassland, and Great Basin conifer woodland. The projected biome shifts in Greater Yellowstone were more extreme than for the Great Northern LCC, with conifer forest types dropping from 82 percent of the area to 26 percent and scrub types increasing from 0 to 48 percent of the area. The Glacier protected area centered ecosystem (PACE) was projected to shift in climate suitability from subalpine conifer to interior cedar-hemlock conifer; Northeast deciduous,

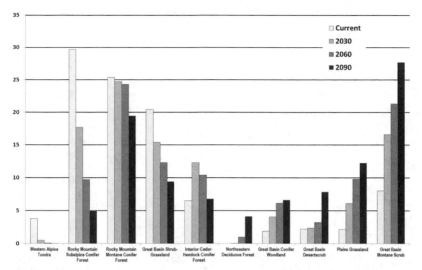

FIGURE 9-2   Projected shift in biome climate suitability for the Great Northern LCC under consensus of the A2 and B1 scenarios and three global circulation models represented by proportion of the study area. (From Hansen and Phillips 2015.)

which includes quaking aspen (*Populus tremuloides*); and Great Basin montane scrub.

All four studies of tree species projected substantial declines in climate suitability for subalpine tree species across the Northern Rocky Mountains. Averaging among the studies, the proportion of the study area with suitable climate for whitebark pine (*Pinus albicaulis*) dropped to 8.8 percent by 2070–2100 under the B1 scenario and to 11 percent under the A2 scenario (fig. 9-3). Remaining suitable climate area by 2100 for Engelmann spruce (*Picea engelmannii*), subalpine fir (*Abies lasiocarpa*), and lodgepole pine (*Pinus contorta*) was 18 to 25 percent under B1 and 16 to 25 percent under A2. Among the montane species—ponderosa pine (*Pinus ponderosa*) and grand fir (*Abies grandis*)—climate-suitable areas were projected to increase substantially. The studies disagreed on Douglas-fir (*Pseudotsuga menziesii*), with some studies projecting expansion and others contraction. Among the tree species now found in the more mesic Rocky Mountain west slope, mountain hemlock was projected to decrease dramatically under both climate scenarios while western redcedar and western hemlock were projected to increase moderately.

The spatial patterns of change in climate suitability revealed projected locations for decline and expansion for the late-century period. For the sub-

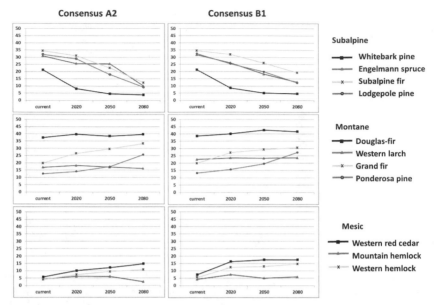

FIGURE 9-3 Projected change in the proportion of the Great Northern LCC with suitable climate for each tree species averaging the results of the four studies considered in Hansen and Phillips (2015) under the B1 and A2 climate scenarios.

alpine species, climate suitability contracted substantially from the reference period to the future period and newly suitable areas were at higher elevations (fig. 9-4). This resulted in major reductions in total area of suitable climate because of the lack of land area on mountaintops in our study area. Montane species generally expanded from mid-elevations to adjacent higher elevations that are currently occupied by subalpine species. Climate suitability for grand fir, western larch, and ponderosa pine, however, expanded in some locations east of the Continental Divide, where they had little suitable habitat in the reference period. Suitable climates for the mesic species were projected by Coops and Waring (2011) to expand from their current westside locations to tens to a few hundreds of miles eastward; however, Crookston et al. (2010) and Gray and Hamann (2013) projected no such shifts to eastside locations. Importantly, both Coops and Waring (2011) and Crookston et al. (2010) project substantial contraction of suitable climate area for mountain hemlock in the western portion of the study area where the species is currently present.

Based on the cardinal ranking of the potential impact scores, whitebark pine was most vulnerable to climate change by a wide margin (fig. 9-5).

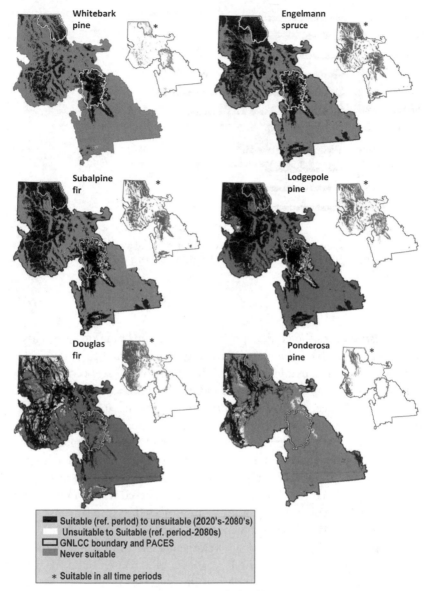

FIGURE 9-4   Change in modeled spatial distribution of climate-suitable areas across the reference and three future time periods under the B1 and A2 climate scenarios based on the consensus results for the studies in Hansen and Phillips (2015).

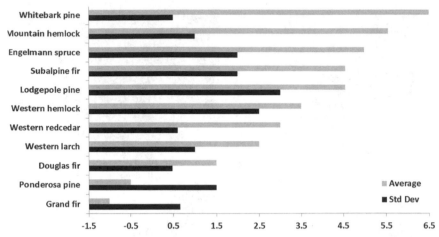

FIGURE 9-5   Results of vulnerability assessment ranking averaged among studies under the A2 scenario. (From Hansen and Phillips 2015.)

Being restricted to the coldest portions of the subalpine, it has relatively little area of suitable habitat currently and was projected to undergo substantial declines in suitable habitat, with little new habitat becoming available. Mountain hemlock had the second highest vulnerability score, largely as a result of its very small area of currently suitable habitat and substantial projected losses of suitable habitat. The other subalpine species (Engelmann spruce, subalpine fir, and lodgepole pine) were also placed in the High vulnerability class because of the large decline in projected suitable area and low gain in newly suitable areas. Western hemlock, western redcedar, western larch, and Douglas-fir were considered moderate in vulnerability. Ponderosa pine and grand fir were projected to gain substantially in the area of suitable habitat and were considered low in vulnerability. Level of agreement among the studies was relatively high for the most vulnerable and least vulnerable species. The lower level of agreement among studies for species ranked moderate in vulnerability suggests that more research is needed on these species (chap. 10).

Among the species currently present in the Greater Yellowstone PACE and the Glacier PACE, the order of vulnerability ranking under the A2 climate scenario was similar to that of the Great Northern LCC, with whitebark pine having the highest vulnerability score and Douglas-fir low scores. The score for mountain hemlock in Glacier was nearly as high as whitebark pine, just as was the case for the Great Northern LCC. The order of vulnerability rankings under the B1 climate scenario were similar to those under A2, but the values were lower than under A2 due to the less extreme climate projections.

Ecological Consequences of Projected Shifts in Climate Suitability

Despite differences in methods among these studies, there was a high level of agreement in terms of dramatic shifts in projected climate suitability in vegetation. The biome analysis and the tree species projections under the higher and lower climate change scenarios suggested substantial reductions in the area of forests across the northern Rockies. Tree species currently in the region are projected to contract in suitable area, and no species from outside the region are projected to gain climate suitability in the region (Crookston et al. 2010). Rehfeldt et al. (2012) suggested the replacement of conifer forests by desert shrub and grassland life-form types.

Such changes in life-form, if realized, would have large implications for evapotranspiration, snowpack, runoff, and habitat for other species. Within the Greater Yellowstone Ecosystem, for example, the area projected to be suitable for the open sage and juniper habitat type expands from 41 percent currently to 71 percent during 2070–2090 while the area of closed forest types declines from 41 percent to 19 percent (based on Piekielek, Hansen, and Chang 2015; also see chap. 10). This change would be expected to reduce snowpack substantially, increase late summer soil drying, and reduce runoff in rivers and streams.

For the areas that remain suitable for forests, composition of tree species is projected to shift, with suitable climates for subalpine species largely moving off the mountain tops and with montane species conditions expanding in the subalpine. Some tree species now largely on the west slope of the Rockies, such as grand fir and ponderosa pine, are projected to have suitable conditions east of the Continental Divide. Actual dispersal and establishment of these species in the newly suitable habitat, however, would likely take several decades.

The ranking of vulnerability among tree species was very similar among studies and among climate change scenarios. Whitebark pine was ranked most vulnerable across the northern Rockies and within the Glacier and Yellowstone PACEs. More recent climate suitability modeling in the Yellowstone PACE (Chang, Hansen, and Piekielek 2014) projected 71 to 98 percent reductions in suitable area by 2100 under RCP 4.5 (similar to B1) and 90 to 99 percent loss under RCP 8.5, with the variation due to differences among climate models. Adaptive capacity of this species is thought to be relatively low because dispersal is fairly limited, the species is often outcompeted by other subalpine conifers, and the species is highly susceptible to mountain bark beetles and the exotic white pine blister rust. Motivation to retain whitebark pine and other conifers in the Greater Yellowstone Ecosystem is high because shifts in climate suitability away from conifer communities and toward desert scrub communities would likely have large negative impacts on snowpack, runoff, wildlife, and aesthetics.

Implications for Research and Management

Several authors (summarized by McKinney et al. 2011) have emphasized that the extent to which individual tree species will actually shift through natural processes interacting with climate change is highly uncertain and involves complex interacting factors, such as edaphic and land use considerations, dispersal ability, genetic controls, carbon dioxide fertilization effects, disturbance patterns, and competitive, predatory, and mutalistic relationships among species. Consequently, efforts to assess species vulnerability under climate change have used an expanding list of predictors beyond climate suitability. These include demographic, life history, genetic, and habitat dynamics factors for past, current, and future periods (Pearson et al. 2014; Iverson et al. 2012; Aubry et al. 2011). Future work should consider these factors beyond potential impact in assessing vulnerability and evaluating management strategies.

The analyses summarized here focused on climate suitability as a first-order predictor of climate vulnerability because of its ecological and management relevance. There is a wealth of evidence that plant species have specific tolerances to climate and that their distributions reflect these tolerances. Thus, strong inference can be drawn that (1) viable populations of a tree species are unlikely to develop or persist where climate is unsuitable, (2) existing populations are likely to decline if climate becomes unsuitable, and (3) locations of newly suitable climate may be able to support future viable populations if several other ecological conditions are met. These inferences are highly relevant to management. Although managers cannot manipulate climate over large landscapes, they can manipulate many of the other factors that influence tree population viability, including seedling establishment, genetic composition, interactions with other species (e.g., completion, parasitism, and disease), and disturbances (e.g., fire).

In other words, knowledge of climate suitability is a critical filter for deciding where to use management actions to protect, restore, or establish tree populations under climate change. Accordingly, Dawson et al. (2011) suggested that climate suitability analysis is an important initial step in integrated science assessment of biodiversity under climate change, with complementary approaches that include the use of paleoecological records, ecophysiological and population models, and experimental manipulations. Accordingly, species that were ranked high in this study in potential impact based on climate envelope modeling are candidates for the additional steps that are often more expensive and sometimes subject to higher levels of uncertainty.

Management strategies for species deemed high priorities for management can be stratified geographically based on the locations of current and projected suitable habitats. Locations where populations are currently present and habitat is projected to remain suitable in the future (known as core habitats) are obviously of high importance, and management strategies should be aimed at maintaining populations in these locations. These strategies might include suppression of fire or other disturbances that could destroy the target tree population, or the use of prescribed fire aimed at maintaining the current population. Within locations projected to become suitable in the future, strategies to facilitate natural colonization on the leading edge of habitat suitability may allow populations to better track changing conditions. Similarly, assisted migration is feasible for newly suitable habitats that are more distant from current populations.

Within locations where populations are present but habitat is projected to become unsuitable, research beyond climate envelope modeling may reveal mechanisms of population viability (Hansen et al., in Review). These are defined as story lines of ecological interactions or geographic settings where some individuals are able to establish, survive, and reproduce at adequate levels to prevent local population extinction. Developing and evaluating hypotheses on such mechanisms may identify management strategies that can enhance natural mechanisms and elevate the probability of population persistence despite deteriorating climate conditions. Examples of additional management strategies of each of these types can be found in Heller and Zavaleta (2009), and examples of coordinated management strategies for maintenance and restoration of whitebark pine can be found in GYCC (2011), McLane and Aitken (2012), and chapters 10 and 15.

A substantial challenge to managing wildlands under climate change relates to issues of land allocation. The tree species and biome types most vulnerable to climate change currently occur in land allocation types where active management may be considered inappropriate or even illegal. Our northern Rockies study area includes unprotected private lands, private lands protected by conservation easements, federal general-use lands (e.g., multiple-use national forest lands and Bureau of Land Management lands), and federal protected lands, such as national parks, wilderness areas, and roadless areas. Suitable habitats for subalpine tree species in the study area lie primarily on the two classes of federal lands.

Under the climate change scenarios, suitable habitats for these species increasingly shift to federal protected lands where enabling legislation (e.g., the Wilderness Act) or current policy dissuades active management. It is important to point out, however, that many of the activities recommended for climate adaptation planning (chap. 2) are appropriate across each of the land allocation types. Research can be used to project potential future response to climate change and reduce uncertainty. Monitoring in fast-changing places provides information on actual rates of change and ecological response to this change. Vulnerability assessments can reveal which species or ecosystems are most at risk, where these are located, and why they are at risk. Education programs for natural resource staff and the public can help in understanding the issues and formulating effective policy. Agency planning documents can incorporate consideration of climate change in order to mitigate undesirable climate change impacts on projects. Passive management such as allowing fires to burn can sometimes favor species vulnerable to climate change.

Finally, a variety of types of active management are being developed and evaluated aimed at protecting existing populations until newly suitable habitats develop, facilitating natural establishment in newly suitable habitats, and assisting migration to suitable habitats. While the debate over active management in wildlands facing climate change will continue, it should be noted that research, monitoring, education, vulnerability assessment, and passive management are all viable options for managers of restricted federal lands.

We close by noting that beyond the northern Rockies study area, the methods described in this chapter can be applied in other geographic locations to conduct vulnerability assessments based on climate suitability. See chapter 8 for a companion study in the Appalachian Mountains. The results of such assessments can help managers prioritize species for more detailed research and climate adaptation planning.

Conclusion

The dramatic patterns of vegetation life forms and species across the complex landscapes of the Rocky Mountains reflect the strong climate gradients of the region. Our synthesis of five previous studies of vegetation response to climate change across western North America reveals high levels of agreement among the studies in projected declines in climate suitability for forested areas, especially subalpine forests, and expansion for desert scrub and grassland communities. Among the tree species analyzed, whitebark pine ranked highest in vulnerability to climate change across the Great Northern LCC and within the Glacier and Yellowstone PACEs. While these projected changes in climate suitability for vegetation will challenge natural resource managers, they also provide a basis for prioritizing management activities. Protection, restoration, and translocation actions can be stratified across planning areas based on projected climate suitability, with the goals of maintaining a species where climate is likely to remain suitable and establishing a species in places where climate is expected to become suitable. Chapters 10 and 14 provide more detailed looks at such climate-based management approaches.

Acknowledgments

Funding was provided by the NASA Applied Sciences Program (10-BIO-CLIM10-0034) and the North Central Climate Sciences Center. We thank the authors of each of the studies included in this synthesis for providing original data. William B. Monahan provided extensive review and suggestions on drafts of the manuscript.

References

Aubry, C., W. Devine, R. Shoal, A. Bower, J. Miller, and N. Maggiulli. 2011. *Climate Change and Forest Biodiversity: A Vulnerability Assessment and Action Plan for National Forests in Western Washington*. Portland, OR: US Forest Service, PNW Region.

Bell, D. M., J. B. Bradford, and W. K. Lauenroth. 2014. Mountain landscapes offer few opportunities for high-elevation tree species migration. *Global Change Biology* 20:1441–51. doi: 10.1111/gcb.12504.

Chang, T., A. J. Hansen, and N. Piekielek. 2014. Patterns and variability of projected bioclimate habitat for *Pinus albicaulis* in the Greater Yellowstone Ecosystem. *PLOS ONE* 9 (11): e111669.

Colwell, R., S. Avery, J. Berger, G. E. Davis, H. Hamilton, T. Lovejoy, S. Malcom, A. McMullen, M. Novacek, R. J. Roberts, R. Tapia, and G. Machlis. 2012. *Revisiting Leopold: Resource Stewardship in the National Parks*. Report. Washington, DC: National Park System Advisory Board Science Committee.

Coops, N. C., and R. H. Waring. 2011. Estimating the vulnerability of fifteen tree species under changing climate in northwest North America. *Ecological Modelling* 222:2119–29.

Crookston, N. L., G. E. Rehfeldt, G. E. Dixon, and A. R. Weiskittel. 2010. Addressing climate change in the forest vegetation simulator to assess impacts on landscape forest dynamics. *Forest Ecology and Management* 260:1198–1211.

Dawson, T. P., S. T. Jackson, J. I. House, I. C. Prentice, and G. M. Mace. 2011. Beyond predictions: Biodiversity conservation in a changing climate. *Science* 332:53–58.

Gray, L. K., and A. Hamann. 2013. Tracking suitable habitat for tree populations under climate change in western North America. *Climate Change* 117:289–303.

Guisan, A., and W. Thuiller. 2005. Predicting species distribution: Offering more than simple habitat models. *Ecology Letters* 8:993–1009.

GYCC (Greater Yellowstone Coordinating Committee). 2011. Whitebark Pine Strategy for the Greater Yellowstone Area. Report.

Hansen, A. J., K. Ireland, K. Legg, R. Keane, E. Barge, M. Jenkis, and M. Pillet. In review. Complex challenges of maintaining whitebark pine in Greater Yellowstone under climate change: A call for innovative research, management, and policy approaches. *Forests*.

Hansen, A. J., and L. B. Phillips. 2015. Which tree species and biome types are most vulnerable to climate change in the US Northern Rocky Mountains? *Forest Ecology and Management* 338:68–83.

Heller, N. E., and E. S. Zavaleta. 2009. Biodiversity management in the face of climate change: A review of 22 years of recommendations. *Biological Conservation* 142 (1): 14–32.

Hunter, M. L., G. L. Jacobson, and T. Webb. 1988. Paleoecology and the coarse-filter approach to maintaining biological diversity. *Conservation Biology* 2 (4): 375–85.

Huntley, B., P. M. Berry, W. Cramer, and A. P. Mcdonald. 1995. Modelling present and potential future ranges of some European higher plants using climate response surfaces. *Journal of Biogeography* 22:967–1001.

Iverson, L. R., S. N. Matthews, A. M. Prasad, M. P. Peters, and G. Yohe. 2012. Development of risk matrices for evaluating climatic change responses of forested habitats. *Climatic Change* 114:231–43.

McKinney, D. W., J. H. Pedlar, R. B. Rood, and D. Price. 2011. Revisiting projected shifts in the climate envelopes of North American trees using updated general circulation models. *Global Change Biology* 17:2720–30.

McLane, S. C., and S. N. Aitken. 2012. Whitebark pine (*Pinus albicaulis*) assisted migration potential: Testing establishment north of the species range. *Ecological Applications* 22:142–53.

Pearson, R. G., and T. P. Dawson. 2003. Predicting the impacts of climate change on the distribution of species: Are bioclimate envelope models useful? *Global Ecology and Biogeography* 12:361–71. doi: 10.1046/j.1466-822X.2003.00042.x.

Pearson, R. G., J. C. Stanton, K. T. Shoemaker, M. E. Aiello-Lammens, P. J. Ersts, N. Horning, D. A. Fordham, C. J. Raxworthy, H. Y. Ryu, J. McNees, et al. 2014. Life history and spatial traits predict extinction risk due to climate change. *Nature Climate Change* 4:217–21.

Piekielek, N., A. J. Hansen, and T. Chang. 2015. Using custom scientific workflow software and GIS to inform protected area climate adaptation planning across Greater Yellowstone. *Ecological Informatics* 30:40–48.

Rehfeldt, G. E., N. L. Crookston, C. Saenz-Romero, and E. M. Campbell. 2012. North American vegetation model for land-use planning in a changing climate: A solution to large classification problems. *Ecological Applications* 22:119–41.

Serra-Diaz, J. M., J. J. Franklin, M. Ninyerola, F. W. Davis, A. D. Syphard, H. R. Regan, and M. Ikegami. 2014. Bioclimatic velocity: The pace of species exposure to climate change. *Diversity and Distributions* 20 (2): 169–80.

Stein, B. A., P. Glick, N. Edelson, and A. Staudt, eds. 2014. *Climate-Smart Conservation: Putting Adaptation Principles into Practice*. Washington, DC: National Wildlife Federation.

Chapter 10

Past, Present, and Future Impacts of Climate on the Vegetation Communities of the Greater Yellowstone Ecosystem across Elevation Gradients

Nathan B. Piekielek, Andrew J. Hansen,
and Tony Chang

The Greater Yellowstone Ecosystem (GYE) poses distinct environmental challenges to plant life. From lower elevations that routinely experience prolonged dry periods during warm summer months to higher elevations that are often covered in snow and below freezing for more than half of the year, plant growth is severely limited by climate in different ways across the GYE. As a result, many plant species found here are survivors—that is, they are better adapted to survive the poor growing conditions in their part of the ecosystem than are other species. This is reflected in distinct vegetation communities that change across zones of elevation along with climate, from generally hot and dry conditions at lower elevations to cold and wet conditions at higher elevations (chap. 4).

The influence of climate on vegetation is reflected in the ebb and flow of boundaries separating vegetation communities, as climate changed through paleologic time. Climate also interacts with other landscape processes, such as soil conditions, natural disturbance, dispersal and establishment, and human management, to determine vegetation community membership and persistence. It is our understanding of how climate shapes current vegetation communities, as well as the evidence in the paleologic record of how vegetation has responded to changing climates in the past, that provides the foundation for understanding how future climate change may impact the vegetation communities of the GYE.

Yellowstone National Park was the world's first protected area created in the contemporary "national park" ethic, and it remains an international icon for environmental preservation. The GYE forms the headwaters of three major river systems (the Missouri, Colorado, and Columbia) and is widely regarded as one of the most intact ecosystems left in the contiguous forty-eight states of the United States. By land area, the GYE is more than half protected for environmental conservation and related uses and contains natural resources of tremendous social, political, and economic interest, including endangered species, the natural source of human pharmaceuticals, and other rare, unique, and important natural and cultural resources (Schrag, Bunn, and Graumlich 2008). Great interest in environmental preservation of the GYE presents important opportunities to demonstrate cutting-edge natural resource management strategies by, for example, better linking science and management (chaps. 2 and 3). The most recent and arguably largest challenge faced by GYE managers stems from the expected ecological impacts of human-induced climate change over the course of the next century.

Because vegetation communities in the GYE are strongly shaped by climate, and climate and associated ecosystem processes are already changing rapidly in the northern Rockies (chaps. 4, 6, and 7), it is feared that some vegetation communities are especially vulnerable to climate change. Changes in GYE vegetation could threaten existing wildlife conservation goals and provide a positive feedback to even more climate and vegetation change through, for example, climate warming resulting in forest loss leading to decreased summer snowpack and further climate warming (Schlaepfer, Lauenroth, and Bradford 2012; Gleason, Nolin, and Roth 2013; but see also Grant, Tague, and Allen 2013).

These and other potential consequences of climate change highlight the need for GYE natural resource managers to better understand vegetation response within long-term past, current, and future contexts. This understanding is a critical first step to developing management response in which human intervention is likely to be successful, as well as to determining where management action either could be counterproductive to environmental preservation goals or is unlikely to succeed and, therefore, not worth the expenditure of limited resources. A better understanding of likely future change will also provide natural resource managers with an opportunity to reevaluate or even rewrite existing goals, which is one of the central tenets of the Climate-Smart Conservation framework (Stein et al. 2014).

The natural resource science community is well positioned to help managers better understand vegetation response to climate and climate change. Specifically, species distribution models (which include bioclimate envelope models but can also use other predictors in addition to climate) have been used to identify and quantify relationships between current species ranges and environmental factors, including climate. Species distribution models are often projected through time to understand how the suitability of future conditions may differ from present conditions (Araújo and Guisan 2006). The practice of projecting future habitat suitability under climate change has become common as a result of increasing demand from the natural resource management community and improved spatial resolution of climate model projections that now produce results at management-relevant scales (Franklin 2009; Knutti and Sedláček 2012; Thrasher et al. 2013).

Climate modeling results have been used to make natural resource management decisions around the world, and through this process we have learned about complexities that can exist when trying to relate habitat suitability to species' distributions (Benito-Garzón et al. 2013; Wang et al. 2006). One of these complexities includes lag effects, whereby distributions respond slowly to changing conditions governed in part by a differential response of species life stages, such as adult survival and seedling establishment (Bell, Bradford, and Lauenroth 2013). The leading edge of a vegetation species' range is thought to be governed primarily by seedling dispersal and establishment, and the trailing edge by adult survival. Each life stage (e.g., dispersal, establishment, and maturation) takes time, meaning that the rate of environmental change can be as important as the magnitude of change in determining complex ecological outcomes of actual shifts in vegetation distributions. Because species distribution models only consider changes in habitat suitability, it can also be useful to consider other sources of information, such as those from the paleoecological record, to better understand how vegetation may respond to future climate change.

The primary objective of this chapter is to examine the potential response of GYE vegetation to climate change using new species distribution models tailored specifically to decision-support needs of GYE natural resource management. Our approach was designed to build on previous analyses of vegetation response to climate change across the US Northern Rocky Mountains (chap. 9) by doing new analyses using finer-scale predictor and response data for tree species in the GYE.

To put these results into a longer-term context, we also summarized paleoecological studies of vegetation and climate change across the GYE since the last glacial advance. In order to make the results most useful to managers, we then suggest new research directions to reduce scientific uncertainty in the context of climate change. Finally, we demonstrate the application of this new science information with a focus on whitebark pine (*Pinus albicaulis*), a keystone species of special conservation concern that is already responding to climate change (Logan, Macfarlane, and Willcox 2010).

The material in this chapter is a synthesis of three published papers. Piekielek, Hansen, and Chang (2015) developed species distribution models for eight vegetation communities in the GYE. Chang, Hansen, and Piekielek (2014) used similar methods and focused on whitebark pine. And Hansen et al. (in review) evaluated the research and management implications of new science results for whitebark pine.

Tree Species

We developed species distribution models for nine vegetation communities across four elevation zones in the GYE (fig. 10-1 and table 10-1). Multiple species of sagebrush occupy the lowest elevations on well-drained soils where there is complete soil recharge in the spring that wets deeper soil layers (Schlapfer, Lauenroth, and Bradford 2012). Moving up in elevation, lower-treeline species are found in often steep and rocky terrain in the transition zone from valley bottoms to midslope settings. Montane forests occupy the best growing conditions where there is often consistent moisture

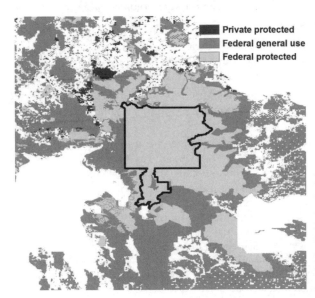

FIGURE 10-1
Study area, including areas managed by federal agencies for environmental protection and general use.

TABLE 10-1. Vegetation communities by elevation zone and the environmental predictors used to model their habitat suitability.

Species	Elevation Zone (meters)	Environmental Predictors	Strength of Model (AUC)
sagebrush	< 1,900	August deficit; April snowpack; solar radiation; rock volume; % sand; June soil moisture	0.731
Lower treeline	1,901–2,200		
juniper		April snowpack; wetness	0.961
limber pine		rock volume; August deficit; June soil moisture; % sand; April snowpack; solar radiation	0.655
Montane	2,201–2,500		
aspen		August deficit; rock volume; April snowpack; % sand; June soil moisture	0.863
Douglas-fir		April snowpack; % sand; August deficit; wetness; solar radiation; June soil moisture; rock volume	0.777
lodgepole pine		% sand; June soil moisture; August deficit; April snowpack; solar radiation; rock volume	0.768
Upper treeline	2,501–3,300		
Engelmann spruce		August deficit; rock volume; % sand; April snowpack; solar radiation; wetness	0.765
subalpine fir		% sand; April snowpack; solar radiation; rock volume; August deficit	0.857
whitebark pine		July maximum temperature; April snowpack; January minimum temperature; March vapor-pressure deficit; April precipitation; July actual evapotranspiration; August potential evapotranspiration; September precipitation	0.940

Note: Predictors are listed in the order of their relative contribution to each species model from high importance to low. Elevation zones are from Despain (1991).

available for plant growth and the growing season is long as a result of limited snow accumulation and moderate temperatures. Soil conditions play a large role in determining montane forest membership, where lodgepole pine (*Pinus contorta*) can be found on sandy soils of poor moisture-holding capacity and nutrient availability, such as those across much of the Yellowstone plateau, and Douglas-fir (*Pseudotsuga menziesii*) and sometimes Engelmann spruce (*Picea engelmannii*) or subalpine fir (*Abies lasiocarpa*) are found on other soil types. Upper treeline species are well adapted to long periods of snow cover and cold where growing seasons are short and moisture is rarely limiting.

To build species distribution models, we used observations of presence and absence from the US Forest Service Inventory and Analysis database, Whitebark and Limber Pine Information System, and long-term monitoring plots established by the Greater Yellowstone Network of the National Park Service Inventory and Monitoring Program.

Environmental Predictors of Tree Species Distributions

To represent climate, thirty-year (1981–2010) monthly averages of temperature and precipitation from the Parameter-elevation Relationship on Independent Slopes Model data set were used as predictors and were input into a dynamic water balance model (in the sense of Lutz, van Wagtendonk, and Franklin 2010). The water balance model accumulated snow and soil moisture to field capacity (or released it as runoff) and depleted soil moisture as a result of estimated rates of plant evapotranspiration based on temperature, with adjustments for latitude, slope, and aspect.

From water balance outputs, we generated monthly averages for soil moisture content, snowpack, actual evapotranspiration, potential evapotranspiration (not limited by the availability of soil moisture), and deficit (potential minus actual evapotranspiration). Soil conditions were represented by the CONUS-SOIL data set of Miller and White (1998), and a US Geological Survey digital elevation model was used to derive a topographic wetness index (hereafter, wetness) and solar radiation that considered shading from adjacent topography as well as the seasonal migration of the sun from lower to higher angles.

To represent the range of possible future climate conditions, we chose nine of the most recent Coupled Model Intercomparison Project Phase 5 models, as well as a low and high future atmospheric carbon dioxide concentration scenario for a total of eighteen climate change scenarios.

We considered three future thirty-year time periods, ending in years 2040, 2070, and 2099.

Tree Species Distribution Models

Species distribution models were constructed using multivariate adaptive regression splines (Leathwick, Elith, and Hastie 2006) and a randomForest model for whitebark pine (Chang, Hansen, and Piekielek 2014). Iterative model construction with a single predictor randomized at each step produced variable importance scores that identified which predictors contributed the most to explaining species distributions. The area under the receiver operating characteristic curve (AUC) is a common model diagnostic that was used to evaluate model fit; a score of 0.5 indicated that a model performed no better than random, while a score of 1.0 indicated perfect model discrimination between presence and absence in withheld portions of training data. AUC scores above 0.7 are generally considered to be good, and above 0.9, excellent (Franklin 2009).

To examine habitat suitability under alternative future climates, we ran species distribution models with future climate as input (both a high and a low carbon dioxide scenario were considered), along with static environmental predictors for such factors as soil properties and incoming solar radiation. Because species distribution models produce probabilities of presence (i.e., habitat suitability) as output, there was a need to reclassify probabilities to "presence" or "absence" based on a threshold suitability and level of consensus among different climate models (out of nine considered). We used a threshold suitability that balanced the predictive accuracy of presences versus absences and identified as suitable habitat areas where there was a simple majority consensus (at least five of nine) among the different future climate models. We also examined the standard deviation of change in suitable area when the threshold level of agreement among climate models was varied from one to nine. Finally, we reported the mean elevation of suitable habitat area for each time period.

Impact of Climate Change on Yellowstone Tree Species

When models were used to calculate change in suitable area for the end of the century and a higher carbon dioxide concentration scenario, upper-treeline species demonstrated the largest reduction in suitable area (mean

= −89 percent; range = −80 to −97 percent) (table 10-2 and fig. 10-2), and montane species followed with a mean decrease of −73 percent (range −60 percent to −85 percent). Lower-treeline species were split, with one species

TABLE 10-2. Area and elevation of projected suitable habitat by species and carbon dioxide emissions scenario.

Species	Current Suitable Area (km²)	2040	2070	2099
Low CO²				
sagebrush	132,252	17% (+/−10)	23% (+/−15)	31% (+/−16)
elevation (m)	1,795	1,879	1,905	1,940
(range)	(897–3,230)	(897–3,608)	(897–3,608)	(897–3,608)
Lower treeline				
juniper	133,727	18% (+/−10)	26% (+/−15)	32% (+/−15)
elevation (m)	1,684	1,757	1,790	1,815
(range)	(893–2,849)	(893–3,195)	(893–3,195)	(893–3,246)
limber pine	104,874	−13% (+/−12)	−8% (+/−19)	−22% (+/−20)
elevation (m)	2,013	2,136	2,184	2,231
(range)	(917–4,015)	(917–4,015)	(917–4,015)	(1,007–4,015)
Montane				
aspen	61,028	−1% (+/− 25)	−5% (+/−32)	−10% (+/−31)
elevation (m)	2,091	2,241	2,301	2,399
(range)	(1,048–3,117)	(1,059–3,512)	(1,135–3,512)	(1,135–3,772)
Douglas-fir	78,229	−35% (+/−16)	−38% (+/−25)	−53% (+/−26)
elevation (m)	2,086	2,283	2,341	2,429
(range)	(992–3,833)	(1,099–3,734)	(1,099–3,577)	(1,110–3,577)
lodgepole pine	54,199	−28% (+/−24)	−42% (+/−36)	−50% (+/−38)
elevation (m)	2,460	2,560	2,602	2,631
(range)	(1,736–3,833)	(1,811–3,867)	(1,896–3,842)	(1,964–3,867)
Upper treeline				
Engelmann spruce	53,843	−46% (+/−24)	−61% (+/−36)	−77% (+/−38)
elevation (m)	2,712	2,864	2,934	3,021
(range)	(1,123–4,015)	(1,123–4,015)	(1,123–4,015)	(1,123–4,015)
subalpine fir	42,144	−43% (+/−30)	−56% (+/−46)	−68% (+/−49)
elevation (m)	2,797	2,929	2,982	3,038
(range)	(1,368–4,015)	(1,354–4,015)	(1,354–4,015)	(1,354–4,015)
whitebark pine	29,251	−44% (+/−27)	−69% (+/−13)	−81% (+/−12)
elevation (m)	2,876	3,020	3,128	3,218
(range)	(2,842–2,895)	(2,938–3,182)	(3,055–3,297)	(3,114–3,471)

TABLE 10-2. (*Continued*)

Species	Current Suitable Area (km²)	2040	2070	2099
High CO²				
sagebrush	132,252	18% (+/–9)	28% (+/–16)	40% (+/–17)
elevation (m)	1,795	1,878	1,929	1,995
(range)	(897–3,230)	(897–3,608)	(897–3,711)	(897–3,771)
Lower treeline				
juniper	133,727	16% (+/–9)	32% (+/–16)	55% (+/–16)
elevation (m)	1,684	1,749	1,813	1,928
(range)	(893–2,849)	(893–3,195)	(893–3,246)	(893–3,608)
limber pine	104,874	–15% (+/–12)	–37% (+/–21)	–29% (+/–21)
elevation (m)	2,013	2,147	2,189	2,307
(range)	(917–4,015)	(917–4,015)	(971–4,015)	(1,071–4,015)
Montane				
aspen	61,028	7% (+/–25)	–1% (+/–23)	–60% (+/–36)
elevation (m)	2,091	2,234	2,382	2560
(range)	(1,048–3,117)	(1,059–3,512)	(1,135–3,772)	(1,356–3,772)
Douglas-fir	78,229	–37% (+/–16)	–63% (+/–27)	–73% (+/–28)
elevation (m)	2,086	2,284	2,394	2,559
(range)	(992–3,833)	(1,099–3,577)	(1,110–3,577)	(1,121–3,714)
lodgepole pine	54,199	–26% (+/–23)	–53% (+/–40)	–85% (+/–41)
elevation (m)	2,460	2,550	2,622	2,758
(range)	(1,736–3,833)	(1,811–3,867)	(1,964–3,842)	(2,130–3,833)
Upper treeline				
Engelmann spruce	53,843	–47% (+/–23)	–77% (+/–40)	–90% (+/–41)
elevation (m)	2,712	2,864	3,016	3,145
(range)	(1,123–4,015)	(1,123–4,015)	(1,123–4,015)	(1,123–4,015)
subalpine fir	42,144	–44% (+/–29)	–66% (+/–51)	–80% (+/–52)
elevation (m)	2,797	2,930	3,036	3,114
(range)	(1,368–4,015)	(1,354–4,015)	(1,354–4,015)	(1,394–4,015)
whitebark pine	29,251	–46% (+/–9)	–82% (+/–11)	–97% (+/–12)
elevation (m)	2,876	3,023	3,226	3,471
(range)	(2,842–2,895)	(2,974–3,061)	(3,116–3,412)	(3,255–3,749)

Note: Low emissions are RCP 4.5 and high are 8.5. Area is presented in square kilometers for current and as percentage change from current for projected future. Percentage change numbers in parentheses refer to the standard deviation of change when threshold levels of agreement between climate models to determine suitability were varied from one to nine.

Age (years Before Present)	Southern Region (summer dry)		Northern Region (summer wet)	
0				
2000	Increasing *Picea, Abies, Pinus* / closed pine forest	Temperature decrease; moisture increase	Increasing *Pseudotsuga, Populus* / Douglas-fir parkland	Moisture decrease
4000				
6000	*Pseudotsuga* / Douglas-fir forest	Temperature greater; moisture lower than present	*Pinus, Juniperus* / Pine-juniper forest	Moisture greater than present
8000				
	North and South Region			
10000	*Pinus* / lodgepole pine			
12000	*Abies* / subalpine forest		Temperature increase; moisture increase	
	Picea engelmannii / Engelmann spruce parkland			
14000	*Artemisia* / subalpine parkland-alpine tundra		Temperature lower; moisture lower than present	
	Deglaciation		Warmer than earlier	

FIGURE 10-2   Summary of past climate and vegetation changes in the Yellowstone region. (Adapted from Whitlock and Bartlein 1993.)

exhibiting a decrease (limber pine, –29 percent) and the other an increase (juniper, 55 percent). Sagebrush was projected to see a 40 percent increase in suitable area.

For all species, change in the elevation of suitable habitat area was to higher elevations than present (table 10-2 and fig. 10-3). The magnitude of elevation shifts was generally positively related to the elevations that the species currently occupy (i.e., low-elevation sagebrush moved up the least), with the exception that montane species exhibited two of the four largest increases in mean elevation. Montane species habitat shifted up in elevation an average of 270 meters (but lodgepole pine only 208 meters); upper-treeline species, an average of 304 meters (whitebark pine, 448 meters); lower-treeline species, an average of 170 meters; and sagebrush, 117 meters. In addition to climate tolerances, elevation shifts were influenced by an interaction between mountain shape and soil properties (Elsen and Tingley 2015).

Species distribution models performed well, as indicated by an average AUC that was greater than 0.8 (range 0.655–0.961) (table 10-1). April snowpack appeared in every model and was often one of the most important predictors across elevation zones. Estimates of July maximum temperature or August soil moisture deficit also appeared in every model and were often of high importance, especially in the models of lower-elevation communities (e.g., sagebrush) and moisture-sensitive species (e.g., quaking aspen [*Populus tremuloides*] and Douglas-fir). Soil predictors (percent sand and rock volume) contributed to seven species models and were the most important predictor in three, including lodgepole pine. Topographic predictors appeared in seven models and generally contributed the least.

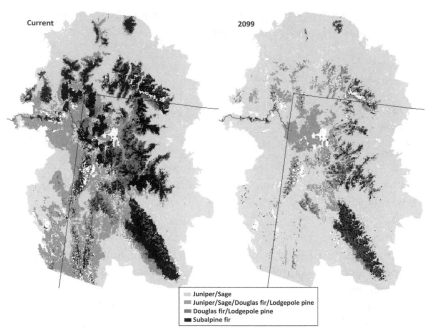

FIGURE 10-3   Current and projected community types across the Greater Yellowstone Ecosystem under the high carbon dioxide concentration scenario. The results of species distribution analysis from Piekielek, Hansen, and Chang (2015) were combined into the community types illustrated in the legend to depict shifts in nonforest and forest types.

Both the high and low future carbon dioxide concentration scenarios agreed on large reductions in future suitable habitat area for most forest species and increasing future suitable habitat area for sagebrush and juniper.

Paleoecological Synthesis

The GYE has undergone numerous vegetation and climate changes over the past fifteen thousand years. Major climate events included the end of the Pleistocene, when many North American glaciers retreated upslope and northward. Whitlock and Bartlein (1993) state that during the late Pleistocene and early Holocene, GYE vegetation shifted rapidly following deglaciation. During the last part of the Pleistocene (12,000 to 11,000 years BP [before present]), much of the GYE region was dominated by sagebrush, with species assemblages that were similar to modern-day shrublands, meadow, and tundra. Many of the coniferous forest communities of present-day North America existed within the lower-latitude southwestern regions of the United States during this time period.

At the end of the Pleistocene and into the early Holocene (circa 12,000 to 6,000 BP), amplification of the seasonal cycle of solar radiation resulted in summer radiation that was approximately 8 percent higher than the present and winter values that were 10 percent lower at 45 degrees N latitude (Iglesias, Krause, and Whitlock 2015). Increased solar radiation was associated with changes in atmospheric circulation that led to higher summer temperatures (about 5 degrees F, or 3 degrees C, higher than present), colder winters (about 3 degrees F, or 2 degrees C, lower than the present), and lower moisture (Alder and Hostetler 2014). This time period marked the migration of subalpine fir and whitebark pine into the GYE, where they displaced communities of sagebrush. Several thousand years later (5,000 BP), a north-south climate division became apparent with warm-dry summers in the southern GYE and warm-wet summers in the north. Northern regions became dominated by montane forest species, including Douglas-fir, aspen, and lodgepole pine, while the southern regions were dominated by upper-treeline species, including Engelmann spruce, subalpine fir, and whitebark pine (fig. 10-2).

Interpreting the Future in Light of the Past

The examination of past climate and vegetation change in the GYE generally supported our interpretation of future species distribution modeling results with a few differences to consider. Species of sagebrush and juniper have thrived in the GYE during paleologic periods of lower precipitation or higher temperatures, such as following deglaciation and in the northern region. In a projected hotter and drier future, the current study's results suggested that area suitable for sagebrush and juniper may increase at the expense of forest species habitat at higher elevations. In contrast to current species distribution modeling results, Douglas-fir appears to have thrived in the southern region of the GYE circa 6,000 BP (which was hotter and drier than the present), when it probably displaced stands of lodgepole pine and subalpine fir. Suitable Douglas-fir habitat over the next century in the present study was limited by generally unsuitable upslope soil conditions (i.e., sandy soils) across much of the mid-elevation Yellowstone plateau that is just upslope of much of its current range. This was reflected by Douglas-fir exhibiting one of the largest increases in mean elevation of suitable habitat area because it had to skip over the Yellowstone Plateau to find suitable upslope conditions in future climates.

The magnitudes of climate change over paleologic time as compared to projected future change are similar; however, the time frame over which this change is expected to take place in the future is much shorter. The transition from a sagebrush-dominated GYE landscape to present-day forest communities took nearly fifteen thousand years, while the current study's projections of sagebrush and juniper suitable habitat expansion onto the Yellowstone plateau spanned about twenty years. The ecological consequences of this difference in rate of change are not well understood. At a minimum, we do not expect that natural rates of in situ adaptation and dispersal will be able to keep pace with changing climate over the next century. The mismatch of expected future rates of change that outpace anything that we have seen in the paleoecological record and the natural abilities of species to respond brings the possibility of human intervention to the forefront of natural resource management debate.

In total, the species distribution modeling projections suggested that climate suitability for forests of the GYE will change substantially in the coming century. Warming temperatures, decreasing springtime snowpack, and decreasing late-season soil moisture would result in a longer, warmer, and drier growing season than present. In general, vegetation communities are projected to shift upward in elevation, while climate suitability for the highest-elevation species, such as whitebark pine, largely moves off the tops of mountains. Upper-treeline forest habitat suitability shifts to the current alpine zone (i.e., above treeline), where in some cases there are only rock outcroppings that may not support forest establishment. Montane species lose suitability across the large mid-elevation portion of the GYE and shift to the current upper-treeline zone in places where soil conditions are suitable. Sagebrush and juniper savanna communities retain climate suitability at current lower treeline but also expand over large portions of the current montane forest zone.

Combining these results into sagebrush, lower-treeline, montane, and upper-treeline community types provides a way of visualizing projected distributions of forest and nonforest types across the GYE (fig. 10-3). The projections for 2099 under the high carbon dioxide scenario are striking in the expansion of habitat suitability for sagebrush and juniper savanna communities and the contraction of closed-canopy conifer forest suitability (fig. 10-3).

With habitat suitability as a basis of understanding, it is also useful to consider how vegetation communities may be affected by the indirect effects of climate change, such as from changing disturbance regimes and pest demography. Based on climate change alone, fire frequency was projected to increase dramatically across all elevations of the GYE (Westerling et al. 2011). However, when feedbacks from vegetation were included (Clark, Keane, and Loehman, in review), fire frequency also increased under future climate scenarios, but the changes were less extreme. Regarding mountain pine beetles, a major agent of mortality for pine species in the GYE, Buotte et al. (in review) projected increasingly favorable climate conditions for beetles.

Such findings lead to the question of how changes in forest habitat suitability, fire regimes, and pest outbreaks will interact to influence patterns of vegetation across the GYE. We speculate that these interacting factors will result in vegetation in the GYE later in the century being more dominated by nonforest communities and the remaining forest communities being earlier in seral stage and lower in canopy cover. Initial modeling studies are largely in agreement with this speculation (Clark, Keane, and Loehman, in review).

Changes in the distributions of the species considered above would likely have large consequences for the provisioning of ecosystem services across the GYE. Loss of coniferous forest cover (especially where fire facilitates the transition) could exacerbate reductions in summer snowpack, with large consequences for streamflows and temperature, coldwater fish populations, and downstream water availability for irrigation and human consumption (chaps. 7, 12, 15, and 16). Habitat quality would be expected to deteriorate for the many species of wildlife now dependent on forest habitats and snow cover. As a consequence, some GYE wildlife conservation goals would need to be modified or rewritten entirely to reflect a potential future without the habitat to support some native species. A revised wildlife conservation goal could include the facilitation of wildlife migration to newly suitable habitat that becomes available outside of the GYE in the future. Finally, implications for the quality of visitor experience and recreational opportunities are poorly understood but are assumed to be substantial.

Research to Reduce Uncertainty

The present study was performed in support of the Climate-Smart Conservation framework for climate adaptation for natural resources (chap. 2). Thus, it is important to evaluate how results can best inform management.

Although species distribution models are highly valuable as first filters for prioritizing research and management under climate change (Hansen and Phillips 2015), they do not consider many factors influencing the viability of a species in newfound conditions. The method assumes that interactions among climate, disturbance, dispersal ability, biotic interactions, and other ecological factors continue in the future as they are today, even when we understand this to be an unlikely outcome.

The list of important ecological factors that are not considered in species distribution models can be used to guide future research and adaptive management, especially for species found to be highly vulnerable to changes in climate suitability. Whitebark pine was projected to have the greatest loss of suitable habitat area in the current study and has already undergone high levels of adult mortality across the GYE from mountain pine beetle infestation (Logan, Macfarlane, and Willcox 2010). Consequently, it is listed as a threatened species by the US Fish and Wildlife Service. The Greater Yellowstone Coordinating Committee's Whitebark Pine Subcommittee has developed a strategy for managing this species (GYCC 2011) and has begun implementation of the plan (chap. 15). The subcommittee is being asked by stakeholders if this management is likely to be effective given the projected loss of suitable climate conditions in the future (although approximately 370 square miles, or 960 square kilometers, are projected to remain suitable; Chang, Hansen, and Piekielek 2014). To address this question, Hansen et al. (in review) identified hypotheses under which whitebark pine populations may remain viable despite projected climate change. These hypotheses led to the following research questions and considerations:

1. Under what conditions can whitebark pine seedlings establish and mature in the alpine zone, and will establishment require human intervention to be successful?

Field studies of close relatives of whitebark pine and in other climate regions suggest that seedling establishment in the alpine zone and north of the current whitebark pine range is possible and may require human intervention to be successful (McLane and Aitken 2012). Further study of Yellowstone whitebark pine ecology would help to guide management of local populations.

2. Will microrefugia provide suitable habitat for whitebark pine?

Some locations projected to become unsuitable by moderate spatial resolution (about 250 acres, or 1 square kilometer) species distribution models may actually contain small patches that remain suitable because of their microsite characteristics. Narrow and steep, north-facing slopes may maintain cooler temperatures and longer winter snowpack than is represented by moderate-resolution climate projections (chap. 6). These sites may serve as microrefugia where whitebark pine could persist even while the surrounding landscape becomes unsuitable.

3. Will some GYE whitebark pine genotypes be able to tolerate future conditions that are projected to be unsuitable by species distribution models?

Within the GYE whitebark pine population, genetic variants may exist that are better able to tolerate the hotter and drier conditions that are projected by future climate models. These variants are expected to be favored by natural selection and may persist in places where average models trained on entire populations (e.g., species distribution models) predict future unsuitable habitat.

4. Would active management to reduce competition with whitebark pine across some of its range provide suitable future habitat in addition to what is projected by species distribution models?

The current distribution of whitebark pine is thought to be strongly limited by competition with lodgepole pine and subalpine fir and may be able to persist in warmer conditions in the absence of competition.

5. Can existing whitebark pine management programs use climate science results to facilitate the establishment of small but vigorous populations of whitebark pine that contribute to the maintenance of the species in the GYE?

Some of the current mortality of whitebark pine is caused by white pine blister rust. Seedlings that are genetically resistant to the rust have been propagated and are being planted. If these seedlings are planted in locations projected to maintain suitable habitat, if competing vegetation is controlled, and if mountain pine beetles do not cause mortality, these seedlings will likely contribute to the maintenance of a viable population in the future.

In the context of the Climate-Smart Conservation framework, these research questions and hypotheses of whitebark pine persistence should be tested in adaptive management experiments. Experiments could include management implementations and controls that are monitored to evaluate outcomes and advance our understanding of whitebark pine ecology in the GYE and the effectiveness of management interventions. In turn, the work on whitebark pine as an early responder to climate change would inform active management of other vegetation species and communities across zones of elevation.

Management Implications

Projected climate change represents a very significant challenge to natural resource managers. There is high uncertainty about the magnitude of climate change, the ecological response to it, the effectiveness of management treatments, and even the appropriateness of active management in some wildland settings (e.g., designated wilderness, roadless areas, and so forth).

Fortunately, approaches to handling these challenges are being developed, tested, and revised, such as the Climate-Smart Conservation framework (chap. 2). Research is being used to project potential future response to climate change and to reduce uncertainty. Monitoring in fast-changing places provides information on actual rates of change and ecological response. Vulnerability assessments can reveal which species or ecosystems are most at risk, where these are located, and why they are at risk. Education programs for natural resource management staff and the public can help promote better understanding of the issues and to formulate effective public policy. Agency planning documents can incorporate consideration of climate change in order to mitigate undesirable future outcomes. Passive management techniques, such as allowing fires to burn naturally, can sometimes favor species vulnerable to climate change. Finally, a variety of types of active management are being developed and evaluated that are aimed at protecting existing populations in situ until newly suitable habitats develop, facilitating natural establishment in adjacent newly suitable habitats and assisting migration to more distant newly suitable habitats.

Climate adaptation planning will be most effective if it is stratified geographically based on the locations of current and projected future suitable climate (Loarie et al. 2008; McLane and Aitken 2012). Hansen and Phillips (2015) recognized three climate suitability zones relevant to climate adaptation. "Core habitats" are locations where populations are currently present and habitat is projected to remain suitable in the future. "Future habitats" are currently unsuitable but are projected to become suitable in the future. "Deteriorating habitats" are those that are currently suitable but are projected to become unsuitable. These zones are illustrated for whitebark pine in the GYE in figure 10-4.

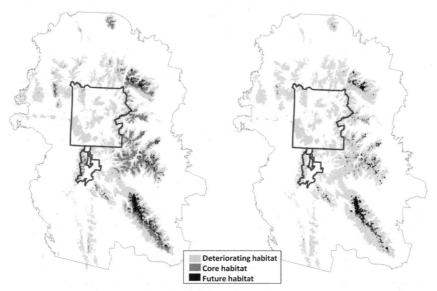

FIGURE 10-4   Distribution of projected deteriorating, core, and future habitat for whitebark pine in the Greater Yellowstone Ecosystem under a moderate warming scenario (low carbon dioxide concentration—left) and under a more extreme warming scenario (high carbon dioxide concentration—right). Habitat types are defined in the chapter text. The period of reference is current to 2099. (From Hansen and Phillips 2015.)

Core habitats for whitebark pine are largely in the current upper-treeline zone. These habitats are of greatest ecological importance because the populations are most likely to remain viable in the future. The primary management goal in these locations could be to maintain population viability (fig. 10-5). Most of the current range of whitebark pine in the GYE is projected to become deteriorating habitat by 2099, especially in the high carbon dioxide concentration scenario. The management goal in these habitats could be to maintain whitebark pine populations as long as possible while new populations establish. Future habitats for whitebark pine in the GYE are mostly located in the alpine zone that is not currently forested. Management strategies to encourage expansion into the alpine zone could include promoting high levels of seed immigration and assisted migration.

In the natural resource science and management communities, there is currently heated debate about whether active management should be part of our climate adaptation response on federal lands where current land use policy precludes its use. The enabling legislations for national parks, roadless areas, and designated or proposed wilderness areas encourage or require minimal human activity and intervention (Long and Biber 2014).

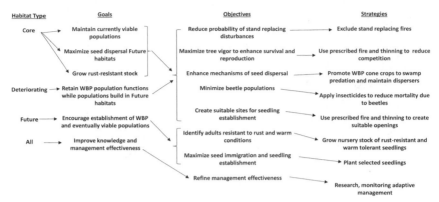

FIGURE 10-5   An example for whitebark pine (WBP) of using future habitat suitability as a framework for management. Habitat types are defined in the chapter text. From Hansen et al. (in review).

All lines of evidence considered in the present study suggest that future suitable forest habitat increasingly shifts from mid-elevation federal lands in the montane zone, where active management is currently well accepted, to higher-elevation lands in the upper-treeline and alpine zones, where it is not. While the debate over active management in wildland settings will likely continue, it should be noted that research, monitoring, education, vulnerability assessment, and passive management are all currently viable options for managers of all federal lands.

Regardless of whether current management policy is changed in the future to allow active management, it remains unclear whether even the most well executed climate adaptation strategy will be sufficient to preserve all contemporary GYE forest species and their associated ecosystem services in the future. It is, however, the explicit consideration of potential futures that do not resemble our world today, as well as the development of management policies that are a reflection of what we would like the future to look like, that forms the basis of any Climate-Smart Conservation response strategy (Stein et al. 2013).

Conclusion

Species distribution models and the results of paleoecological studies both suggest that forest habitat suitability in the GYE will deteriorate over the course of the next century under expected climate change. This was found to be true across zones of elevation and was accompanied by an expansion of suitable habitat for sagebrush communities and juniper woodlands. Upper-treeline species in the GYE, such as whitebark pine, appear to be most at risk of climate change—for example, whitebark pine's preferred habitat is projected to shift into the alpine zone (or off the tops of mountains), where soil conditions may not be suitable for seedling establishment. Projected loss of habitat suitability combined with an anticipated rate of change that greatly exceeds anything that we have seen before poses a host of challenges for natural resource managers who are interested in preserving GYE forest communities.

Research, monitoring, adaptive management, and the Climate-Smart Conservation framework for climate adaptation planning all provide useful responses to these challenges. In particular, management response to climate change will most likely be successful if it is stratified geographically based on the locations of current and projected future suitable habitat. This chapter provided an example of applying the Climate-Smart Conservation framework to forest vegetation communities in the GYE and made recommendations for future research to reduce scientific uncertainty.

Acknowledgments

Funding was provided by the NASA Applied Sciences Program (10-BIO-CLIM10-0034) and the North Central Climate Sciences Center. We thank Linda B. Phillips for her assistance with figure preparation, and William B. Monahan for his comments and suggestions on this chapter.

References

Alder, J., and S. Hostetler. 2014. Global climate simulations at 3000 year intervals last 21 000 years with the GENMOM coupled atmosphere–ocean model. *Climate of the Past Discussions* 10:2925–78.

Araújo, M. B., and A. Guisan. 2006. Five (or so) challenges for species distribution modelling. *Journal of Biogeography* 33:1677–88.

Bell, D. M., J. B. Bradford, and W. K. Lauenroth. 2013. Early indicators of change: Divergent climate envelopes between tree life stages imply range shifts in the western United States. *Global Ecology and Biogeography* 23 (2): 1–13.

Benito-Garzán, M., M. Ha-Duong, N. Frascaria-Lacoste, and J. Fernández-Manjarrés. 2013. Habitat restoration and climate change: Dealing with climate variability, incomplete data, and management decisions with tree translocations. *Restoration Ecology* 21 (5): 530–36.

Buotte, P. C., J. A. Hicke, H. K. Preisler, J. T. Abatzoglou, K. F. Raffa, and J. A. Logan. In review. Historical and future climate influences on mountain pine beetle outbreaks in whitebark pine forests of the Greater Yellowstone Ecosystem. *Ecological Applications*.

Chang, T., A. J. Hansen, and N. Piekielek. 2014. Patterns and variability of projected bioclimate habitat for *Pinus albicaulis* in the Greater Yellowstone Ecosystem. *PLOS ONE* 9 (11): e111669.

Clark, J., R. E. Keane, and R. A. Loehman. In review. Climate changes and wildfire alter forest composition of Yellowstone National Park, but forest cover persists. *Climate Change*.

Despain, D. 1991. *Yellowstone Vegetation: Consequences of Environment and History in a Natural Setting*. Lanham, MD: Roberts Rinehart.

Elsen, P. R, and M. W. Tingley. 2015. Global mountain topography and the fate of montane species under climate change. *Nature Climate Change* 5:772–76.

Franklin, J. 2009. *Mapping Species Distributions: Spatial Inference and Prediction*. Cambridge: Cambridge University Press.

Gleason, K. E., A. W. Nolin, and T. R. Roth. 2013. Charred forests increase snowmelt: Effects of burned woody debris and incoming solar radiation on snow ablation. *Geophysical Research Letters* 40:1–8.

Grant, G. E., C. L. Tague, and C. D. Allen. 2013. Watering the forest for the trees: An emerging priority for managing water in forest landscapes. *Frontiers in Ecology and Environment* 11 (6): 314–21.

Greater Yellowstone Coordinating Committee (GYCC), Whitebark Pine Subcommittee. 2011. Whitebark Pine Strategy for the Greater Yellowstone Area. Report.

Hansen, A. J., K. Ireland, K. Legg, R. E. Keane, E. Barge, M. Jenkins, and M. Pillet. In review. Complex challenges of maintaining whitebark pine in Greater Yellowstone under climate change: A call for innovative research, management, and policy approaches. *Forests*.

Hansen, A. J., and L. B. Phillips. 2015. Which tree species and biome types are most vulnerable to climate change in the US Northern Rocky Mountains? *Forest Ecology and Management* 338:68–83.

Iglesias, V., T. R. Krause, and C. Whitlock. 2015. Complex response of pine to past environmental variability increases understanding of its future vulnerability. *PLOS ONE* 10 (4): e0124439.

Knutti, R., and J. Sedláček. 2012. Robustness and uncertainties in the new CMIP5 climate model projections. *Nature Climate Change* 3:369–73.

Leathwick, J. R., J. Elith, and T. Hastie. 2006. Comparative performance of generalized additive models and multivariate adaptive regression splines for statistical modelling of species distributions. *Ecological Modelling* 199:188–96.

Loarie, S. R., B. E. Carter, K. Hayhoe, S. McMahon, R. Moe, C. A. Knight, and D. D. Ackerly. 2008. Climate change and the future of California's endemic flora. *PLOS ONE* 3 (6): e2502.

Logan, J. A., W. W. Macfarlane, and L. Willcox. 2010. Whitebark pine vulnerability to climate-driven mountain pine beetle disturbance in the Greater Yellowstone Ecosystem. *Ecological Applications* 20:895–902.

Long, E., and E. Biber. 2014. The Wilderness Act and climate change adaptation. *Environmental Law* 44:632–91.

Lutz, J. A., J. W. van Wagtendonk, and J. F. Franklin. 2010. Climatic water deficit, tree species ranges, and climate change in Yosemite National Park. *Journal of Biogeography* 37:936–50.

McLane, S. C., and S. N. Aitken. 2012. Whitebark pine (*Pinus albicaulis*) assisted migration potential: Testing establishment north of the species range. *Ecological Applications* 22:142–53.

Miller, D. A., and R. A. White. 1998. A conterminous United States multi-layer soil characteristics data set for regional climate and hydrology modeling. *Earth Interactions* 2.

Piekielek, N., A. J. Hansen, and T. Chang. 2015. Using custom scientific workflow software and GIS to inform protected area climate adaptation planning across the Greater Yellowstone. *Ecological Informatics* 30:40–48.

Schlaepfer, D. R., W. K. Lauenroth, and J. B. Bradford. 2012. Consequences of declining snow accumulation for water balance of mid-latitude dry regions. *Global Change Biology* 18:1988–97.

Schrag, A. M., A. G. Bunn, and L. J. Graumlich. 2008. Influence of bioclimatic variables on tree-line conifer distribution in the Greater Yellowstone Ecosystem: Implications for species of conservation concern. *Journal of Biogeography* 35:698–710.

Stein, B. A., P. Glick, N. Edelson, and A. Staudt, eds. 2014. *Climate-Smart Conservation: Putting Adaptation Principles into Practice*. Washington, DC: National Wildlife Federation.

Thrasher, B., J. Xiong, W. Wang, F. Melton, A. Michaelis, and R. Nemani. 2013. Downscaled climate projections suitable for resource management. *Eos, Transactions American Geophysical Union* 94:321–23.

Wang, T., A. Hamann, A. Yanchuk, G. A. O'Neill, and S. N. Aitken. 2006. Use of response functions in selecting lodgepole pine populations for future climates. *Global Change Biology* 12 (12): 2404–16.

Westerling, A. L., M. G. Turner, E. A. Smithwick, W. H. Romme, and M. G. Ryan. 2011. Continued warming could transform Greater Yellowstone fire regimes by mid-21st century. *Proceedings of the National Academy of Sciences of the United States of America* 108:13165–70.

Whitlock, C., and P. J. Bartlein. 1993. Spatial variations of Holocene climatic change in the Yellowstone region. *Quaternary Research* 39 (2): 231–38.

Chapter 11

Vulnerability of Tree Species to Climate Change in the Appalachian Landscape Conservation Cooperative

Brendan M. Rogers, Patrick Jantz, Scott J. Goetz, and David M. Theobald

Forests of the Appalachian Landscape Conservation Cooperative provide critical ecological and management functions. The moist climate of the eastern United States fosters productive stands that store relatively high amounts of carbon; for example, the Appalachian Landscape Conservation Cooperative (Appalachian LCC) accounts for only 7.6 percent of the contiguous United States but contains 18.8 percent of its aboveground forest biomass (derived from Kellndorfer et al. 2012). The Appalachian Mountains create substantial topographic and microclimatic diversity, and forests in the southern Appalachian LCC have some of the highest levels of endemic mammal, bird, amphibian, reptile, freshwater fish, and tree species biodiversity in the conterminous United States (Jenkins et al. 2015). Forest types vary from commercial pine plantations in the south to temperate hardwoods in the central Appalachians to high-elevation spruce-fir forests in the north.

Appalachian LCC forests are managed for a variety of objectives by private landowners and state and federal agencies. These objectives include production forestry, low-intensity harvesting, wildlife habitat, native vegetation, water quality, aesthetics and recreation, and, increasingly, climate change mitigation through carbon sequestration. However, forest health and ecosystem service management can be challenging in the Appalachian LCC. Compared to the western United States, these forests tend to be fragmented and privately owned by a large number of stakeholders. A dated property tax system, based on a land's "highest and best use," undervalues ecosystem services and does not promote forest preservation (Beecher 2013). Suburban and exurban growth tends to fragment forests near metropolitan areas (Brown et al. 2005), while natural gas drilling, wind farms, and mining disturb remote and higher-elevation forests (Dunscomb et al. 2014). Land is increasingly divided between more parties ("parcelization"), limiting regional coherence in management objectives and approaches.

Many of these forests are also fundamentally transformed from their presettlement state as a result of altered fire regimes, deforestation and agricultural abandonment, and invasive species. These land use activities have tended to decrease biodiversity, replacing tracts of fire-adapted woodlands and open forests with relatively mesic even-aged and closed-canopy forests (Nowacki and Abrams 2008). This, along with exotic introductions and warming, has increased susceptibility to insects and diseases. In many affected areas, regeneration is hindered by deer herbivory (Eschtruth and Battles 2008).

Climate change will add additional stressors to current management challenges. As shown in chapter 5, temperatures have risen over the last several decades (especially in the northern Appalachian LCC) and are projected to increase by a further 4 to 5.4 degrees F (2.2 to 3 degrees C) by 2055 and 4.8 to 8.3 degrees F (2.7 to 4.6 degrees C) by 2080 in the Appalachian LCC. As a rule of thumb, to keep pace with warming, tree species will need to migrate roughly thirty-five miles for every 1 degree F of warming (Melillo et al. 1990). Woodall et al. (2009) provide evidence that many tree species in the eastern United States are beginning to shift their ranges northward at a rate that could approach sixty miles per century, although recent analyses of the same data have challenged this finding (e.g., Zhu, Woodall, and Clark 2012). Nonetheless, numerous studies indicate that tree migration will not be able to track the velocity of climate change under even the most optimistic of dispersal and climate scenarios (e.g., Iverson, Schwartz, and Prasad 2004; Nathan et al. 2011).

Because of this information, there is a growing appreciation that climate change needs to be included in management plans. Some preliminary implementations are promising. For example, working with the US Forest Service's Northern Institute of Applied Climate Science, various federal, state, and private stakeholders have incorporated climate and habitat projections into their management activities. Climate-smart activities include improving core, matrix, and connecting habitat; selective thinning and plantings that favor native tree species projected to thrive in future climates; cutting poorly performing species that are not projected to do well; and improving resiliency to pests and pathogens.

Some of the best and most widely used tools for informing climate-smart management are species distribution models (SDMs). As detailed in chapters 8, 9, and 10, SDMs statistically relate climate and other geospatial data to observed species occurrence or abundance. When projected to future climates, SDMs suggest where habitat suitability may change. By themselves, SDMs may provide enough information for certain climate-focused management activities, especially those involving more intensive approaches (e.g., selective thinnings or plantings). However, for more passively managed or unmanaged forests, additional information must be coupled with SDM results to assess vulnerability. This is accomplished by characterizing a system's adaptive capacity, which, for forests, typically involves factors affecting regeneration, migration, and tolerance to various disturbances and stressors.

In this chapter, we present a vulnerability analysis for individual tree species and forest types in the Appalachian LCC and discuss management implications. As an illustration, we provide a case study for the Delaware Water Gap National Recreation Area and protected area centered ecosystem (PACE). For context, we also present results for the entire eastern United States. We used habitat suitability models (SDMs) for forty tree species from chapter 8 as our starting point to define potential impacts. We then coupled this with information on seed source strength, dispersal potential, and forest fragmentation to define adaptive capacity and to estimate vulnerability. Our approach to vulnerability differs from many past efforts (e.g., Matthews et al. 2011; Butler et al. 2015) in that it is entirely spatially explicit. Thus, while we necessarily neglect some types of information, we are able to provide data layers across a broad domain at a management-relevant resolution (0.5 mile, or 800 meter). We believe this can add significant value to ongoing and future activities, particularly those requiring spatial estimates of adaptive capacity and vulnerability.

Defining Vulnerability

We operated within the framework presented in Glick, Stein, and Edelson (2011) to calculate vulnerability. This involved combining climate exposure, inherent sensitivity, and adaptive capacity. The SDM output of changes in habitat suitability from chapter 8 effectively represented the combination of exposure and sensitivity as "potential impact," which we formalize here into a quantitative index. For simplicity, we focused on results obtained with climate projections from the representative concentration pathway

(RCP) 8.5 model ensemble at 2055 (discussed in chapter 8), although we present comparisons to RCP 4.5 and other future time periods.

Potential Impact

For a given species, 0.5-mile (0.25-square-mile) pixel, and point in time, habitat suitability was calculated as the fraction of random forest trees (1,500 for each species, described in chapter 8) that predicted presence of a species. We calculated potential impact as the difference in suitability be-tween a future scenario and the historical time period (thirty-year window centered at 1995) (fig. 11-1). We incorporated both absolute and relative differences in suitability:

$$P_a = S_h - S_f$$

$$P_r = \sqrt{\frac{s_h - s_f}{s_h + s_f}}$$

$$P = \frac{P_a + P_r}{2}$$

where P_a, P_r, and P = absolute, relative, and combined potential impact, respectively, and s_h and s_f = historical and future suitability, respectively. Potential impact is on a scale of −1 to 1. To align with our vulnerability metric, positive values are associated with increasingly deleterious changes in suitability (i.e., higher impact), and negative values indicate increasingly favorable changes (i.e., lower, or no, impact). Note that all vulnerability components are calculated on scales of −1 to 1 (or 0 to 1 in the case of forest fragmentation, propagule pressure, and adaptive capacity) and subsequently multiplied by 100 for ease of interpretation.

Adaptive Capacity

Although potential impact provides information on changes in habitat suitability, it does not inform whether trees will be resilient in the face of unfavorable change or whether they will be able to capitalize on newly suitable locations. To address this, we characterized adaptive capacity. Estimates of adaptive capacity are inherently subjective and dependent on the particular

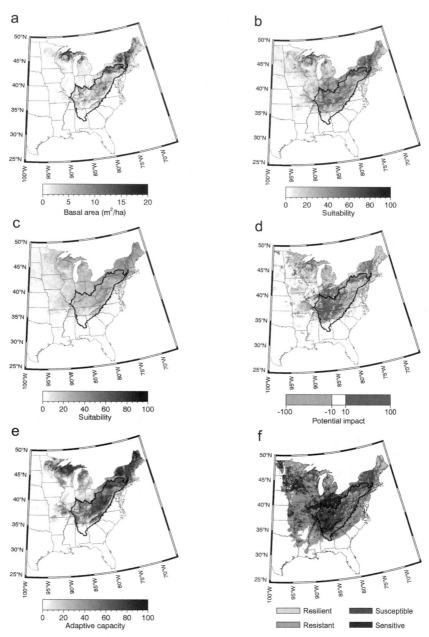

FIGURE 11-1 Example vulnerability layers for sugar maple using RCP 8.5 ensemble model projection for the year 2055. Panels show (a) historical basal area, (b) historical modeled suitability, (c) future modeled suitability, (d) potential impact, (e) adaptive capacity, and (f) vulnerability categories.

elements of vulnerability in question. We chose an approach that incorporates attributes of species' survival and migration that can be quantified spatially with current knowledge and data sets. These attributes included forest fragmentation, seed source strength, and dispersal potential. We did not consider other potentially influential sources of information, such as resistance to disturbance or seed viability, as these properties are uncertain and difficult to quantify. Our approach allowed us to provide spatially explicit estimates in a consistent manner across forty tree species.

Fragmented landscapes negatively impact biodiversity, genetic exchange, and species movements (e.g., Jump and Penuelas 2005). One component of fragmentation is human modification, which introduces impervious surfaces, managed vegetation, agriculture, and other land cover unfavorable to natural vegetation. As an estimate of human modification we used layers from Theobald (2013) projected to future time periods using a method similar to Theobald (2010). Human modification does not represent unfavorable patches created by natural features, such as water bodies, rock outcroppings, bogs, and open grasslands. These, however, can be captured by estimates of forest cover. For example, Iverson, Schwartz, and Prasad (2004) used forest cover as a metric of sink strength in their dispersal model. To calculate forest fragmentation, we therefore averaged human modification and (1 – tree cover) (Carroll et al. 2011), where tree cover is scaled between 0 and 1, capped at a maximum of 70 percent (fig. 11-2). The resulting index ranges from 0 to 1. For future time periods, we assumed that any increases in human modification were associated with similar decreases in tree cover (which is consistent with Jantz, Goetz, and Jantz's (2005) analysis of past change using satellite observations).

The level of seed rain from surrounding forest patches, or "propagule pressure," positively affects adaptive capacity by increasing resilience (e.g., regeneration after disturbances or mortality) or allowing a species to colonize newly suitable areas. We leveraged the results of Iverson, Schwartz, and Prasad (2004), who calculated propagule pressure as the product of importance value and forest cover. Importance value for a given species is the sum of percent basal area and tree density relative to site total. As a proxy for this, we used gridded basal area for every species from Wilson, Lister, and Riemann (2012). Iverson et al. also found that propagule pressure ceased to limit migration when importance values were greater than approximately two and forest cover at a source pixel was greater than 75 percent (product of 150). We applied this result conservatively, such that our final metric of propagule pressure was the product of percent basal area

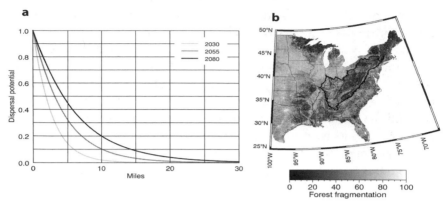

FIGURE 11-2 Dispersal kernels for (a) future time periods and (b) historical (1995) forest fragmentation.

and tree cover, capped at 100 (equivalent to a value of 200 from Iverson's approach) and scaled from 0 to 1.

We also incorporated an estimate of dispersal potential in our adaptive capacity metric. Tree dispersal and migration rates have a long history of active research and debate, and they remain highly uncertain. Estimates from paleoecological literature, current observations, mechanistic models, and statistical models frequently disagree. Maximum migration rates are ultimately thought to be controlled by rare long-distance dispersal events (Clark 1998), which are difficult to capture with observations. Glacially altered landscapes, potential refugia, varying rates of climate change, and data uncertainties limit the direct application of paleoecological studies. Because of this, we chose a simplified but often-used approach that characterizes dispersal potential as a negative exponential kernel, as in Crossman, Bryan, and Summers (2012):

$$D = e^{-\theta d}$$

where D = dispersal potential, d = distance, and θ = the dispersal constant. We chose a dispersal constant of 0.0001 for a time scale of eighty-five years (1995–2080), which generally mimics the colonization probability curves from Iverson's model for several tree species in the eastern United States (Iverson, Schwartz, and Prasad 2004; Prasad et al. 2013) (fig. 11-2) and lies in the middle of the range for previous estimates of tree migration in North America. Because there is little evidence to suggest that life history traits or dispersal mechanisms (e.g., wind versus bird) influence migration rates (Higgins et al. 2003), we apply this same function to every tree species.

For a given species and pixel, adaptive capacity was defined by the product of propagule pressure and (1 – forest fragmentation) of all the surrounding pixels within a thirty-one mile (50-kilometer) radius (fig. 11-1). In this case, forest fragmentation between a given pixel and any surrounding pixel was calculated as the mean fragmentation of the intervening landscape (i.e., matrix habitat), defined by a 30-degree arc extending from the pixel of interest. To incorporate migration, the contributions from surrounding pixels were weighted by their dispersal potentials divided by their distance to the center (to account for the two-dimensional nature of dispersal).

Vulnerability

Our overall metric of vulnerability was calculated as:

$$V = \begin{cases} P \times (1-A) & P \geq 0 \\ P \times A & P < 0 \end{cases}$$

where V = vulnerability, P = potential impact, and A = adaptive capacity. Note that this allows higher adaptive capacity to decrease vulnerability when positive (i.e., harmful impacts) and to increase the magnitude of negative vulnerability (i.e., improved conditions). Unless otherwise stated, overall vulnerability for a given species was calculated by weighting pixels by their historical basal area, thus emphasizing areas in the core habitat. Similarly, we present landscape vulnerability as the mean from all species present at a given pixel, weighted by their basal areas.

Although this metric is helpful to rank species and evaluate spatial patterns of vulnerability, it collapses potentially useful information. We therefore also present management-relevant categories of vulnerability based on the representation of potential impact and adaptive capacity in two-dimensional space (figs. 11-1 and 11-3). Similar to categories used in Mazziotta et al. (2015), *sensitive* is defined by positive (i.e., harmful) potential impacts and low adaptive capacity, whereas *susceptible* indicates positive potential impacts and high adaptive capacity. Susceptible pixels tend to occur within a species' core range; sensitive pixels, closer to the fringes or with rare species. For areas with improving suitability (i.e., negative potential impact),

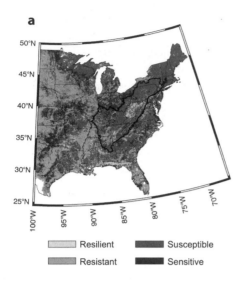

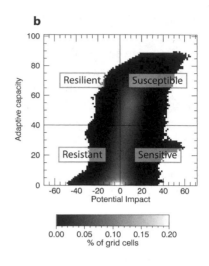

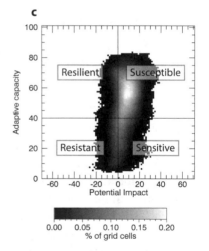

FIGURE 11-3 Landscape-level categories of vulnerability considering all forty species weighted by their historical basal area (a) and density plots of pixels in vulnerability categories for the entire eastern United States (b) and the Appalachian LCC (c). As potential impact increases, habitat suitability for a given species declines. As adaptive capacity increases, the potential for a given species to adapt to either favorable or unfavorable climate change increases.

resistant pixels exhibit low adaptive capacity and *resilient* exhibit high adaptive capacity (fig. 11-3). Resilient is the optimal scenario, suggesting that a species will thrive in a given area, whereas species may not be able to utilize newly suitable area in resistant pixels.

Pests and pathogens are a priority concern for a variety of species—a concern that is projected to be exacerbated by climate warming. To address this concern, we include vulnerability layers that incorporated gridded projections of species-specific basal area loss due to pests and pathogens from Krist, Sapio, and Tkacz (2010). We used these projections in two ways: (1)

by decreasing adaptive capacity directly by the percent loss in basal area, thereby representing the negative impact of site-level mortality, and (2) by reducing the basal area of the surrounding pixels used to define propagule pressure. Note that these estimates are limited to 2030.

Landscape Vulnerability Assessment

As an overarching response to warming, species were projected to do poorer in their core historical ranges and better upslope and to the north. As an illustration of this, mean potential impact across all species and pixels was −4.3 (improving conditions, scale of −100 to +100). When limited to locations with historical basal area or nonzero adaptive capacity for each species, this increased to −3.8; when weighted by species' historical basal areas, the mean jumped to 7.9. Despite an overall increase in suitability across the Appalachian LCC, many newly favorable areas had low adaptive capacity that limited the species' abilities to utilize new habitat. For example, mean vulnerability across all species and pixels was 0.9 (as opposed to a potential impact of −4.3) and increased to 2.7 when weighted by historical basal area. This can be visualized in figure 11-3, which shows that areas with high adaptive capacity tended to have high potential impacts, whereas most areas with negative potential impacts had low adaptive capacity. Conditions also tended to worsen later in the twenty-first century and under the higher emissions scenario: mean vulnerability was 1.0, 1.8, and 2.3 by 2030, 2055, and 2080 under RCP 4.5, and 1.4, 2.7, and 3.5 under RCP 8.5. Note that these are averages across the domain, which includes a large amount of spatial heterogeneity (figs. 11-1 and 11-3). This generally supports previous claims that many tree species will not be able to keep pace with climate change, and it highlights the importance of coupling suitability models with estimates of dispersal.

Compared to the entire eastern United States, adaptive capacity was significantly higher in the Appalachian LCC; thus, most of the Appalachian LCC was categorized as susceptible at the landscape level (fig. 11-3). Notable exceptions included the Appalachian foothills in Virginia and West Virginia, which exhibited some of the most consistently resilient landscapes, and parts of the southwestern Appalachian LCC (Illinois, Indiana, Kentucky, Tennessee, and Alabama) with lower adaptive capacity that were categorized as mixtures of sensitive and resistant landscapes. This was also true of most areas west of the Appalachian LCC, which contained limited adaptive capacity due to high levels of human modification (mostly agriculture) or low levels of forest cover.

Species Vulnerability Assessment

Although analyses at the landscape scale are useful for describing overarching patterns, they arise from variable species-level patterns that may be more useful in many management contexts. General responses to warming were modulated by species-specific sensitivities, associations with static landscape attributes (e.g., soils, solar insolation, and so forth), and precipitation, which tended to increase in climate projections (chap. 5).

An example for one of the most widespread species, sugar maple (*Acer saccharum*), can be seen in figure 11-1. Suitability in much of sugar maple's core habitat decreased by 2055 due to warming. High adaptive capacity limited the deleterious impacts in most of its core habitat, especially in the Great Lakes and greater Appalachians regions. However, relatively low adaptive capacity in the Midwest and southwestern Appalachian LCC resulted in sensitive areas. Suitability increased in many high-elevation and northern landscapes that had high adaptive capacity (i.e., resilient) and therefore may be able to support increased sugar maple populations without direct human intervention, especially parts of Appalachia and Maine. Despite warming, a number of areas to the west and east of sugar maple's current core range also became more suitable as a result of increases in precipitation and, especially in the Midwest, decreases in the seasonality of precipitation. However, these areas were classified as resistant because of low adaptive capacity and are therefore unlikely to see major increases in sugar maple without active management, according to our analysis.

In general, northern and high-elevation species tended to fare considerably worse than southern and low-elevation to mid-elevation species (table 11-1; fig. 11-4). In areas of complex terrain, habitat suitability moved upslope for most species with relatively high levels of adaptive capacity (due to intact forests and large temperature gradients with low climate velocity), indicating the potential for continued upslope migration, as has been observed in other temperate mountain regions (e.g., Kelly and Goulden 2008). However, species already residing at the highest elevations, such as balsam fir (*Abies balsamea*) and red spruce (*Picea rubens*), will likely have nowhere nearby to migrate to and are highly vulnerable.

When categorized by forest type across the Appalachian LCC (described in chapter 8), spruce-fir, northern hardwoods, northern pines, central/mountain pines, and maple-beech-birch forests are consistently the most vulnerable (fig. 11-4). Southern species, particularly oaks, hickories, and pines, were either more resistant to changes in their core habitats or

TABLE 11-1. Mean historical basal area and vulnerability for forty tree species across the Appalachian LCC and entire eastern United States in 2055 under the RCP 8.5 scenario.

	Appalachian LCC			Eastern United States		
Rank	Species	Basal Area (ft² ac⁻¹)	Mean Vulnerability	Species	Basal Area (ft² ac⁻¹)	Mean Vulnerability
1	Red spruce (*Picea rubens*)	0.48	16.2 (0.0)	Pitch pine	0.29	16.4 (1.3)
2	Yellow birch (*Betula alleghaniensis*)	1.54	11.1	Balsam fir	2.31	13.1 (6.6)
3	Balsam fir (*Abies balsamea*)	0.04	9.7 (0.0)	Striped maple	1.63	9.9
4	Longleaf pine (*Pinus palustris*)	0.13	8.6 (0.0)	Red spruce	1.05	9.4 (3.2)
5	Striped maple (*Acer pensylvanicum*)	0.44	8.4	Yellow birch	1.21	9.2
6	Sugar maple (*Acer saccharum*)	16.78	7.1 (0.7)	Quaking aspen	3.04	7.9 (0.0)
7	Quaking aspen (*Populus tremuloides*)	0.66	7.0 (0.0)	Mountain maple	0.03	7.2
8	Eastern white pine (*Pinus strobus*)	7.7	6.3 (0.2)	Red maple	9.31	6.3 (0.0)
9	Red maple (*Acer rubrum*)	29.59	6.0 (0.0)	Eastern white pine	2.82	6.2 (0.2)
10	Virginia pine (*Pinus virginiana*)	4.98	5.2 (0.2)	Blackjack oak	0.38	5.5 (0.0)
11	Pitch pine (*Pinus rigida*)	0.94	5.2 (0.0)	Sugar maple	4.96	5.5 (0.9)
12	Yellow poplar (*Liriodendron tulipifera*)	19.88	4.9	Virginia pine	1.01	5.5 (0.2)
13	Eastern hemlock (*Tsuga canadensis*)	6.52	4.5 (1.5)	Shagbark hickory	0.78	5.5
14	American beech (*Fagus grandifolia*)	8.1	4.1 (1.0)	Yellow poplar	3.86	5.2
15	Fraser fir (*Abies fraseri*)	0.01	3.9 (0.0)	Eastern hemlock	2.19	4.7 (1.2)
16	Black cherry (*Prunus serotina*)	10.07	3.2	Black cherry	2.29	4.4
17	Red hickory (*Carya ovalis*)	0.03	2.7	American beech	2.18	4.0 (0.9)
18	Shagbark hickory (*Carya ovata*)	3.01	2.3	Slash pine	2.37	3.9 (0.0)

TABLE 11-1. (Continued)

		Appalachian LCC		Eastern United States		
Rank	Species	Basal Area ($ft^2\ ac^{-1}$)	Mean Vulnerability	Species	Basal Area ($ft^2\ ac^{-1}$)	Mean Vulnerability
20	Yellow buckeye (*Aesculus flava*)	0.77	1.8	Winged elm	0.79	3.6
21	Mountain maple (*Acer spicatum*)	0.01	1.7	Post oak	2.65	3.2 (0.0)
22	American elm (*Ulmus americana*)	2.04	1.5 (0.5)	Black hickory	0.41	3.1
23	Slash pine (*Pinus elliottii*)	0.004	1.2 (0.0)	Loblolly pine	14.08	3.1 (0.0)
24	Table Mountain pine (*Pinus pungens*)	0.22	1.2	Sweetgum	4.33	2.8
25	Pignut hickory (*Carya glabra*)	6.65	0.8	American elm	1.28	2.7 (0.3)
26	Winged elm (*Ulmus alata*)	0.85	0.5	Red hickory	0.01	1.9
27	American basswood (*Tilia americana*)	2.54	0.2	Black oak	2.5	1.8 (0.0)
28	White ash (*Fraxinus americana*)	8.19	0.1 (0.9)	Chestnut oak	2.47	1.8 (0.0)
29	Silver maple (*Acer saccharinum*)	0.62	−0.6	Yellow buckeye	0.09	1.8
30	Blackjack oak (*Quercus marilandica*)	0.11	−0.7 (0.0)	Pignut hickory	1.17	1.2
31	Black walnut (*Juglans nigra*)	1.93	−0.8	Table Mountain pine	0.02	1.2
32	White oak (*Quercus alba*)	18.18	−1.0 (0.7)	White oak	5.19	1.1 (0.6)
33	Shortleaf pine (*Pinus echinata*)	1.6	−1.1 (0.0)	Black walnut	0.65	1
34	Loblolly pine (*Pinus taeda*)	7.2	−1.3 (0.0)	American basswood	0.96	0.9
35	Sweetgum (*Liquidambar styraciflua*)	3.2	−1.6	Shortleaf pine	1.83	0.7 (0.1)
36	Northern red oak (*Quercus rubra*)	14.25	−1.6 (0.8)	White ash	1.93	0.4 (0.7)
37	Mockernut hickory (*Carya alba*)	3.69	−1.7	Mockernut hickory	1.02	0.4
19	Chestnut oak (*Quercus prinus*)	20.63	1.8 (0.0)	Fraser fir	0.001	3.8 (0.0)

TABLE 11-1.   *(Continued)*

| | Appalachian LCC | | | Eastern United States | | |
Rank Species	Basal Area (ft² ac⁻¹)	Mean Vulnerability	Species	Basal Area (ft² ac⁻¹)	Mean Vulnerability
38 Black oak (*Quercus velutina*)	8.29	−2.0 (0.0)	Northern red oak	3.56	0.1 (0.7)
39 Black hickory (*Carya texana*)	0.02	−4.3	Longleaf pine	0.69	−0.8 (0.1)
40 Post oak (*Quercus stellata*)	1.42	−5.6 (0.0)	Silver maple	0.48	−0.8

Note: Values in parentheses indicate the increase in vulnerability when accounting for pest- and pathogen-induced losses in basal area (in this case, calculated only for the 2030 scenario). Vulnerability was calculated by weighting grid cells by their historical basal area.

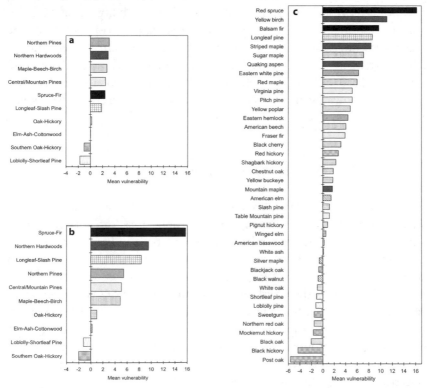

FIGURE 11-4   Vulnerability rankings by forest type (a, b) and species (c) in 2055 for RCP 8.5. Rankings in (a) were calculated by taking the average across each species' range, and rankings in (b) and (c) were weighted by each species' historical basal area. Bar patterning in (c) associates species with their respective forest types in (a) and (b). Note that mean vulnerability increases when grid cells are weighted by their historical abundance.

better able to utilize newly suitable areas (typically to the north and east). Northern species also tended to be more vulnerable to pests and pathogens, although many of the highly affected areas were in the northern Great Lakes regions and northeast United States, outside the Appalachian LCC (table 11-1). The most affected species included balsam fir (due to balsam woolly adelgid and eastern spruce budworm), red spruce (due to root diseases and brown longhorned spruce beetle), pitch pine (*Pinus rigida*, due to *Sirex* woodwasp, southern pine beetle, and engraver beetles), eastern hemlock (*Tsuga canadensis*, due to hemlock woolly adelgid), sugar maple (due to Asian longhorned beetle), and American beech (*Fagus grandifolia*, due to beech bark disease).

We note that vulnerability rankings are in some cases subject to the idiosyncrasies of domain definition. For example, longleaf pine (*Pinus palustris*) emerged as highly vulnerable in the Appalachian LCC. However, this neglected considerable areas of improving suitability classified as resilient and resistant on the Gulf and southern Atlantic coasts (longleaf pine was the second least vulnerable across the entire eastern United States). Moreover, many of the northern species we included have ranges that extend into Canada. Because of the general northward trend in suitable area, we are undoubtedly missing favorable landscapes in Canada, particularly given the region's high levels of intact forest.

Management Implications

Spatially explicit vulnerability analyses can add significant value to climate-smart management and planning. For example, areas categorized as sensitive for a particular species have the worst prognosis. These could be designated as locations for intensive, active management if the species in question is of high priority. Alternatively, it may be better to set these areas aside and focus on others with more favorable projections or higher adaptive capacity. Susceptible landscapes are projected to be unfavorable but have higher adaptive capacity. Depending on the level of potential impact and management context, these areas may justify increases in monitoring or preventative measures (e.g., increasing resistance to pests and diseases, selective thinning and planting, and so forth).

Resilient areas are the most favorable but the least common. They may become increasingly important for species persistence and forest health and may potentially warrant increased monitoring and protection as reserves. Finally, resistant areas are projected to experience higher suitability for a given species but may lack adequate adaptive capacity. These areas may be candidates for more aggressive measures, such as assisted migration, given appropriate motivation and resources. However, such activities should be carefully evaluated against the potential for unintended impacts on native communities (Schwartz et al. 2012).

Our landscape-level analysis may also be useful for broader-scale, holistic activities, such as improving matrix habitat and connectivity. For example, higher connectivity between sensitive and resistant/resilient patches may increase adaptive capacity. Ultimately, management plans and activities are context dependent, varying by land ownership, management priorities, and available resources, but may benefit from vulnerability analyses such as this. Our layers can be considered climate change "threat maps," on which a structured decision-making process can be based and used to prioritize conservation actions (Tulloch et al. 2015).

Case Study: The Delaware Water Gap

Of all the eastern PACEs considered, including the Delaware Water Gap, the Great Smoky Mountains, and Shenandoah, the Delaware Water Gap PACE had the highest mean potential impact (11.4 versus 9.3 for Great Smoky Mountains and 2.2 for Shenandoah). However, Delaware Water Gap also had the highest mean adaptive capacity (64.4 versus 57.7 for Great Smoky Mountains and 48.4 for Shenandoah), resulting in a mean vulnerability (2.8) that was slightly lower than Great Smoky Mountains (3.1), the Appalachian LCC as a whole (3.2), and the entire eastern United States (3.4), but significantly higher than Shenandoah (1.4). The Delaware Water Gap PACE also contains complex terrain (fig. 11-5,) and is therefore fairly representative of many northern forests in the eastern United States. Here, we use the Delaware Water Gap National Recreation Area and PACE as a case study for assessing vulnerability and management implications at a local to regional scale.

Figure 11-5 provides a sample of species vulnerability on a management-relevant domain. We focus on four important tree species for the region whose habitat suitability exhibited similar qualitative responses to climate change (upslope movement) but whose vulnerability assessments varied. These species include red spruce, eastern white pine (*Pinus strobus*), white oak (*Quercus alba*), and eastern hemlock. Red spruce occupies the highest-elevation sites in the Delaware Water Gap PACE; hence, there were no areas of improved suitability, and the species can be considered

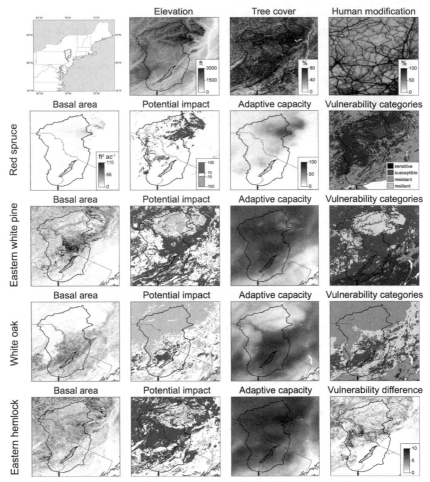

FIGURE 11-5   Example inputs (top row) and vulnerability layers (second through fifth rows; legends shown in second row) for the Delaware Water Gap National Recreation Area and PACE for RCP 8.5 in 2055. Note that because of data limitations, the vulnerability difference map for eastern hemlock (bottom right panel) has been calculated for 2030.

highly vulnerable in the region. Eastern white pine occurs throughout but is most prevalent at mid-elevation sites. Despite unfavorable future conditions at low to mid-elevation, our analysis suggests that white pine has a relatively high chance of colonizing and increasing abundance at higher elevations (categorized here as resilient) due to its high adaptive capacity throughout. White oak, on the other hand, had low adaptive capacity at high elevations and may have a more difficult time migrating upslope. Eastern hemlock displayed patterns similar to white pine but may experience substantial mortality from hemlock woolly adelgid (HWA), which increased its vulnerability.

Regarding the latter, hemlock woolly adelgid is a major concern for the Delaware Water Gap National Recreation Area. HWA was first reported in the eastern United States in the 1950s but was not detected in the park until 1989. Because HWA populations are strongly affected by cold winter temperatures, climate warming has and will likely continue to exacerbate HWA infestations. HWA defoliation can have a quick and devastating impact on growth and mortality: it is estimated that every 10 percent increase in HWA infestation causes a 7 percent decline in new growth the following year (Evans 2010). Since 1993, when permanent monitoring plots were established in the park, total hemlock mortality in many stands is between 30 and 40 percent.

Hemlock forests are recognized as "Outstanding Natural Features" with "high intrinsic or unique values" in the park (National Park Service 1987). Hemlock mortality adversely affects a number of associated species and ecosystem and park functions by disrupting microclimates, which allows for increased deer herbivory and invasion of exotic plants; negatively impacting fish and invertebrates in riparian areas from increased stream temperatures; causing hazardous trees near visitor use areas; and impairing aesthetics and recreational opportunities (Evans 2010). The park is actively managing for this threat through a variety of measures, including cutting hazardous trees, installing boardwalks to avoid soil compaction, applying insecticides, installing deer exclosure fences, suppressing invasive plants, planting native trees, and releasing biological control agents (Evans 2015; Evans, personal communication).

Our analysis indicates that hemlock is considerably more vulnerable to climate change in the northeast portion of the park. In the south, higher elevation buffers against temperature increases, and vulnerability is positive but small. Similarly, most of the projected basal area loss due to HWA is concentrated in the north of the park. This information has the potential to inform management plans. For example, limited resources may dictate prioritizing less vulnerable areas in the south for protective measures, such as chemical injections, removing exotic plants, and hemlock plantings after mortality. Alternatively, park managers might consider actively transitioning diseased hemlock stands in the north to species that can serve somewhat comparable ecosystem and management services, such as white pine, eastern red cedar (*Juniperus virginiana*), or, especially in riparian areas, evergreen shrubs such as American rhododendron (*Rhododendron maximum*) and mountain laurel (*Kalmia latifolia*) (Beecher 2013).

Conclusion

There are both benefits and limitations to conducting an entirely spatially explicit vulnerability analysis at 0.5-mile (0.25-square-mile) resolution. On the one hand, the data layers are relevant for landscapes that managers are familiar with and actively manage. On the other hand, our analysis does not incorporate qualitative expert assessment metrics, such as shade or drought tolerance, seedling establishment and other regeneration characteristics, resistance and resilience to disturbance and pollution, community associations, or other competitive interactions. Although some of these are implicitly included in our SDM approach, others are not. There are also uncertainties associated with the SDMs and our relatively simplistic treatment of dispersal, propagule pressure, and fragmentation. Nonetheless, our approach clearly identifies vulnerable and resilient species and landscapes, and can help inform a variety of management activities within and across the Appalachian LCC and the eastern United States.

Acknowledgments

This work was funded by the NASA Climate and Biological Response Program (10-BIOCLIM10-0034). We would like to thank Woody Turner for support, William Monahan for helpful comments, and Tina Cormier and Kevin Guay for providing data layers.

References

Beecher, S. 2013. *Adapting to a Changing Climate: Risks and Opportunities for the Upper Delaware River Region*, edited by G. Griffith, T. Thaler, T. Crossett, and R. Rasker. Sagle, ID: Model Forest Policy Program in association with Common Waters Partnership, Pinchot Institute for Conservation and the Cumberland River Compact and Headwaters Economics.

Brown, D. G., K. M. Johnson, T. R. Loveland, and D. M. Theobald. 2005. Rural land-use trends in the conterminous United States, 1950–2000. *Ecological Applications* 15:1851–63.

Butler, P. R., L. Iverson, F. R. Thompson, et al. 2015. *Central Appalachians Forest Ecosystem Vulnerability Assessment and Synthesis*. A Report from the Central Appalachians Climate Change Response Framework Project. Newton Square, PA: US Forest Service, Northern Research Station.

Carroll, M., J. Townshend, M. Hansen, C. DiMiceli, R. Sohlberg, and K. Wurster. 2011. MODIS vegetative cover conversion and vegetation continuous fields. In *Land Remote Sensing and Global Environmental Change*, edited by B. Ramachandran, C. O. Justice, and M. J. Abrams, 725–45. New York: Springer.

Clark, J. S. 1998. Why trees migrate so fast: Confronting theory with dispersal biology and the paleorecord. *American Naturalist* 152:204–24.

Crossman, N. D., B. A. Bryan, and D. M. Summers. 2012. Identifying priority areas for reducing species vulnerability to climate change. *Diversity and Distributions* 18:60–72.

Dunscomb, J. K., J. S. Evans, J. M. Strager, M. P. Strager, and J. M. Kiesecker. 2014. *Assessing Future Energy Development across the Appalachian Landscape Conservation Cooperative*. Charlottesville, VA: The Nature Conservancy.

Eschtruth, A. K., and J. J. Battles. 2008. Deer herbivory alters forest response to canopy decline caused by an exotic insect pest. *Ecological Applications* 18: 360–76.

Evans, R. 2010. Hemlock woolly adelgid and hemlock ecosystems at Delaware Water Gap National Recreation Area. *Proceedings of the Fifth Symposium on Hemlock Woolly Adelgid in the Eastern US*. Asheville, NC.

Glick, P., B. A. Stein, and N. A. Edelson, eds. 2011. *Scanning the Conservation Horizon: A Guide to Climate Change Vulnerability Assessment*. Washington, DC: National Wildlife Federation.

Higgins, S. I., J. S. Clark, R. Nathan, T. Hovestadt, F. Schurr, J. M. V. Fragoso, M. R. Aguiar, E. Ribbens, and S. Lavorel. 2003. Forecasting plant migration rates: Managing uncertainty for risk assessment. *Journal of Ecology* 91:341–47.

Iverson, L. R., M. W. Schwartz, and A. M. Prasad. 2004. How fast and far might tree species migrate in the eastern United States due to climate change? *Global Ecology and Biogeography* 13:209–19.

Jantz, P., S. Goetz, and C. Jantz. 2005. Urbanization and the loss of resource lands in the Chesapeake Bay watershed. *Environmental Management* 36:808–25.

Jenkins, C. N., K. S. Van Houtan, S. L. Pimm, and J. O. Sexton. 2015. US protected lands mismatch biodiversity priorities. *Proceedings of the National Academy of Sciences of the United States of America* 112:5081–86.

Jump, A. S., and J. Penuelas. 2005. Running to stand still: Adaptation and the response of plants to rapid climate change. *Ecology Letters* 8:1010–20.

Kellndorfer, J., W. Walker, E. LaPoint, J. Bishop, T. Cormier, G. Fiske, M. Hoppus, K. Kirsch, and J. Westfall. 2012. NACP Aboveground Biomass and Carbon Baseline Data (NBCD 2000), USA, 2000. Oak Ridge, TN: ORNL DAAC.

Kelly, A. E., and M. L. Goulden. 2008. Rapid shifts in plant distribution with recent climate change. *Proceedings of the National Academy of Sciences of the United States of America* 105:11823–26.

Krist, F. J., F. J. Sapio, and B. Tkacz. 2010. A multicriteria framework for producing local, regional, and national insect and disease risk maps. In *Advances in Threat Assessment and Their Application to Forest and Rangeland Management*, edited by J. M. Pye, H. Rauscher, M. Sands, et al., 621–36. Portland, OR: US Department of Agriculture, Forest Service, Pacific Northwest and Southern Research Stations.

Matthews, S. N., L. R. Iverson, A. M. Prasad, M. P. Peters, and P. G. Rodewald. 2011. Modifying climate change habitat models using tree species-specific assessments of model uncertainty and life history-factors. *Forest Ecology and Management* 262:1460–72.

Mazziotta, A., M. Trivino, O.-P. Tikkanen, J. Kouki, H. Strandman, and M. Monkkonen. 2015. Applying a framework for landscape planning under climate change for the conservation of biodiversity in the Finnish boreal forest. *Global Change Biology* 21:637–51.

Melillo, J. M., R. V. Callaghan, F. I. Woodward, E. Salati, and S. K. Sinha. 1990. Effects on ecosystems. In *Climate Change: The IPCC Scientific Assessment*, edited by J. T. Houghton, F. J. Jenkins, and J. J. Ephraums, 131–72. Cambridge: Cambridge University Press.

Nathan, R., N. Horvitz, Y. He, A. Kuparinen, F. M. Schurr, and G. G. Katul. 2011. Spread of North American wind-dispersed trees in future environments. *Ecology Letters* 14:211–19.

National Park Service. 1987. General Management Plan for Delaware Water Gap National Recreation Area. Report. US Department of the Interior, National Park Service.

Nowacki, G. J., and M. D. Abrams. 2008. The demise of fire and "mesophication" of forests in the eastern United States. *BioScience* 58:123–38.

Prasad, A. M., J. D. Gardiner, L. R. Iverson, S. N. Matthews, and M. Peters. 2013. Exploring tree species colonization potentials using a spatially explicit simulation model: Implications for four oaks under climate change. *Global Change Biology* 19:2196–2208.

Schwartz, M. W., J. J. Hellmann, J. M. McLachlan, D. F. Sax, J. O. Borevitz, J. Brennan, A. E. Camacho, G. Ceballos, J. R. Clark, H. Doremus, et al. 2012. Managed relocation: Integrating the scientific, regulatory, and ethical challenges. *BioScience* 62:732–43.

Theobald, D. M. 2010. Estimating natural landscape changes from 1992 to 2030 in the conterminous US. *Landscape Ecology* 25:999–1011.

Theobald, D. M. 2013. A general model to quantify ecological integrity for landscape assessments and US application. *Landscape Ecology* 28:1859–74.

Tulloch, V. J. D., A. I. T. Tulloch, P. Visconti, B. S. Halpern, J. E. M. Watson, M. C. Evans, N. A. Auerbach, M. Barnes, M. Beger, I. Chades, et al. 2015. Why do we map threats? Linking threat mapping with actions to make better conservation decisions. *Frontiers in Ecology and the Environment* 13:91–99.

Wilson, B. T., A. J. Lister, and R. I. Riemann. 2012. A nearest-neighbor imputation approach to mapping tree species over large areas using forest inventory plots and moderate resolution raster data. *Forest Ecology and Management* 271:182–98.

Woodall, C. W., C. M. Oswalt, J. A. Westfall, C. H. Perry, M. D. Nelson, and A. O. Finley. 2009. An indicator of tree migration in forests of the eastern United States. *Forest Ecology and Management* 257:1434–44.

Zhu, K., C. W. Woodall, and J. S. Clark. 2012. Failure to migrate: Lack of tree range expansion in response to climate change. *Global Change Biology* 18:1042–52.

Chapter 12

Likely Responses of Native and Invasive Salmonid Fishes to Climate Change in the Rocky Mountains and Appalachian Mountains

Bradley B. Shepard, Robert Al-Chokhachy,
Todd Koel, Matthew A. Kulp, and Nathaniel Hitt

Effects of climate change over the next century will have important consequences for freshwater fish distributions and abundance. A fish's body temperature closely mirrors that of its environment. Consequently, the physiology, ontogeny, and life histories of freshwater fishes are regulated by the timing and magnitude of streamflow and temperature regimes, which are directly influenced by climatic conditions.

Here, we focus on the likely effects of climate change on native trout in the eastern and western portions of the United States. Over the past two decades, concerns have been growing about the effects of climate change on fishes in the salmonid family (Keleher and Rahel 1996). Such concerns stem from the strong relationships between salmonids and streamflow (Latterell et al. 1998; Hakala and Hartman 2004) and water temperature (Bulkley and Benson 1962; Dunham, Schroeter, and Rieman 2003).

Native trout species across North America have experienced dramatic reductions in distribution and abundance. Declines of native cutthroat trout subspecies (westslope and Yellowstone cutthroat trout, *Oncorhynchus clarkii lewisi* and *O. c. bouvieri*) and bull trout (*Salvelinus confluentus*) in western North America and of native brook trout (*S. fontinalis*) in eastern North America (Flebbe, Roghair, and Bruggink 2006; Hudy et al. 2008) have largely been attributed to anthropogenic-related factors (figs. 12-1 and 12-2; Shepard et al. 1997; Shepard, May, and Urie 2005; US Fish and Wildlife Service 2008; Gresswell 2011). Disease pathogens, especially exotic

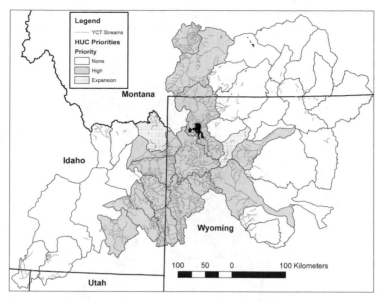

FIGURE 12-1   High-priority river basins (gray shading) and two river basins where opportunities for expansion make them high-priority basins (dot stippling) within the historical range of Yellowstone cutthroat trout (YCT). Gray lines show current distributions of designated conservation populations of YCT. (Draft from Yellowstone Cutthroat Trout Conservation Work Group, Helena, Montana.)

pathogens, have been shown to impact native trout (Koel et al. 2006). Overharvest was likely an important historical threat, but this threat has been much reduced through restrictive harvest regulations throughout the ranges of native trout (Shepard, May, and Urie 2005). Habitat modifications by humans have dramatically reduced coldwater habitats throughout the United States (Hudy et al. 2008; Torterotot et al. 2014), and water withdrawals for agricultural, industrial, and domestic uses have considerably reduced base flows in many streams and rivers (Walters, Bartz, and McClure 2013).

Nonnative fishes, particularly nonnative salmonids that have been introduced to provide recreational sport fisheries, threaten many extant populations of native trout (Larson and Moore 1985; Krueger and May 1991; Quist and Hubert 2005) through both hybridization (Rhymer and Simberloff 1996) and competition (Shepard 2010). Hybridization can compromise entire genomes, while loss of allelic diversity is a risk many isolated populations currently face (Humston et al. 2012). The added im-

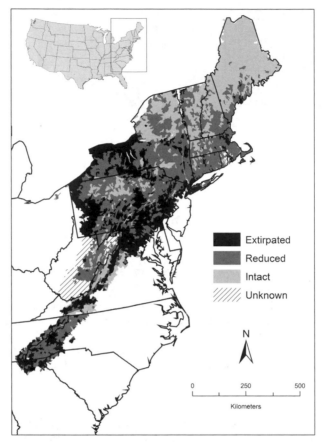

FIGURE 12-2   Range loss of eastern brook trout based on 2008 rangewide assessment. (From http://easternbrooktrout.org.)

pacts of acid precipitation and mine drainage have likely also contributed to declines of brook trout (fig. 12-2; Neff et al. 2009; Jastram et al. 2013).

Native trout in the western and northeastern United States evolved over tens of thousands of years, and current distributions resulted from these trout recolonizing habitats as continental glaciers receded eight to twelve thousand years ago (Behnke 1992). Native trout in the southeastern United States were not exposed to continental glaciers in the last two to three hundred thousand years and therefore have been isolated for considerably longer periods than species in the West and Northeast (Rash, Lubinski, and King 2014). As the continental glaciers receded, climatic conditions fluctuated with periods of time that were both much colder and much warmer than the present. Native trout populations persisted throughout this time period, even in the face of these climatic shifts. However, the threats discussed above that are related to human activities have increased the risk of extinction for many populations of native trout. Additional stresses from a rapidly changing climate may tip the balance for many of these species unless conservation actions are taken to secure them. We discuss how climate may interact with these current threats to native trout in the eastern and western United States and what national parks may offer for conserving the coldwater habitats these species require.

Effects of Climate Change on Aquatic Environments

Climate change is anticipated to alter current water temperature and hydrologic regimes. Long-term data clearly indicate that stream and river temperatures track air temperatures and that warming of rivers has occurred over the last fifty to one hundred years (Mosheni, Stephan, and Eaton 2003; Webb and Nobilis 2007). Some hydrologic features—such as glaciers, headwater lakes, and groundwater springs—can mediate temperatures, often resulting in lower variability in thermal regimes. However, climate changes may more rapidly affect these features, leading to higher variability in thermal regimes. The effects of a changing climate will vary by region, but most climate predictions provide strong evidence for an increase in summer air temperatures and, concomitantly, summer water temperatures (Eaton and Scheller 1996). Thermally suitable salmonid habitat in streams of the eastern United States may be more spatially dispersed than in streams of the western United States as a result of the dominance of snowmelt or glacial hydrology in the West versus the dominance of groundwater–surface water interactions in the East (Trumbo et al. 2014). The presence and volume of groundwater–surface water interfaces will likely influence the thermal resiliency of native trout habitat.

While there is evidence that climate change has influenced, and will continue to influence, hydrologic regimes, these effects are less certain than those for temperature. Changes have already been demonstrated for large portions of the United States (Leppi et al. 2012). Effects of climate change on streamflows will likely vary more by region than effects for temperature. Current climate models indicate that summer base flows will be lower in most areas, thus reducing the amount of available habitat (space) for fish (Ruesch et al. 2012). It is likely that snowmelt peak flows will occur earlier in most areas, but uncertainty exists for how magnitudes and durations of peak streamflows will change. Rain-on-snow winter peak flow events will likely become more frequent. Along with climatic influences on flows, human demand for water is increasing (Haddeland et al. 2014), which is likely to have direct implications for native coldwater fish (Van Kirk and Benjamin 2001).

Climate Change Effects on Native Trout

Water temperature regimes are changing, and seasonal variation in temperature is important for salmonid population dynamics. Temperature affects survival and growth of trout, especially younger fish, and can influence reproductive success.

Summer water temperatures are now staying warmer for longer periods of time, with concomitant increases in peak water temperatures (Kanno et al. 2015; Rice and Jastram 2015). Extending a fish's potential growth season by having water temperatures warm earlier in the spring and remain warm later into the fall will provide longer growing seasons (Al-Chokhachy et al. 2013). Xu, Letcher, and Nislow (2010) showed increasing brook trout growth with increasing water temperatures but also found an interactive effect of temperature with flow. However, higher maximum annual water temperatures may exceed a trout's thermal tolerance, resulting in direct mortality events, or place additional stress on fish that would lead to slower growth, poorer reproduction success, and mortality caused as stressors accumulate over time.

Fish, particularly salmonids, will move to more thermally suitable microhabitats or macrohabitats (i.e., local or regional thermal refuges) if they are available. At the macrohabitat scale, this can occur across latitude or elevation gradients. Because thermal differences are more subtle across latitudes than elevations, fish would need to move farther in a northern direction than up gradient in elevation to achieve the same magnitude of cooling. There is currently little opportunity for fish to make long movements north, but there are some opportunities to move to higher elevations (Al-Chokhachy et al. 2013).

Barriers to fish movement, especially those caused by humans in the last two centuries, have severely restricted opportunities for movement. This fragmentation of aquatic networks is particularly severe in the eastern United States, where brook trout have been shown to move between coldwater main stem and tributary habitats to track flow and thermal refuges (Petty, Lamothe, and Mazik 2005). Unfortunately, most native brook trout streams have lost this connectivity (Kanno et al. 2014; Whiteley et al. 2013). Humans may sometimes have to act as dispersal agents to move fish into thermally suitable habitats above fish barriers (Shepard, May, and Urie 2005).

Thermally suitable microhabitats are often associated with groundwater inputs, stratified lake habitats, vegetative or orthographic shading, or elevation gradients along streams, where fish can select water temperatures across thermal gradients (Aunins et al. 2015). A stream's sensitivity to changes in air temperatures is influenced by groundwater upwelling (Sinokrot et al. 1995), but groundwater temperatures will also respond to changes in air temperatures—Kurylyk, MacQuarrie, and Voss (2014) found that groundwater temperatures are close to annual mean air temperatures. Groundwater upwelling areas cause patchiness in thermal habitat for fish as well as spatial structure in how stream reaches will respond to increasing air temperature. Groundwater upwelling may increase the time lag for thermal habitat to become unsuitable for native trout but may increase population fragmentation. In winter months, groundwater upwelling zones are generally warmer than ambient stream temperatures, which may increase survival of salmonids, especially during early life history stages (Baxter and McPhail 1999).

Water temperatures and flows provide cues that trigger critical behaviors in salmonids, such as spawning, hatching, emergence, movements, and feeding. Changes in hydrologic regimes, particularly lower base flows and more variable peak flows, will likely change timing of these important life history events. Connections between tributaries and main stem river habitats are often dependent on high flows that scour stream mouth delta sediments and on higher base flows that allow surface flow to persist in lower reaches of tributary streams (National Park Service 2010). Winter precipitation is expected to increase in the northeastern United States (Horton et al. 2014). Higher winter and peak spring flows could reduce embryo survival of fall and early spring spawning fish by scouring streambed gravels and crushing or dislodging incubating embryos (Kondolf et al. 1991; Kanno et al. 2015). Shorter-duration peak flows coupled with extended periods of lower base flows during summer months could reduce growth and survival by delivering less food to fish (Xu, Letcher, and Nislow 2010; Kanno et al. 2015).

Indirect Effects of Climate Change on Native Trout

Changing climatic conditions are likely to indirectly influence native salmonids by creating conditions that are more favorable for nonnative species

(McMahon et al. 2007; Wenger et al. 2011). Hitt et al. (2003) observed a stepping-stone model of rainbow trout (*Oncorhynchus mykiss*) introgression into westslope cutthroat trout via admixed population dispersal that was not constrained by thermal gradients (cold water) or stream gradients. Muhlfeld et al. (2014) showed that increasing water temperature facilitates this introgression. Nonnative species can also impact native trout by gametic wastage (DeHaan, Schwabe, and Ardren 2010), competition (Hasegawa and Maekawa 2006), and predation (Martinez et al. 2009). In many instances, nonnative species have broader or higher thermal tolerances than natives (Bear, McMahon, and Zale 2007; Wenger et al. 2011). Such differences in thermal tolerances and metabolic rates are likely to favor nonnative species for both suitable habitat (Rahel and Olden 2008) and growth rates (Al-Chokhachy et al. 2013).

Changing thermal regimes are expected to increase the vulnerability of salmonids to the invasions of cool-water and warmwater fishes (Sanderson, Barnas, and Rub 2009). Species such as smallmouth bass (*Micropterus dolomieu*), a cool-water predator, are continuing to expand their distribution (Lawrence et al. 2014). Extant populations of native coldwater species that are proximate and accessible to source populations of cool-water nonnative species are particularly vulnerable.

Reduced summer base flows and increasing temperatures under changing climatic conditions are also likely to increase the exposure of fishes to disease. Disease resistance for coldwater fishes such as salmonids decreases with increased thermal stress (Marcogliese 2001). Some diseases or their hosts may be limited by cold temperatures, and changing thermal regimes may increase the within and among year vulnerability to infection (Koel et al. 2006).

Climate Change Impacts on Important Coldwater Sport Fisheries

We recognize that nonnative trout provide important sport fisheries in many areas, and we are not advocating replacing most nonnative trout populations with native trout. In natural systems native fish species provide ecological, evolutionary, and socioeconomic values, but in disrupted systems nonnative species may represent the only opportunity to provide these socioeconomic and ecological values (Fausch et al. 2009). Thus, in highly disrupted systems, evolutionary values provided by native fish may be lost and nearly impossible to recover.

Quist and Hubert (2004) discuss the trade-offs between the conservation of native cutthroat trout by removing nonnative trout in terms of ecological function, socioeconomic costs and benefits, and loss of evolutionary capacity. They suggest that economic benefits may often not justify costs. Managers and the public will need to make hard choices regarding the allocation of shrinking coldwater habitats for important sport fisheries supported by nonnative species versus conservation of native species. Providing native trout sport fishing opportunities may be one alternative to this dilemma.

Uncertainty Associated with Response of Trout to Climate Change

We acknowledge that there are uncertainties related to how trout might respond to changes in climate. Because of their genetic and behavioral plasticity, salmonids have a relatively high capacity to respond to their environment. Brook trout have demonstrated responses to variable environmental conditions by changing their feeding behaviors, growth rates, and body shape (Hutchings 1996). Cutthroat trout can shift life history behaviors between resident and migratory forms (Johnson et al. 2010). Bull trout have been shown to respond to changes in density through changes in growth, survival, and reproductive life-history characteristics (Johnston and Post 2009). This plasticity will likely allow these salmonids to respond to climate changes over short and long time frames (Crozier et al. 2008).

Many native salmonid populations persist in headwater areas above fish barriers (Shepard et al. 1997; Shepard, May, and Urie 2005). While these isolated populations are at a higher risk for extinction due to stochastic demographic or environmental processes, they will be less vulnerable to warming because of their locations in higher, colder habitats. We caution that spatial variability in how climate change will affect aquatic environments, plasticity in fish responses, invasions by exotic species, and interactions of many biotic and abiotic parameters will increase the uncertainty of how climate change might influence a specific fish species in a specific location.

Climate Adaptation or Mediation Measures

Prioritizing conservation for native fish in the context of climate change can provide a framework for management priorities across a species' range.

This type of prioritization effort puts conservation actions in perspective at a variety of spatial and temporal scales and can be used to allocate scarce conservation resources to locations and actions that will have the highest likely return. While climate change often will not be the primary criteria for prioritizing conservation areas or actions, these potential effects should be considered. The highest priority should be given to securing extant populations, especially groups of connected populations (Hitt and Roberts 2012). Securing extant populations in national parks should be a high priority because this is already a mission of the National Park Service. Actions to secure extant populations may include the following:

- eliminating risks posed by nonnative species by preventing their movements into habitats occupied by native species
- restoring occupied habitats to more natural conditions
- maintaining or increasing genetic diversity

The next priority should be expanding extant populations into connected, unoccupied habitats. Finally, translocations of fish into suitable, vacant habitats within their historical range should be considered (Minckley 1995). Vacant habitats could be naturally vacant or a result of physical or chemical removal of nonnative fish. We caution that habitats that historically did not support fish populations may support unique aquatic communities, and managers should conduct a careful biological assessment for any translocation of fish but especially into these habitats (George et al. 2009).

Extant groups of populations offer the best option for long-term persistence of native communities. For these connected populations, limited human intervention is needed because natural processes of local extinction and recolonization can operate over a broad landscape that may buffer the potential effects of climate change. Eliminating threats posed by nonnative species and providing adequate habitat protection are usually the biggest challenges to preserving larger groups of connected populations. Commonly, a barrier is necessary to secure these populations from invasion by nonnative species (Torterotot et al. 2014). National parks can provide core, high-elevation strongholds for coldwater species (figs. 12-1 and 12-2), but these habitats often need to be linked to lower-elevation, larger, and more productive habitats located outside of parks. Conservation of large groups of connected populations often requires managing across management jurisdictions, thus presenting additional challenges, but benefits from these types of projects include leveraging of funding and sharing field personnel to complete large, logistically complex projects.

Where nonnative species pose a risk of invading habitats occupied by native species and no natural barriers are present, a decision needs to be made whether to construct a fish barrier (Fausch et al. 2009). Fish barriers have commonly been used to secure native trout in headwater reaches from invasion by nonnative fish, especially those that pose a risk of hybridization (Simmons, Lavretsky, and May 2010). Peterson et al. (2014) used empirical evidence to conclude that cutthroat trout populations persisted in relatively small habitat patches (1 to 9 miles, or 2 to 14 kilometers) above barriers for decades to a century, depending on the quality of the local habitat. Since headwater habitats currently support many of the extant populations of native trout and are some of the coldest habitats remaining, these areas are likely to remain thermally suitable in the face of climate change, so conserving native trout in these headwater areas should be a priority, even if it means isolating these areas with barriers. However, isolation may result in the loss of genetic variability, and human-assisted dispersal of a few genetically pure individuals may be warranted (Frankham 2010).

When nonnative fish occur together with native fish, removal or suppression of these nonnative fish is usually necessary to conserve the native fish. In small streams and many lakes, this may be accomplished using physical removal techniques, including netting, angling, and electrofishing (Moore, Larson, and Bromfield 1986; Kulp and Moore 2000; Shepard et al. 2014). Unfortunately, physical techniques are often ineffective in larger waters and piscicides, such as rotenone, are usually necessary to eradicate nonnative species. Projects to suppress or eradicate nonnative fish species have been implemented in Yellowstone (Gresswell 1991; Koel et al. 2005), Crater Lake (Buktenica et al. 2013), and Great Smoky Mountains national parks (Moore, Larson, and Bromfield 1986).

Where nonnative species have been eliminated and habitats have been degraded, habitat restoration can increase the security of native fish populations. Large-scale restorations of anadromous salmonid habitats have been completed in Olympic National Park and at Point Reyes National Seashore (Duda, Freilich, and Schreiner 2008; Miller 2010). Furthermore, novel approaches, such as groundwater upwelling manipulations, may be necessary to provide more suitable thermal conditions for salmonids (Kurylyk et al. 2015).

Translocations of native species into vacant habitats within their historical ranges have frequently been accomplished, and selection of translocation sites should consider climate change (Cooney et al. 2005). These projects often require extensive collaboration. For example, biologists from Yellowstone National Park, Gallatin National Forest, Turner Enterprises, and Montana Fish, Wildlife and Parks recently worked together in Grayling Creek, located in western Yellowstone National Park, to construct a barrier near its mouth and remove hybridized and nonnative trout above this barrier. These efforts will provide a large headwater refuge for westslope cutthroat trout and native fluvial Arctic grayling (*Thymallus arcticus*).

Angling regulations and special angling closures may be necessary to protect both native and nonnative coldwater fish species as water temperatures increase. Boyd et al. (2010) evaluated thermal effects on catch-and-release angling mortality for several species of salmonids and found relatively high angling mortality for mountain whitefish at higher water temperatures.

Fish Management and Conservation in Yellowstone National Park

Yellowstone National Park has a core of high-priority basins for Yellowstone cutthroat trout (YCT; fig. 12-1). The park contains hundreds of lakes and thousands of kilometers of flowing waters. While native fish were widely distributed throughout the park, nearly half of the park's waters were historically fishless. However, soon after the park was established, its aquatic species composition began to change. The US Fish Commission and park managers began planting native fish in fishless waters in 1881 and started bringing nonnative species into the park by 1889 (Varley 1981). The purpose of these releases was to meet a public demand for recreational and subsistence fisheries. Most of the nonnative fish introductions were trout species, most notably brook trout, brown trout (*Salmo trutta*), and rainbow trout. Native YCT of Yellowstone Lake were stocked extensively, both within and outside their native range. From the early 1880s to the mid-1950s, almost all park waters, including most remote backcountry locations, were stocked. By the 1950s, nonnative sport fish were established in most major waters in the park.

The 1950s marked a paradigm shift in fisheries management in Yellowstone as stocking for recreational purposes was abandoned in favor of wild fish management and native species conservation. By then, over 300 million fish had been stocked in park waters, and nonnative species were firmly established in most lakes, rivers, and streams. Of the 400 miles (about 640 kilometers) of river habitat that originally supported native fish when the park was established, only 30 miles (about 50 kilometers) currently support only native fish that are genetically unaltered, while 460 miles (about 740 kilometers) now contain native species along with nonnative or hybridized trout. At present, river dwelling Arctic grayling are completely gone from park waters, westslope cutthroat trout currently remain in only a few streams, and YCT face serious threats in the few waters where they remain.

The Yellowstone Lake drainage above Upper Yellowstone Falls represents the largest remaining undisturbed habitat for a genetically pure YCT population with high ecological, economical, and social values. YCT are a valuable food source for several species of birds and mammals, including grizzly bears, otters, eagles, white pelicans, and osprey. YCT are the basis for a valuable sport fishing economy in communities surrounding the park, attracting anglers from across the world.

Lake trout were discovered in Yellowstone Lake in 1994. They are highly predatory and have virtually eliminated cutthroat trout in other lakes in the western United States where they have been introduced (Martinez et al. 2009). In 1995, a panel of fisheries experts projected that without control of lake trout in Yellowstone Lake, native YCT would be reduced to a mere fraction of their historical levels or functionally eliminated. Efforts to restore cutthroat trout and the ecology of the Yellowstone Lake focused on the suppression of nonnative lake trout using gill and trap nets. Physical removal of lake trout from 1995 to 2014 has increased cutthroat trout abundance, especially as removal efforts increased from 2012 to 2014. More than 877,000 (51 percent) of the 1.7 million lake trout killed since 1994 were captured during 2012 to 2014. Catch per effort declined from 6.3 in 2012 to 4.8 in 2013 and to 3.7 in 2014, indicating that the population of lake trout decreased during this period.

Hybridization of cutthroat trout as a result of rainbow trout range expansion continues to be the greatest threat to native fish populations in waters outside the Yellowstone Lake ecosystem. Two important cutthroat trout strongholds, Slough and Soda Butte creeks in the Lamar River watershed, have been invaded by rainbow trout during just the past ten years, and hybridization of the native cutthroat trout in those creeks has begun.

Over the past decade, Yellowstone National Park has restored, or is currently restoring, native trout to 51 miles (82 kilometers) of four stream systems and to 50 acres (20 hectares) of four lakes using rotenone. Native trout are also being preserved by suppressing nonnatives via electrofishing in Slough and Soda Butte creeks in the Lamar River drainage.

Since most of Yellowstone Park is at the highest elevations within the historical ranges of native YCT, westslope cutthroat trout, and Arctic grayling, conservation and restoration of these species in this park should be a high priority because this area should be relatively resistant to climate change effects. In particular, the Yellowstone Lake ecosystem is critically important for the conservation of YCT because this large lake can buffer climate effects; however, the invading lake trout must be suppressed.

Fish Management and Conservation in Great Smoky Mountains National Park

Although once thought to occupy up to about 5,000 miles (about 8,000 kilometers) of coldwater habitat throughout their native range in the eastern United States and Canada, today brook trout occupy about 72 percent of this range and only 62 percent of their historic southern Appalachian range (Hudy, Roper, and Gillespie 2007). In Great Smoky Mountains National Park, the range of brook trout has declined approximately 70 percent since 1900. Concomitantly, range expansion of rainbow trout into streams, which previously only supported brook trout, equaled the range loss (Larson and Moore 1985). From 1976 to 1998, the range of southern Appalachian brook trout did not change (Strange and Habera 1998).

Initial losses of brook trout from much of their historical range were attributed to logging and resultant water quality degradation (King 1937). By 1910, brook trout were virtually eliminated from areas of streams below about 3,000 feet (about 900 meters) in elevation. Local residents began putting pressure on local managers and politicians because they had no fish to catch. To meet their demands for recreational angling, logging companies—followed by the National Park Service—stocked about 1.4 million nonnative rainbow trout and 800,000 northern-strain brook trout between 1934 and 1975 (Kulp and Moore 2000). Until 1974, Great Smoky Mountains National Park staff saw no harm in stocking rainbows and believed that, as reforestation occurred in logged-over areas, brook trout would reclaim their lost range. However, distribution surveys in the 1970s found brook trout were extirpated from 45 percent of the range that they had exclusively occupied previously (Kelly, Griffith, and Jones 1980) and that this decline was directly attributed to the expansion of rainbow trout (Larson and Moore 1985).

Native brook trout were restricted to marginal headwater streams, and numerous populations that were previously considered secure had declined or been extirpated due to low pH associated with acid deposition (Neff et al. 2009). Many streams that had pH values from 5.5 to 6.5 during base flow conditions experienced a significant drop in their pH (1.0–2.0 units) when intense storms increased flows, which led to elevated proton and aluminum concentrations toxic to brook trout (Neff et al. 2009, 2013). These episodic events were believed to be the dominant cause of lethal and sublethal toxicity to trout, leading to the extirpation of native brook trout in six Great Smoky Mountains National Park headwater streams over the past two decades (Neff et al. 2009). These findings increased the urgency of restoring mid- to low-elevation stream segments identified in the Great Smoky Mountains National Park Fishery Management Plan to provide long-term refuges for native brook trout.

Research concluded that successfully restoring brook trout to larger streams required the use of piscicides to ensure that nonnative rainbow trout were eradicated (Moore et al. 2005). In 1996, because of increased concerns related to acid deposition and headwater range loss, Great Smoky Mountains National Park fishery staff evaluated the piscicide Fintrol® (antimycin) for eradicating rainbow trout in larger streams (Moore et al. 2005). Twenty streams covering 40 miles (65 kilometers) met the following selection criteria for restoration: (1) a suitable barrier to reinvasion, (2) a historical record of brook trout in the stream segment, and (3) a feasible size that would not impact a large number of nongame fish species. Brook trout have been restored in three of these streams totaling 15 miles (24 kilometers) since 2000 (Moore et al. 2005). All of these streams were reopened to fishing.

To date, Great Smoky Mountains National Park has restored brook trout to 27 miles (44 kilometers) of eleven streams (Kulp and Moore 2000). Roughly 11 miles (18 kilometers) of the park's streams remain for potential restoration. Future brook trout restoration priorities in Great Smoky Mountains National Park will focus on stream segments at low elevations in order to minimize projected acid deposition impacts (Ellis et al. 2013), in subwatersheds where genetically robust source stocks are available (Whiteley et al. 2013), and where the potential for establishing connected populations is highest. Future restoration efforts may also focus on genetic rescue in order to improve genetic fitness and sustain healthy populations rather than moving isolated populations with low allelic richness (Whiteley et al. 2013). Park staff have also worked with federal and state agency partners to restore southern Appalachian brook trout in two streams in Tennessee and two in South Carolina.

As climate change effects continue, there will be increased pressure on National Park Service managers to take bolder steps to preserve and improve habitat currently and formerly occupied by brook trout. Such bold steps could include large-scale liming of high-elevation watersheds and recommendations for more restrictive air quality regulations. Another step could be the implementation of more radical programs, such as a Trojan Y Chromosome (TYC) program for rainbow trout, in which hatchery-produced genetically YY male fish (known as supermales) could be regularly released into an undesired population over time, skewing the population toward 100 percent males over time, theoretically resulting in population extirpation. Although some programs, including TYC, have not been fully vetted, new and innovative approaches should be considered in order to protect and preserve native brook trout.

Conclusion

Our climate is changing, and these changes will undoubtedly affect native fish distributions, timing of life history events, survival rates, and invasion success of nonnative fish. In some cases, habitats that are currently only marginally suitable for native fish because they are too cold will become suitable, but in most cases habitats that are now suitable or marginally suitable because they are too warm will become less suitable or uninhabitable. Human impacts have already reduced distributions and abundances of many native fish species, and these human impacts will also increase over time as the human population grows. The synergistic effects of these human impacts and climate change dictate that we take aggressive conservation actions if we wish to preserve our native fish species, particularly coldwater species such as trout. National parks offer us an opportunity to preserve native fish under the reality of climate change because native fish conservation is a mandate for the National Park Service and many parks are located at the highest elevations in many regions of the United States.

References

Al-Chokhachy, R., J. Alder, S. Hostetler, R. Gresswell, and B. Shepard. 2013. Thermal controls of Yellowstone cutthroat trout and invasive fishes under climate change. *Global Change Biology* 19:3069–81.

Aunins, A. W., J. T. Petty, T. L. King, M. Schilz, and P. M. Mazik. 2015. River mainstem thermal regimes influence population structuring within an Appalachian brook trout population. *Conservation Genetics* 16 (1): 15–29.

Baxter, J. S., and J. D. McPhail. 1999. The influence of redd site selection, groundwater upwelling, and over-winter incubation temperature on survival of bull trout (*Salvelinus confluentus*) from egg to alevin. *Canadian Journal of Zoology* 77:1233–39.

Bear, E. A., T. E. McMahon, and A. V. Zale. 2007. Comparative thermal requirements of westslope cutthroat trout and rainbow trout: Implications for species
interactions and development of thermal protection standards. *Transactions of
the American Fisheries Society* 136 (4): 1113–21.

Behnke, R. J. 1992. *Native trout of Western North America*. Bethesda, MD: American
Fisheries Society.

Boyd, J. W., C. S. Guy, T. B. Horton, and S. A. Leathe. 2010. Effects of catch-and-
release angling on salmonids at elevated water temperatures. *North American
Journal of Fisheries Management* 30:898–907.

Buktenica, M. W., D. K. Hering, S. F. Girdner, B. D. Mahoney, and B. D. Rosenlund. 2013. Eradication of nonnative brook trout with electrofishing and Anti-
mycin-A and the response of a remnant bull trout population. *North American
Journal of Fisheries Management* 33:117–29.

Bulkley, R. V., and N. Benson. 1962. Predicting Year-Class Abundance of Yellowstone Lake Cutthroat Trout. Research Report 59. US Fish and Wildlife Service.

Cooney, S. J., A. P. Covich, P. M. Lukacs, A. L. Harig, and K. D. Fausch. 2005.
Modeling global warming scenarios in greenback cutthroat trout (*Oncorhyn-
chus clarkii stomias*) streams: Implications for species recovery. *Western North
American Naturalist* 65:371–81.

Crozier, L. G., A. P. Hendry, P. W. Lawson, T. P. Quinn, N. J. Mantua, J. Bat-
tin, R. G. Shaw, and R. B. Huey. 2008. Potential responses to climate change
in organisms with complex life histories: Evolution and plasticity in Pacific
salmon. *Evolutionary Applications* 1:252–70.

DeHaan, P. W., L. T. Schwabe, and W. R. Ardren. 2010. Spatial patterns of hybridization between bull trout, *Salvelinus confluentus*, and brook trout, *Salvelinus
fontinalis*, in an Oregon stream network. *Conservation Genetics* 11 (3): 935–49.

Duda, J. J., J. E. Freilich, and E. G. Schreiner. 2008. Baseline studies in the Elwha
River ecosystem prior to dam removal: Introduction to the special issue.
Northwest Science 82 (special issue): 1–12.

Dunham, J., R. Schroeter, and B. Rieman. 2003. Influence of maximum water
temperature on occurrence of Lahontan cutthroat trout within streams. *North
American Journal of Fisheries Management* 23:1042–49.

Eaton, J., and R. M. Scheller. 1996. Effects of climate warming on fish thermal habitat in streams of the United States. *Limnology and Oceanography* 41:1109–15.

Ellis, R. A., D. J. Jacob, M. P. Sulprizio, L. Zhang, C. D. Holmes, B. A. Schichtel,
T. Blett, E. Porter, L. H. Pardo, and J. A. Lynch. 2013. Present and future
nitrogen deposition to national parks in the United States: Critical load
exceedances. *Atmospheric Chemistry and Physics* 13 (17): 9083–95.

Fausch, K. D., B. E. Rieman, J. B. Dunham, M. K. Young, and D. P. Peterson.
2009. Invasion versus isolation: Trade-offs in managing native salmonids with
barriers to upstream movement. *Conservation Biology* 23:859–70.

Flebbe, P. A., L. D. Roghair, and J. L. Bruggink. 2006. Spatial modeling to project
southern Appalachian trout distribution in a warmer climate. *Transactions of the
American Fisheries Society* 135:1371–82.

Frankham, R. 2010. Challenges and opportunities of genetic approaches to biological conservation. *Biological Conservation* 143:1919–27.

George, A. L., B. R. Kuhajda, J. D. Williams, M. A. Cantrell, P. L. Rakes, and J. R. Shute. 2009. Guidelines for propagation and translocation for freshwater fish conservation. *Fisheries* 34:529–45.

Gresswell, R. E. 1991. Use of antimycin for removal of brook trout from a tributary of Yellowstone Lake. *North American Journal of Fisheries Management* 11:83–90.

Gresswell, R. E. 2011. Biology, status, and management of the Yellowstone cutthroat trout. *North American Journal of Fisheries Management* 31:782–812.

Haddeland, I., J. Heinke, H. Biemans, S. Eisner, M. Florke, N. Hanasaki, M. Konzmann, F. Ludwig, Y. Masaki, J. Schewe, et al. 2014. Global water resources affected by human interventions and climate change. *Proceedings of the National Academy of Sciences of the United States of America* 111:3251–56.

Hakala, J. P., and K. J. Hartman. 2004. Drought effect on stream morphology and brook trout (*Salvelinus fontinalis*) populations in forested headwater streams. *Hydrobiologia* 515:203–13.

Hasegawa, K., and K. Maekawa. 2006. The effects of introduced salmonids on two native stream-dwelling salmonids through interspecific competition. *Journal of Fish Biology* 68 (4): 1123–32.

Hitt, N. P., C. Frissell, C. C. Muhlfeld, and F. W. Allendorf. 2003. Spread of hybridization between native westslope cutthroat trout, *Oncorhynchus clarkii lewisi*, and nonnative rainbow trout, *Oncorhynchus mykiss*. *Canadian Journal of Fisheries and Aquatic Sciences* 60:1440–51.

Hitt, N. P., and J. H. Roberts. 2012. Hierarchical spatial structure of stream fish colonization and extinction. *Oikos* 121:127–37.

Horton, R., G. Yohe, W. Easterling, R. Kates, M. Ruth, E. Sussman, A. Whelchel, D. Wolfe, and F. Lipschultz. 2014. Northeast. Chap. 16 in *Climate Change Impacts in the United States: The Third National Climate Assessment*, edited by J. M. Melillo, T. C. Richmond, and G. W. Yohe. US Global Change Research Program.

Hudy, M., B. Roper, and N. Gillespie. 2007. Large scale assessments: Lessons learned from native trout management. In *Sustaining Wild Trout in a Changing World: Proceedings of Wild Trout IX Symposium*, edited by R. F. Carline and C. LoSapio, 223–35. October 9–12, West Yellowstone.

Hudy, M., T. M. Thieling, N. Gillespie, and E. P. Smith. 2008. Distribution, status, and land use characteristics of subwatersheds within the native range of brook trout in the eastern United States. *North American Journal of Fisheries Management* 28:1069–85.

Humston, R., K. Bezold, N. D. Adkins, R. J. Elsey, Huss, B. A. Meekins, P. R. Cabe, and T. L. King. 2012. Consequences of stocking headwater impoundments on native populations of brook trout in tributaries. *North American Journal of Fisheries Management* 32 (1): 100–108.

Hutchings, J. A. 1996. Adaptive phenotypic plasticity in brook trout, *Salvelinus fontinalis*, life histories. *Ecoscience* 3:25–32.

Jastram, J. D., C. D. Snyder, N. P. Hitt, and K. C. Rice. 2013. *Synthesis and Interpretation of Surface-Water Quality and Aquatic Biota Data Collected in Shenandoah National Park, Virginia, 1979–2009*. Scientific Investigations Report 2013-5157. Reston, VA: US Geological Survey.

Johnson, J. R., J. Baumsteiger, J. Zydlewski, J. M. Hudson, and W. Ardren. 2010. Evidence of panmixia between sympatric life history forms of coastal cutthroat trout in two lower Columbia River tributaries. *North American Journal of Fisheries Management* 30:691–701.

Johnston, F. D., and J. R. Post. 2009. Density-dependent life-history compensation of an iteroparous salmonid. *Ecological Applications* 19:449–67.

Kanno, Y., B. H. Letcher, J. A. Coombs, K. H. Nislow, and A. R. Whiteley. 2014. Linking movement and reproductive history of brook trout to assess habitat connectivity in a heterogeneous stream network. *Freshwater Biology* 59 (1): 142–54.

Kanno, Y., B. H. Letcher, N. P. Hitt, D. A. Boughton, J. E. B. Wofford, and E. F. Zipkin. 2015. Seasonal weather patterns drive population vital rates and persistence in a stream fish. *Global Change Biology* 21 (5): 1856–70.

Keleher, C. J., and F. J. Rahel. 1996. Thermal limits to salmonid distributions in the Rocky Mountain region and potential habitat loss due to global warming: A geographic information system (GIS) approach. *Transactions of the American Fisheries Society* 125:1–13.

Kelly, G. A., J. S. Griffith, and R. D. Jones. 1980. Changes in Distribution of Trout in Great Smoky Mountains National Park, 1900–1977. US Fish and Wildlife Service Technical Papers 102.

King, W. 1937. Notes on the distribution of native speckled and rainbow trout in the streams at Great Smoky Mountains National Park. *Journal of the Tennessee Academy of Science* 12 (4): 351–61.

Koel, T. M., P. E. Bigelow, P. D. Doepke, B. D. Ertel, and D. L. Mahony. 2005. Nonnative lake trout result in Yellowstone cutthroat trout decline and impacts to bears and anglers. *Fisheries* 30 (11): 10–19.

Koel, T. M., D. L. Mahony, K. L. Kinnan, C. Rasmussen, C. J. Hudson, S. Murcia, and B. L. Kerans. 2006. *Myxobolus cerebralis* in native cutthroat trout of the Yellowstone Lake ecosystem. *Journal of Aquatic Animal Health* 18 (3): 157–75.

Kondolf, G. M., G. F. Cada, M. J. Sale, and T. Felando. 1991. Distribution and stability of potential salmonid spawning gravels in steep boulder-bed streams of the eastern Sierra Nevada. *Transactions of the American Fisheries Society* 120:177–86.

Krueger, C. C., and B. May. 1991. Ecological and genetic effects of salmonid introductions in North America. *Canadian Journal of Fisheries and Aquatic Sciences* 48:66–77.

Kulp, M. A., and S. E. Moore. 2000. Multiple electrofishing removals for eliminating rainbow trout in a small southern Appalachian stream. *North American
Journal of Fisheries Management* 20 (1): 259–66.

Kurylyk, B. L., K. T. B. MacQuarrie, T. Linnansaari, R. A. Cunjack, and R. A. Curry.
2015. Preserving, augmenting, and creating cold-water thermal refugia in
rivers: Concepts derived from research on the Miramichi River, New Bruns-
wick (Canada). *Ecohydrology* 8 (6): 1095–1108.

Kurylyk, B. L., K. T. B. MacQuarrie, and C. I. Voss. 2014. Climate change impacts
on the temperature and magnitude of groundwater discharge from shallow,
unconfined aquifers. *Water Resources Research* 50:3253–74.

Larson, G. L., and S. E. Moore. 1985. Encroachment of exotic rainbow trout into
stream populations of native brook trout in the southern Appalachian mountains. *Transactions of the American Fisheries Society* 114:195–203.

Latterell, J. J., K. D. Fausch, C. Gowan, and S. C. Riley. 1998. Relationship of trout
recruitment to snowmelt runoff flows and adult trout abundance in six Colorado mountain streams. *Rivers* 6:240–50.

Lawrence, D. J., B. Stewart-Koster, J. D. Olden, A. S. Ruesch, C. E. Torgersen, J. J.
Lawler, D. P. Butcher, and J. K. Crown. 2014. The interactive effects of climate
change, riparian management, and a nonnative predator on stream-rearing
salmon. *Ecological Applications* 24 (4): 895–912.

Leppi, J. C., T. H. DeLuca, S. W. Harrar, and S. W. Running. 2012. Impacts of
climate change on August stream discharge in the Central-Rocky Mountains.
Climatic Change 112 (3–4): 997–1014.

Marcogliese, D. J. 2001. Implications of climate change for parasitism of animals in
the aquatic environment. *Canadian Journal of Zoology* 79 (8): 1331–52.

Martinez, P. J., P. E. Bigelow, M. A. Deleray, W. A. Fredenberg, B. S. Hansen,
N. J. Horner, S. K. Lehr, R. W. Schneidervin, S. A. Tolentino, and A. E. Viola.
2009. Western lake trout woes. *Fisheries* 34 (9): 424–42.

McMahon, T. E., A. V. Zale, F. T. Barrows, J. H. Selong, and R. J. Danehy. 2007.
Temperature and competition between bull trout and brook trout: A test of the
elevation refuge hypothesis. *Transactions of the American Fisheries Society* 136
(5): 1313–26.

Miller, G. 2010. In central California, coho salmon are on the brink. *Science*
327:512–13.

Minckley, W. L. 1995. Translocation as a tool for conserving imperiled fishes: Experiences in Western United States. *Biological Conservation* 72:297–309.

Moore, S. E., M. A. Kulp, J. Hammonds, and B. Rosenlund. 2005. Restoration of
Sams Creek and an Assessment of Brook Trout Restoration Methods: Great
Smoky Mountains National Park. National Park Service Technical Report/NPS/
NRWRD/NRTR-2005/342.

Moore, S. E., G. L. Larson, and R. Bromfield. 1986. Population control of exotic
rainbow trout in streams of a natural area park. *Environmental Management*
10:215–19.

Mosheni, O., H. G. Stephan, and J. Eaton. 2003. Global warming and potential changes in fish habitat in U.S. streams. *Climatic Change* 59:389–409.

Muhlfeld, C. C., R. P. Kovach, L. A. Jones, R. Al-Chokhachy, M. C. Boyer, R. F. Leary, W. H. Lowe, G. Luikart, and F. W. Allendorf. 2014. Invasive hybridization in a threatened species is accelerated by climate change. *Nature Climate Change* 4:620–24.

National Park Service. 2010. *National Park Service Climate Change Response Strategy*. Fort Collins, CO: National Park Service Climate Change Response Program.

Neff, K. J., J. S. Schwartz, T. B. Henry, R. B. Robinson, S. E. Moore, and M. A. Kulp. 2009. Physiological stress in native brook trout during episodic stream acidification in the Great Smoky Mountains National Park. *Archives of Environmental Contamination and Toxicology* 57:366–76. doi: 10.1007/s00244-008-9269-4.

Neff, K. J., J. S. Schwartz, S. E. Moore, and M. A. Kulp. 2013. Influence of basin characteristics on baseflow and stormflow chemistry in the Great Smoky Mountains National Park, USA. *Hydrological Processes* 27 (14): 2061–74.

Peterson, D. P., B. E. Rieman, D. L. Horan, and M. K. Young. 2014. Patch size but not short-term isolation influences occurrence of westslope cutthroat trout above human-made barriers. *Ecology of Freshwater Fish* 23:556–71.

Petty, J. T., P. J. Lamothe, and P. M. Mazik. 2005. Spatial and seasonal dynamics of brook trout populations inhabiting a central Appalachian watershed. *Transactions of the American Fisheries Society* 134:572–87.

Quist, M. C., and W. A. Hubert. 2004. Bioinvasive species and the preservation of cutthroat trout in the western United States: Ecological, social, and economic issues. *Environmental Science and Policy* 7:303–13.

Quist, M. C., and W. A. Hubert. 2005. Relative effects of biotic and abiotic processes: A test of the biotic-abiotic constraining hypothesis as applied to cutthroat trout. *Transactions of the American Fisheries Society* 134:676–86.

Rahel, F. J., and J. D. Olden. 2008. Assessing the effects of climate change on aquatic invasive species. *Conservation Biology* 22 (3): 521–33.

Rash, J. M., B. A. Lubinski, and T. L. King. 2014. Genetic analysis of North Carolina's brook trout *Salvelinus fontinalis* with emphasis on previously uncharacterized collections. In *Proceedings of the Wild Trout XI Symposium*, 180–89. September. http://www.wildtroutsymposium.com/ proceedings-11.pdf.

Rhymer, J. M., and D. Simberloff. 1996. Extinction by hybridization and introgression. *Annual Review of Ecology and Systematics* 27:83–109.

Rice, K. C., and J. D. Jastram. 2015. Rising air and stream-water temperatures in Chesapeake Bay region, USA. *Climatic Change* 128:127–38.

Ruesch, A. S., C. E. Torgersen, J. J. Lawler, J. D. Olden, E. E. Peterson, C. J. Volk, and D. J. Lawrence. 2012. Projected climate-induced habitat loss for salmonids in the John Day River network, Oregon, USA. *Conservation Biology* 26:873–82.

Sanderson, B. L., K. A. Barnas, and A. M. W. Rub. 2009. Nonindigenous species of the Pacific Northwest: An overlooked risk to endangered salmon? *BioScience* 59 (3): 245–56.

Shepard, B. B. 2010. Evidence of niche similarity between cutthroat trout (*Oncorhynchus clarkii*) and brook trout (*Salvelinus fontinalis*): Implications for displacement of native cutthroat trout by nonnative brook trout. PhD dissertation. Montana State University, Bozeman.

Shepard, B. B., B. E. May, and W. Urie. 2005. Status and conservation of westslope cutthroat trout within the western United States. *North American Journal of Fisheries Management* 25:1426–40.

Shepard, B. B., L. M. Nelson, M. L. Taper, and A. V. Zale. 2014. Factors influencing successful eradication of nonnative brook trout from four small Rocky Mountain streams using electrofishing. *North American Journal of Fisheries Management* 34:988–97.

Shepard, B. B., B. Sanborn, L. Ulmer, and D. C. Lee. 1997. Status and risk of extinction for westslope cutthroat trout in the upper Missouri River Basin, Montana. *North American Journal of Fisheries Management* 17:1158–72.

Simmons, R. E., P. Lavretsky, and B. May. 2010. Introgressive hybridization of redband trout in the Upper McCloud River watershed. *Transactions of the American Fisheries Society* 139:201–13.

Sinokrot, B. A., H. G. Stefan, J. H. McCormick, and J. G. Eaton. 1995. Modeling of climate change effects on stream temperatures and fish habitats below dams and near groundwater inputs. *Climatic Change* 30:181–200.

Strange, R. J., and J. W. Habera. 1998. No net loss of brook trout distribution in areas of sympatry with rainbow trout in Tennessee streams. *Transactions of the American Fisheries Society* 127:434–40.

Torterotot, J. B., C. Perrier, N. E. Bergeron, and L. Bernatchez. 2014. Influence of forest road culverts and waterfalls on the fine-scale distribution of brook trout genetic diversity in a boreal watershed. *Transactions of the American Fisheries Society* 143 (6): 1577–91.

Trumbo, B. A., K. H. Nislow, J. Stallings, M. Hudy, E. P. Smith, D. Kim, B. Wiggins, and C. A. Dolloff. 2014. Ranking site vulnerability to increasing temperatures in southern Appalachian brook trout streams in Virginia: An exposure-sensitivity approach. *Transactions of the American Fisheries Society* 143:173–87.

US Fish and Wildlife Service (USFWS). 2008. *Bull Trout* (Salvelinus confluentus) *5-Year Review: Summary and Evaluation*. Portland, OR: USFWS.

Van Kirk, R. W., and L. Benjamin. 2001. Status and conservation of salmonids in relation to hydrologic integrity in the Greater Yellowstone Ecosystem. *Western North American Naturalist* 61:359–74.

Varley, J. D. 1981. *A History of Fish Stocking Activities in Yellowstone National Park between 1881 and 1980*. Mammoth, WY: US Department of the Interior, National Park Service, Yellowstone National Park.

Walters, A. W., K. K. Bartz, and M. M. McClure. 2013. Interactive effects of water diversion and climate change for juvenile Chinook salmon in the Lemhi river basin (USA). *Conservation Biology* 27 (6): 1179–89.

Webb, B. W., and F. Nobilis. 2007. Long-term changes in river temperature and the influence of climatic and hydrological factors. *Hydrological Sciences Journal* 52:74–85.

Wenger, S. J., D. J. Isaak, C. H. Luce, H. M. Neville, K. D. Fausch, J. B. Dunham, D. C. Dauwalter, M. K. Young, M. M. Elsner, B. E. Rieman, et al. 2011. Flow regime, temperature, and biotic interactions drive differential declines of trout species under climate change. *Proceedings of the National Academy of Sciences of the United States of America* 108 (34): 14175–80.

Whiteley, A. R., J. A. Coombs, M. Hudy, Z. Robinson, A. R. Colton, K. H. Nislow, and B. H. Letcher. 2013. Fragmentation and patch size shape genetic structure of brook trout populations. *Canadian Journal of Fisheries and Aquatic Sciences* 70 (5): 678–88.

Xu, C., B. H. Letcher, and K. H. Nislow. 2010. Context-specific influence of water temperature on brook trout growth rates in the field. *Freshwater Biology* 55 (11): 2253–64.

PART 4

Managing under Climate Change

Parts 2 and 3 of the book are largely based on climate and ecological sciences. Large data sets derived from both satellite and ground-based sensors are used to parameterize sophisticated simulation and statistical models to understand how global stressors are experienced locally by managers of protected areas. These analyses are carefully crafted to address spatial and temporal scales that would best provide insights into local dynamics in the context of regional to continental trends. Interestingly, as complicated as these analyses and results may be, they are much more tractable than the steps of the Climate-Smart Conservation framework that involve management. As clearly elucidated in this final part of the book, resource management is the product of a complicated social system that involves science, education, policy, budgets, economics, interagency cooperation, communication, and complex stakeholder partnerships. Thus, it is not surprising that progress on the science components of the Climate-Smart Conservation framework has outpaced that of the management components.

Be this as it may, the infrastructure and capacity for climate adaptation is in a state of rapid evolution. Chapter 13 overviews the advances that have been made by federal land management agencies over the past five-year period of activities that we cover in this book, and it highlights several specific case studies illustrating recent progress in climate adaptation. Chapter 14 covers similar topics but from the perspective of resource specialists in Rocky Mountain National Park. This chapter might be considered a "must read" for managers elsewhere who are just beginning to work in the climate adaptation planning arena. One conclusion from chapter 14 is that at least ten years of team and capacity building are needed to begin to handle the extreme climate and disturbance events that are becoming more frequent under climate change.

Chapter 15 focuses on one such extreme event in the Greater Yellowstone Ecosystem: the massive mortality of mature whitebark pine, the keystone subalpine species in that ecosystem. It is in chapter 15 that all the steps of the Climate-Smart Conservation framework are demonstrated. Despite challenges to park planning, interpretive programs, budgets, and interagency collaboration, park resource managers have developed, implemented, and evaluated active management to restore this species in the face of climate change. The prognosis for success is unknown at this time. However, the approach that has been pioneered for this "early responder" to climate change provides a road map for active management for the many other species that will be responding to climate change in the years and decades ahead.

The final chapter in this part returns to the overarching mission in conservation that was introduced early in the book: sustaining the ecological integrity of ecosystems. The Climate-Smart Conservation framework is an approach for attempting to sustain elements of ecosystems that are most vulnerable to climate change. Chapter 16 asks a fundamental question of the ecological science conducted to date in the Greater Yellowstone Ecosystem: how well are we sustaining ecological integrity in this iconic wildland ecosystem within a humanizing planet? The answer is perhaps surprising, given that this is one of our best studied wildland ecosystems. Essentially, we observe that it is very difficult to know how well ecological integrity is being sustained, largely because of a lack of monitoring on the private lands throughout the ecosystem.

This realization is a wake-up call for the need to foster broader partnerships among federal, state, and private lands and to better use our vast remote sensing and scientific analyses capabilities to quantify the condition of our wildland ecosystem and communicate the results to stakeholders and decision makers in ways they find meaningful.

Chapter 13

Approaches, Challenges, and Opportunities for Achieving Climate-Smart Adaptation

S. Thomas Olliff, William B. Monahan,
Virginia Kelly, and David M. Theobald

Only five years ago, Halofsky et al. (2011) wrote: "For climate change adaptation, there is no recipe, no road map, and yet no time to lose; science and management partners must tackle the climate change issue in a timely way, despite uncertainty." Later that same year, the National Wildlife Federation, in partnership with authors from federal, state, university, and other nongovernmental organizations, published one piece of a road map: *Scanning the Conservation Horizon: A Guide to Climate Change Vulnerability Assessment* (Glick, Stein, and Edelson 2011). The four steps described in *Scanning*, introduced in chapter 2, are an excellent introduction to developing and using climate change vulnerability assessments in planning and management of public lands.

However, as managers try to implement climate change adaptation actions, the four steps from *Scanning* serve the role of high-level strategy. More development and definition are needed to make decisions actionable. That development and definition are contained in the follow-up document *Climate-Smart Conservation: Putting Adaptation Principles into Practice* (Stein et al. 2014). *Climate-Smart Conservation* serves as an implementation handbook for putting the concepts developed in *Scanning* into action.

This chapter discusses the approaches, challenges, and opportunities with using climate change vulnerability assessments in planning and managing public lands, specifically addressing steps 4 to 6 of the Climate-Smart Conservation framework (see chapter 2 for more information on Climate-Smart Conservation). We focus on the progress made since the US Department of the Interior, the US Department of Agriculture, the National Park Service, the US Fish and Wildlife Service, and the US Forest Service developed climate change strategies in 2010, and we highlight some of the key programs and workshops that have contributed to this progress, including ones highlighted in other "case study" chapters of this section.

Approaches for Adaptation Options

Climate-Smart Conservation is the intentional and deliberative consideration of climate change in resource management, realized through setting forward-looking goals and linking actions to key climate impacts and vulnerabilities (Stein et al. 2014). The seven steps of the Climate-Smart Conservation framework provide guidance to federal managers who are beginning to gauge the effectiveness of current management approaches under future climate change. Previous chapters have focused on defining planning purpose and objectives (step 1), assessing climate impacts and vulnerabilities (step 2), and reviewing/revising conservation goals and objectives to ensure they are climate informed and forward looking (step 3). This chapter focuses on step 4 (identifying possible adaptation options), step 5 (evaluating and prioritizing adaptation actions), and step 6 (implementing priority adaptation actions).

We draw from experience with the Landscape Climate Change Vulnerability Project (LCCVP), both from our engagement in LCCVP activities and from leadership and participation in other collaborative efforts, such as the Great Northern Landscape Conservation Cooperative (Great Northern LCC), the Crown Managers Partnership, the Greater Yellowstone Coordinating Committee (GYCC), the Northern Rockies Adaptation Partnership (NRAP), and the Yale Framework. We also rely on published literature to analyze how federal land managers are identifying, evaluating, and implementing adaptation actions on federal lands and discuss some of the challenges and opportunities that arise when managing for climate change adaptation.

Identify Possible Adaptation Options

Several frameworks that outline climate change adaption options have been proposed since the US Climate Change Science Program (CCSP) published Science Assessment Product 4.4 in 2008 (SAP4.4, summarized in West et al. 2009). The SAP4.4 reasons that many of the best management practices for existing stressors—land use change, invasive species spread, air and water pollution—can help reduce the exacerbation that climate change will have on already stressed resources. Seven "adaptation approaches" (sometimes called the seven "Rs") are suggested to maximize ecosystem resilience to climate change: (1) promote Resilience by protecting key ecosystem features, (2) Reduce anthropogenic stressors, (3) promote Representation, (4) promote Replication, (5) practice Restoration, (6) establish Refugia, and (7) facilitate Relocation (Todd et al. 2013).

The Yale Framework team (Schmitz et al. 2015) framed adaptation options to protect biodiversity at three levels of ecological organization (species, ecosystems, and landscapes). The options include the following: (1) protect current patterns of biodiversity, (2) protect large, intact natural landscapes and ecological processes, (3) protect the geophysical setting, (4) identify and appropriately manage areas that will provide future climate space for species expected to be displaced by climate change, (5) protect climate refugia, and (6) maintain and restore ecological connectivity.

The National Fish, Wildlife and Plant Adaptation Strategy (NFWP-CAS Steering Committee 2012), developed by over ninety scientists and natural resource managers from federal, state, and tribal agencies across the country, takes a holistic view of protecting resources, increasing knowledge, and building capacity. NFWPCAS goals include the following: (1) conserve and connect habitat, (2) manage species and habitat, (3) enhance management capacity; (4) support adaptive management, (5) increase knowledge and information, (6) increase awareness and motivate action, and (7) reduce nonclimate stressors. The twenty-two landscape conservation cooperatives have adopted NFWPCAS into their strategic and science plans.

The frameworks described above are particularly powerful when applied together. Because the Yale Framework was specifically constructed so that adaptation approaches are applied across three levels of ecological organization, this framework forms the foundation of such a crosswalk, with the SAP4.4 and the NFWPCAS filling in action items and strategic gaps (Todd et al. 2013) (table 13-1). With geographic extent, planning goals, and audience in mind, managers can use this crosswalk to identify the types of strategies and actions that will conserve resources in a rapidly changing climate at different scales. Working at the landscape scale, across ecosystems, and across jurisdictional and political boundaries is critical given the scope of climate change impacts (McKinney, Scarlett, and Kemmis 2010).

TABLE 13-1. Framework for Identifying Climate Adaptation Actions, based on the Yale Framework, Science Assessment Product 4.4 (SAP 4.4), and National Fish, Wildlife and Plant Adaptation Strategy (FWP)

	Levels of Ecological Analysis		
	Species and Populations	Ecosystems	Landscapes
	Strengthen current conservation efforts		
Adaptation Objectives	Management Actions		
Protect current patterns of biodiversity	• Reduce anthropogenic stressors (4.4, FWP) • Protect key ecosystem features (4.4) • Maintain representation (4.4) • Replication (4.4) • Prevent, control invasive species (FWP)	• Reduce anthropogenic stressors (4.4, FWP) • Protect key ecosystem features (4.4) • Maintain representation (4.4) • Replication (4.4) • Restore (4.4) • Prevent, control invasive species (FWP)	• Reduce anthropogenic stressors (4.4, FWP) • Protect key ecosystem features (4.4) • Maintain representation (4.4) • Prevent, control invasive species (FWP)
Protect large, intact, natural landscapes and ecological processes	• Reduce anthropogenic stressors (4.4, FWP)	• Slow and reverse habitat loss and fragmentation (FWP) • Prevent, control invasive species (FWP)	• Slow and reverse habitat loss and fragmentation (FWP) • Prevent, control invasive species (FWP)
Protect the geophysical setting	• Protect key ecosystem features (4.4)	• Protect key ecosystem features (4.4)	
	Anticipate and respond to future conditions		
	Management Actions		
Identify and appropriately manage areas that will provide future climate space for species expected to be displaced by climate change	• Relocate organisms (4.4) • Develop science to detect and describe climate impacts on fish, wildlife, and plants and ecosystems (FWP) • Identify knowledge gaps and define collaborative research priorities (FWP)	• Identify refugia (4.4) • Develop science to detect and describe climate impacts on fish, wildlife, and plants and ecosystems (FWP) • Identify knowledge gaps and define collaborative research priorities (FWP)	• Identify refugia (4.4) • Develop science to detect and describe climate impacts on fish, wildlife, and plants and ecosystems (FWP) • Identify knowledge gaps and define collaborative research priorities (FWP)

TABLE 13-1.   (*Continued*)

	Levels of Ecological Analysis		
	Species and Populations	Ecosystems	Landscapes
	Strengthen current conservation efforts		
Adaptation Objectives	Management Actions		
Identify and protect climate refugia	• Identify refugia (4.4)	• Identify refugia (4.4)	• Identify refugia (4.4)
Maintain and restore ecological connectivity	• Protect key ecosystem features (4.4) • Slow and reverse habitat loss and fragmentation (FWP) • Identify, protect, restore, and establish new areas for an ecologically connected network of conservation areas (FWP)	• Protect key ecosystem features (4.4) • Slow and reverse habitat loss and fragmentation (FWP) • Identify, protect, restore, and establish new areas for an ecologically connected network of conservation areas (FWP)	• Slow and reverse habitat loss and fragmentation (FWP) • Identify, protect, restore, and establish new areas for an ecologically connected network of conservation areas (FWP)

For the most part, federal land managers are just becoming aware of these frameworks; the frameworks typically have not yet been incorporated into day-to-day management decisions or even climate change planning. Importantly, although scientists and workshop facilitators understand and try to incorporate the frameworks into steps 4 and 5 to identify and prioritize adaptation actions, we have found in the workshops we have hosted, attended, or reviewed that managers often generate adaption options from their own experience without the aid of frameworks. This is true in workshops sponsored by the Great Northern LCC to identify adaptation options for coldwater salmonids and aquatic ecosystems (http://www.eco adapt.org/workshops/GNLCC-adaptation-workshop) and NRAP (http://adaptationpartners.org/nrap/); in the report *Adapting to Climate Change in the Olympic National Forest and Olympic National Park* (Halofsky et al. 2011); and in case studies associated with climate-smart training sessions focusing on Bandelier and Pecos national monuments (USFWS Conservation Training Center 2015).

Lastly, the Adaptation for Conservation Targets (ACT) framework (Cross et al. 2012) explicitly relies on local managers' knowledge to generate adaptation options; in fact, the ACT framework is based on the premise that adaptation to climate change can rely on local knowledge of an ecosystem and does not necessarily require detailed projections of climate change or its effects. Whether managers identify adaptation options by using an explicit framework, drawing from their own experience, or relying on previous efforts, the goal of step 4 of the Climate-Smart Conservation framework is to generate the broadest array of possible adaptation options for consideration in step 5 and implementation in step 6 (Stein et al. 2014).

Evaluate and Prioritize Adaptation Actions

After potential adaptation options (management actions) have been identified, they need to be narrowed down and prioritized by their relative feasibility and desirability. Such decisions may be based on assessments that include costs, potential for successful conservation, uncertainty of outcome, consistency with agency policy, time frame of impacts, and public support. The chasm between identifying adaptation options and evaluating whether those options are truly feasible can be vast. Lemieux et al. (2011) surveyed Canadian scientists and managers and found a significant gap between the perceptions of the two groups in what constituted feasible adaptation options. In this study, a panel of scientists identified 165 adaptation options that they deemed to be desirable and feasible. A separate team of senior park agency decision makers reduced this list to the 56 most desirable adaptation options but determined that only two were "definitely implementable," with constraints due largely to fiscal and capacity limitations.

The ACT framework has a strong focus on, and is a good model for, evaluating feasibility of adaptation options. The feasibility matrix includes economic costs, regulatory constraints, social conflict, potential unintended consequences, synergies with other management objectives, potential for removal of the management action (if necessary), consistency with current management practice, and robustness to uncertainty in future climate projections. This approach was used successfully with state and federal fish and wildlife managers and scientists to evaluate the costs and benefits associated with three potential actions for managing Yellowstone River flows under a warmer and dryer climate (Cross et al. 2012).

Lemieux et al. (2011) developed an ordinal scale to rank feasibility criteria, including affordability, ease of implementation, institutional capacity, and capacity to sustain actions over time. Nelson (2014) further developed this approach to have managers rank different adaptation options for resources vulnerable to climate change in the Crown of the Continent area of Montana, British Columbia, and Alberta, defining four ordinal categories to rank options for effectiveness, feasibility (affordability), feasibility (legal, political, institutional, social barriers), and implementation status (from "currently being implemented" to "implementation not possible").

Managers we have worked with often assess feasibility using cost as a first filter: agency budgets are generally flat to shrinking and, in the words of a federal biologist, "a project without funding is just talk." Other issues that constrain feasibility are whether an action falls within current or projected agency policy, the level of uncertainty about projected resource impact, and perceived stakeholder buy-in to adaption actions. Some managers feel that areas managed for pristine conditions, such as wilderness or national parks, should not be actively managed and thus fulfill the role of acting as natural laboratories for monitoring and studying environmental change.

Implement Priority Adaption Actions

The ultimate goal of identifying, evaluating, and prioritizing adaptation options is to implement specific actions through agency land management plans and on-the ground decisions. Implementing climate change adaptation actions is the most important step of the Climate-Smart Conservation framework, yet it is the step where, at least to this point, most efforts have stalled. Halofsky, Peterson, and Marcinkowski (2015) reviewed climate change adaption efforts within federal agencies during 2013–2014 and found that successful adaption efforts on federal lands were the result of motivated managers and successful science–management partnerships, but they caution: "Mainstreaming of climate-smart practices in agencies has been slow to develop, probably because it has not been required at local to regional scales and because systems of accountability are rare." Managers contend with a multitude of challenges when trying to intentionally add climate change actions to the management matrix (many of these are discussed in chapter 3).

Climate change is mandated to be considered by agency managers in high-level unit plans, such as US Forest Service forest plans (Code of Federal Regulations 2012), National Park Service foundation documents (National Park Service 2012), Bureau of Land Management land management plans, and US Fish and Wildlife Service refuge comprehensive conservation plans (Czech et al. 2014). Thinking about climate change in these high-level plans is helping managers to understand the language, concepts, models, and tools of climate change adaptation. The US Forest Service has taken a step beyond high-level guidance by developing the Climate Change Performance Scorecard for national forests (US Department of Agriculture 2011). The scorecard requires that each national forest conduct management actions that reduce the vulnerability of resources and places to climate change.

Although guidance has been developed and is beginning to be used, few projects have actually moved beyond planning and into climate adaptation action on federal lands (Halofsky, Peterson, and Marcinkowski 2015). Many managers have not yet moved forward—some feel that the information available on climate change projections and resource vulnerability is simply not site specific enough to be actionable. In other cases, agencies have not yet developed specific guidance on climate change adaptation for managers to follow when developing environmental assessments or environmental impact statements (National Park Service 2012).

While adaption implementation on federal lands seems to be moving forward slowly, adaptation actions at local districts and field units are clearly underreported. For example, in recent meetings, we learned that staff are (1) using climate change projections from NorWest (Isaak et al. 2011) to inform fisheries strategy on the Gallatin National Forest; (2) using a Greater Yellowstone Ecosystem Watershed Vulnerability Assessment to prioritize replacing culverts on the Bridger-Teton National Forest, thus reducing sedimentation in streams and improving water quality; (3) treating trees killed by a climate-induced beetle epidemic at Timpanogos Cave National Monument; and (4) working with The Nature Conservancy and the Wildlife Conservation Society to install beaver mimicry structures to restore riparian areas and wetlands on private and some public lands in the Madison and Centennial valleys in Montana.

However, these local efforts do not roll up into regional or national reports or databases describing on-the-ground adaptation actions. To cite one example, only seven of eighty-two parks responded to a request for information about actions taken by parks to implement the National Park Service Climate Change Action Plan. Managers simply did not have time to respond, felt like the request was confusing, or didn't believe local adaptation efforts were significant enough to warrant including in a report. Agencies are taking steps to capture these types of management actions in relation to climate change—for example, the National Park Service is creating an adaptation database for parks that will help local managers report their successes as well as discover other projects. In addition, the GYCC

Adaptation Subcommittee is developing one-page briefing statements to capture examples of on-the-ground adaptation.

Challenges and Opportunities

Interest from federal agencies in mitigating and adapting to climate change impacts is increasing. With that interest, a suite of adaptation options have been identified (e.g., West et al. 2009; Cross et al. 2012; Schmidtz et al. 2015; NFWPCAS Steering Committee 2012) and high-level strategies have been developed by several federal agencies (e.g., National Park Service 2010; US Fish and Wildlife Service 2010; US Forest Service 2011; Council on Climate Preparedness and Resilience 2014).

Examples of managers implementing adaptation actions at the unit level are sparse, and some managers' understanding of climate change adaptation practices is limited (National Research Council 2010a, 2010b; Archie et al. 2012; Lemieux et al. 2013; Halofsky, Peterson, and Marcincowski 2015). Lemieux et al. (2013) surveyed managers and found that the greatest gap between perceived importance of climate change and implementing climate change actions was lack of adequate adaptation policies and lack of climate change adaptation activities that are well defined at the management unit scale. Archie et al. (2012) found that the most common barriers to adaptation planning were lack of information at relevant scales, budget constraints, lack of specific agency direction, and lack of useful information. The challenges that managers face when considering actions based on climate change vulnerability are well documented in chapter 3.

Despite these challenges, the US Government Accountability Office (GAO) reported that federal agencies have made some progress in planning and implementing climate change adaptation since they began tracking in 2007. In 2007, GAO found that natural resource managers in federal agencies lacked specific guidance for incorporating climate change into their planning efforts and management actions, and in 2009, GAO reported that climate change is a complex, interdisciplinary issue and that government-wide adaptation planning and collaboration could assist climate change adaptation efforts. By 2013, GAO reported that the Departments of Agriculture and the Interior had developed various mechanisms to address climate change impacts on natural resources. These mechanisms include climate change adaptation strategies within the departmental strategic plans, adaptation-related policies and guidance, and national interagency collaborative initiatives. In addition, the departments have both, in various ways, informed their respective agencies to begin addressing climate change adaptation (GAO 2007, 2009, 2013).

The magnitude and expanse of climate change impacts will far exceed the ability of any one federal unit, agency, or organization to effectively respond in isolation. Collaboration between scientists and resource managers from different organizations and agencies will greatly assist in guiding the development of appropriate climate change adaptation activities (Halofsky, Peterson, and Marcincowski 2015). The LCCVP project has advanced a model of pairing cutting-edge science with management engagement—both during the project proposal stage and by staying engaged throughout the life of the project. Climate change adaptation works best when scientists and managers collaborate to develop cross-jurisdictional conservation plans to mitigate the impacts of climate change at the scale at which they occur.

Several cross-jurisdictional partnerships are well established at the ecosystem level in the northern Rockies, yet climate change has further blurred agency and jurisdictional boundaries. For example the Crown Managers Partnership (CMP), established in 1990, seeks to demonstrate leadership in addressing the environmental management challenges in the Crown of the Continent ecosystem (anchored by Glacier National Park but including lands managed by the US Forest Service as well as areas of Alberta and British Columbia) by adopting transboundary collaborative approaches to environmental management. The CMP hosted a Climate Change Scenario Planning workshop in March 2010 and followed up by identifying specific adaptation actions in a 2014 forum. The Great Northern LCC has partnered with the CMP to develop ecosystem health indicators and to enable partners to collaborate on climate change adaptation, work effectively across jurisdictions, and share data and utilize a common science template. The overall goal is to achieve amplified management outcomes that address these shared conservation threats.

Specifically, the CMP is working to prioritize and implement shared climate adaptation strategies in a coordinated effort with nongovernment organizations and community stakeholders. These strategies collectively advance the CMP's priorities for managing to a desired condition across the Crown of the Continent ecosystem while providing for significant strategic alignment with the Great Northern LCC's conservation priorities.

In the Greater Yellowstone Area, the National Park Service and the Forest Service formed the GYCC in the 1960s to pursue opportunities of mutual cooperation and coordination in the management of core federal lands in the Greater Yellowstone Area. The GYCC now includes the

Greater Yellowstone national wildlife refuges and Bureau of Land Management lands, and the committee's working groups partner with county and state agencies, universities, and nongovernmental organizations on a range of ecosystem topics. The GYCC, Montana State University and Yellowstone National Park hosted a Science Agenda workshop in November 2009 and subsequently published an ecosystem science agenda in response to large landscape stressors, including climate change (Olliff et al. 2010). The GYCC currently sustains an interagency and interdisciplinary Climate Change Adaptation Subcommittee, which addresses employee education, ways to fill data gaps, synthesis of ecosystem climate information, ecosystem-scale future climate projections, and vulnerability assessments to inform land management. The LCCVP projections of future climate and impacts to vegetation (chaps. 4, 9, and 10) have been a fundamental resource available to GYCC managers and Great Northern LCC staff on the subcommittee.

These ecosystem-based cooperatives have recently been augmented by organizations created to operate at the larger landscape scale. In 2009, the Department of the Interior created LCCs that are intended to provide cutting-edge science and to work with federal, state, tribal, and local governments, private landowners, and nongovernmental organizations to "develop landscape-level strategies for understanding and responding to climate change impacts" and to help managers sustain the continent's natural and cultural resources (Secretarial Order 3289). Although a self-directed steering committee guides each LCC, they all (1) use applied conservation science in collaboration with partners within a geographically defined area, (2) function as a fundamental management partnership that will help frame decisions made at the unit level in a larger landscape context, and (3) provide a national (and international) network of land, water, wildlife and cultural resource managers, and interested public and private organizations to respond to landscape-scale stressors, such as climate change and land use change (Austen 2011).

Also established in 2009, climate science centers deliver basic climate change impact science to field managers within their respective regions, including physical and biological research, ecological forecasting, and multiscale modeling. Climate science centers prioritize delivery of fundamental science, data, and decision support to meet the needs of the LCCs and their partner organizations. This includes providing climate change impact information on natural and cultural resources and developing adaptive management and other decision support tools for managers (Beard, O'Malley, and Robertson 2011).

By the end of 2013, the full suite of twenty-two LCCs and eight climate science centers were operational. In addition, in 2013 the US Department of Agriculture launched seven "Regional Hubs for Risk Adaptation and Mitigation to Climate Change" (climate hubs) to develop and deliver science-based, region-specific information and technologies to agriculture and natural resource managers and communities.

Perhaps especially important for climate change adaptation is the role of LCCs as bridging organizations, since adaptation needs to be implemented at large scales and across multiple jurisdictions. As bridging organizations, LCCs foster adaptive cooperative governance by providing platforms for communication, relationship building, and stakeholder engagement, and they have great potential to facilitate conservation of rapidly changing social-ecological systems by providing structure and incentives for collaboration and shared learning (Jacobson and Robinson 2012; McDowell 2012). The international network of twenty-two LCCs has over three hundred active partners and since 2010 has provided funding to more than four hundred science partners to help managers conserve resources in light of not only climate change but also other landscape stressors, such as land use change and invasive species.

Case Studies

Chapters 14 to 16 offer case studies of Climate-Smart Conservation from the perspectives of Rocky Mountain National Park and the Greater Yellowstone Ecosystem (GYE). Here we set the stage for these chapters by offering vignettes of science–management partnerships that have similarly made progress on steps 4 to 6 of the Climate-Smart Conservation framework. We also discuss some of the case studies in chapters 14 to 16 through the lens of steps 4 to 6.

In April 2015, the LCCVP team hosted a workshop for managing vegetation communities in the GYE under predictions of vast changes caused by climate change. The workshop focused on steps 4 and 5: identifying climate change adaptation options, weighing each option against measures of feasibility, and discussing how the feasible options fit together in a landscape strategy. The workshop included thirty managers and scientists from three federal land management agencies, nongovernmental organizations, and universities.

The overall goal of the workshop was to build on the recent Northern Rockies Adaptation Project (NRAP) and lay the conceptual groundwork for integrated assessment and management to sustain healthy and resilient vegetation communities across the GYE under climate change. Participants identified and evaluated climate adaptation options for high-priority vegetation types, including alpine, whitebark pine, low-elevation woodlands, grasslands, and shrublands. For each vegetation type, and for different goals, participants articulated adaption options (defined as an integrated set of management actions aimed at achieving resource goals across a planning area) as well as particular management actions (a specific treatment—for example, planting or prescribed burning). For each action, they rated the feasibility and the likely effectiveness of the action and discussed integrating management actions across the planning area.

This workshop had several unique characteristics. First, the exercises were based on recent science that was conducted specifically to support the workshop exercises and that projected vegetation change under different climate scenarios for the GYE planning area. Second, adaptation options and management actions focused on three different zones (chapter 9): core (areas where the climate is suitable now and will continue to be suitable for species in the future); deteriorating (areas where the climate is suitable now but will no longer be suitable in the future); and future (areas where the climate is not presently suitable but is projected to be suitable in the future).

Third, managers strongly identified several aspects of adaptive capacity and focused adaptation options on supporting that adaptive capacity, especially in the whitebark pine (*Pinus albicaulis*) breakout group. For example, whitebark pine is known to occur predominantly in the subalpine zone but occasionally is found elsewhere, especially on dry sites, even at low elevations. One suggested adaption option was to take advantage of whitebark pine's adaptive capacity by supporting these low-elevation stands through suppressing competition and planting to aid regeneration. LCCVP work identified a similar strategy for a related species of five-needle pine (i.e., limber pine, *Pinus flexilis*) in Rocky Mountain National Park (Monahan et al. 2013). Although the other two elements of vulnerability—exposure and sensitivity—can be modeled and assessed from literature, managers with extensive field experience were often best at articulating the adaptive capacity of a species.

Finally, only a few of the workshop participants had been involved with LCCVP over the duration of our project (four years), but these participants had a deeper understanding of the models and the model results than those whose first exposure was the April 2015 workshop.

Although federal managers can conduct steps 1 through 5 collaboratively, the next step (implementation) is carried out individually for each park, forest, wildlife refuge, or Bureau of Land Management district, because each agency must comply with its own laws, regulations, and policies and carry out its own National Environmental Policy Act analysis to proceed with implementation. The Greater Yellowstone Area Whitebark Pine Strategy is an excellent example of this collaborative planning and analysis with unit-based implementation. As a result of the collective vision and cooperation on the Greater Yellowstone Area Whitebark Pine Strategy, these early adopters are already implementing climate-robust actions on several forests and parks. For other Greater Yellowstone Area vegetation types, fundamental questions remain about desired future condition, how to incorporate consideration of the future while addressing the high-priority current issues, and the methods for monitoring, early change detection, and adaptive management.

NRAP is a science–management partnership between the Forest Service's regional managers (fifteen national forests and the Pacific Northwest and Rocky Mountain research stations), the National Park Service, the Great Northern and Plains and Prairie Potholes LCCs, the North Central Climate Science Center, the Greater Yellowstone Coordinating Committee, Oregon State University, and EcoAdapt. The partnership provides a foundation for climate change adaptation work on national forests, national parks, and adjacent lands. NRAP is developing a climate change vulnerability assessment (step 2 of the Climate-Smart Conservation framework) that addresses hydrology and roads, fisheries, wildlife, vegetation and disturbance, recreation, and socioeconomic conditions.

As part of the assessment, NRAP convened five workshops in Idaho, Montana, and South Dakota in fall 2014, at which two hundred scientists and managers identified the most significant vulnerabilities to climate change throughout the region, brainstormed specific adaptation strategies, and assessed the feasibility of those strategies. NRAP is developing guidance for managers to implement adaptation strategies at various levels of management operation (from project design to forest planning) and is compiling all information into a peer-reviewed, published report. This guidance will effectively inform steps 3 to 5 of the Climate-Smart Conservation framework. Many of the forests in the northern Rockies will use this process as the vulnerability assessment required by the Climate Change Report Card and are planning to use the prioritized adaption actions identified in steps 4 and 5 in their revised forest plans, which will serve as a foundation for implementation (step 6).

The trout study described in chapter 12, as well as similar work presented in the same chapter for Great Smoky Mountains National Park, uses a feasibility framework adapted from a financial portfolio concept aimed at maximizing species persistence in the face of adversity (Schindler et al. 2010; Haak and Williams 2012). This "3-R" framework creates a diverse management portfolio through increasing representation (protecting and restoring diversity), resilience (having sufficiently large populations and intact habitats to facilitate recovery from rapid environmental change), and redundancy (saving a sufficient number of populations so that some can be lost without jeopardizing the species).

In the Yellowstone region, this study integrates existing information on Yellowstone cutthroat trout status (*Oncorhynchus clarkii bouvieri*) and limiting factors to develop a spatially explicit conservation priority system. State and federal partners and collaborators developed criteria and a framework for prioritizing populations of Yellowstone cutthroat trout with respect to risk from climate change and then applied this framework with population-specific ranking of limiting factors and climate risks (steps 4 and 5). This ranking identified and prioritized conservation actions to enhance resilience under a changing climate and to identify areas for potential reintroduction of Yellowstone cutthroat trout into streams that are likely more resilient to regional changes in climate. To reach step 6 (implementation), scientists are working with the Geographic Management Unit teams under the auspices of the Western Native Trout Initiative. Again, each individual unit will implement these common strategies for building resilience in coldwater fisheries under climate change.

The whitebark pine study outlined in chapter 15 takes a unique quantitative approach to defining adaptation options. As mentioned previously, federal land managers of the Whitebark Pine Subcommittee are going a step beyond implementing climate-robust strategies by working with scientists to develop climate projections that will inform the siting of on-the-ground management actions defined in the Greater Yellowstone Area Whitebark Pine Strategy. The intent is to understand which combinations of actions have the greatest impact on preserving whitebark pine in the long term (the next hundred years) in response to the projected decrease in abundance and distribution as a result of climate change. The core LCCVP team has conducted much of the science for this project, which is augmented by funding from the North Central Climate Science Center and supported by the Great Northern LCC. Three spatially explicit management alternatives are being evaluated: No Active Management, Full Implementation of the Whitebark Pine Strategy, and a No Constraints alternative. The project is also using social surveys to understand the attitudes of residents in the northern Rockies toward management. The subcommittee is poised to implement the management actions identified, in large part because of the well-coordinated and thorough work they have conducted.

As these case studies show, managers have come light years from their basic understanding of climate change in 2011: they are now obtaining and conducting vulnerability assessments, rethinking their goals to incorporate climate change, identifying and prioritizing adaptation actions, and beginning the early stages of implementing these actions.

Conclusion

The road maps for climate change adaptation—the initial *Scanning the Conservation Horizon: A Guide to Climate Change Vulnerability Assessment* (Glick, Stein, and Edelson 2011) and the refinements provided by *Climate-Smart Conservation: Putting Adaptation Principles into Practice* (Stein et al. 2014)—have now been developed and are beginning to be used by natural resource managers at federal agencies. Although managers are beginning to better understand, through vulnerability assessments, the impacts that climate change is projected to have on natural resources, there is a long road ahead before they routinely identify, evaluate, and implement climate change adaption options. The science of climate change adaptation is nascent, so current managers have to overcome language, conceptual, and science barriers in addition to acquiring resources and developing policies to implement adaptation options. While improvements have been made, more work is needed to make adaptation strategies more actionable through specific management actions. Chapters 14 to 16 offer case studies of Climate-Smart Conservation from the perspectives of Rocky Mountain National Park and the Greater Yellowstone Ecosystem that illustrate some of this progress.

Acknowledgments

The authors would like to thank the staff and collaborators of the Great Northern LCC (Yvette Converse, Sean Finn, Matt Heller, Mary McFadzen, Anne Carlson, John Pierce, Molly Cross, and Regan Nelson); leaders of the NRAP (Linh Huang and Dave Peterson); staff from the National Park Service (NPS) Climate Change Response Program (Leigh Welling, Cat Hawkins-Hoffman, Nick Fisichelli, Gregor Schuurman, and Patrick Gonzales) and the NPS Greater Yellowstone Inventory and Monitoring Program (Kristin Legg and David Thoma); the GYCC Climate Change

Subcommittee (Scott Barndt, Karri Cari) and NPS Intermountain Region climate change coordinator Pam Benjamin; and members of the Crown Managers Partnership (Erin Sexton, Ian Dyson, and Mary Riddle) for contributing their ideas and experience to this chapter. Dr. Glenn Plumb, NPS, reviewed the chapter and provided helpful comments.

References

Archie, K. M., L. Dilling, J. B. Milford, and F. C. Pampel. 2012. Climate change and western public lands: A survey of U.S. federal land managers on the status of adaptation efforts. *Ecology and Society* 17 (4): 20. http://dx.doi.org/10.5751/ES-05187-170420.

Austen, D. J. 2011. Landscape conservation cooperatives: A science-based network in support of conservation. *Wildlife Professional* 5 (3): 12–15.

Beard, T. D., R. O'Malley, and J. Robertson. 2011. New research on climate's front lines. *Wildlife Professional* 5 (3): 26–30.

Czech, B., S. Covington, T. M. Crimmins, J. A. Ericson, C. Flather, M. Gale, K. Gerst, M. Higgins, M. Kaib, E. Marino, et al. 2014. *Planning for Climate Change on the National Wildlife Refuge System*. Washington, DC: US Fish and Wildlife Service, National Wildlife Refuge System.

Code of Federal Regulations, 36 CFR Part 219. 2012. *USDA Forest Service National Forest System Land Management Planning Rules and Regulations* 77 (68), April 9.

Council on Climate Preparedness and Resilience, Climate and Natural Resources Working Group. 2014. Priority Agenda: Enhancing the Climate Resilience of America's Natural Resources. Report, September. http://www.whitehouse.gov/sites/default/files/docs/enhancing_climate_resilience_of_americas_natural_resources.pdf.

Cross, M., E. S. Zavaleta, D. Bachelet, M. L. Brooks, C. A. F. Enquist, E. Fleishman, L. J. Graumlich, C. R. Groves, L. Hannah, L. Hansen, et al. 2012. The Adaptation for Conservation Targets (ACT) Framework: A tool for incorporating climate change into natural resource management. *Environmental Management* 50 (3): 341–51. doi: 10.1007/s00267-012-9893-7.

GAO (US Government Accountability Office). 2007. Climate Change: Agencies Should Develop Guidance for Addressing the Effects on Federal Lands and Water Resources. GAO-07-863. Washington, DC, August 7.

GAO (US Government Accountability Office). 2009. Climate Change Adaptation: Strategic Federal Planning Could Help Government Officials Make More Informed Decisions. GAO-10-113. Washington, DC, October 7.

GAO (US Government Accountability Office). 2013. Climate Change: Various Adaptation Efforts Are Under Way at Key Natural Resource Management Agencies. GAO-13-253. Washington, DC, June 20.

Glick, P., B. A. Stein, and N. A. Edelson, eds. 2011. *Scanning the Conservation Horizon: A Guide to Climate Change Vulnerability Assessment*. Washington, DC: National Wildlife Federation.

Haak, A. L., and J. E. Williams. 2012. Spreading the risk: Native trout management in a warmer and less-certain future. *North American Journal of Fisheries Management* 32:387–401.

Halofsky, J. E., D. Peterson, and K. W. Marcinkowski. 2015. Climate Change Adaptation in United States Federal Natural Resource Science and Management Agencies: A Synthesis. USGCRP Climate Change Adaptation Interagency Working Group.

Halofsky, J. E., D. Peterson, K. O'Halloran, and C. Hawkins Hoffman, eds. 2011. *Adapting to Climate Change at Olympic National Forest and Olympic National Park*. Gen. Tech. Rep. PNW-GTR-844. Portland, OR: US Department of Agriculture, Forest Service, Pacific Northwest Research Station.

Isaak, D. J., S. J. Wenger, E. E. Peterson, J. M. Ver Hoef, S. Hostetler, C. H. Luce, J. B. Dunham, J. Kershner, B. B. Roper, D. Nagel, et al. 2011. NorWeST: An Interagency Stream Temperature Database and Model for the Northwest United States. US Fish and Wildlife Service, Great Northern Landscape Conservation Cooperative Grant. http://www.fs.fed.us/rm/boise/AWAE/projects/NorWeST.html.

Jacobson, C., and A. Robinson. 2012. Landscape conservation cooperatives: Bridging entities to facilitate adaptive co-governance of social–ecological systems. *Human Dimensions of Wildlife* 17:333–43. doi: 10.1080/10871209.2012.709310.

Lemieux, C. J., and D. J. Scott. 2011. Changing climate, challenging choices: Identifying and evaluating climate change adaptation options for protected areas management in Ontario, Canada. *Journal of Environmental Management* 48 (4): 675–90.

Lemieux, C. J., J. L. Thompson, J. Dawson, and R. M. Schuster. 2013. Natural resource manager perceptions of agency performance on climate change. *Environmental Management* 114:178–89. http://dx.doi.org/10.1016/j.jenvman.2012.09.014.

McDowell, G. 2012. The Role of Bridging Organizations in Facilitating Socioecological Transformation: A Case Study of the Great Northern Landscape Conservation Cooperative. Thesis, master's of science in environmental change and management, University of Oxford, Environmental Change Institute. http://greatnorthernlcc.org/sites/default/files/documents/mcdowell_msc_thesis.pdf.

McKinney, M., L. Scarlett, and D. Kemmis. 2010. *Large Landscape Conservation: A Strategic Framework for Policy and Action*. Cambridge, MA: Lincoln Institute of Land and Policy.

Monahan, W. B., T. Cook, F. Melton, J. Connor, and B. Bobowski. 2013. Forecasting distributional responses of limber pine to climate change at management-relevant scales in Rocky Mountain National Park. *PLOS ONE* 8 (12): e83163. doi:10.1371/journal.pone.0083163.

National Park Service. 2010. *National Park Service Climate Change Response Strategy*. Fort Collins, CO: National Park Service, Climate Change Response Program.

National Park Service. 2012. *Climate Change Action Plan*. Fort Collins, CO: National Park Service, Climate Change Response Program.

National Research Council. 2010a. *Adapting to the Impacts of Climate Change*. Washington, DC: National Academies Press.

National Research Council. 2010b. *Informing an Effective Response to Climate Change*. Washington, DC: National Academies Press.

Nelson, R. 2014. *A Climate Change Adaptation Gap Analysis for the Crown of the Continent*. Commissioned and published by the Crown of the Continent Conservation Initiative.

NFWPCAS (National Fish, Wildlife, and Plants Climate Adaptation Strategy) Steering Committee. 2012. *The National Fish, Wildlife, and Plants Climate Adaptation Strategy*. http://www.wildlifeadaptationstrategy.gov.

Olliff, T., G. Plumb, J. Kershner, C. Whitlock, A. Hansen, M. Cross, and S. Bischke. 2010. A science agenda for the Greater Yellowstone Area: Responding to landscape impacts from climate change, land use change, and invasive species. *Yellowstone Science* 18 (2): 14–22.

Schindler, D. E., R. Hilborn, B. Chasco, C. P. Boatright, T. P. Quinn, L. A. Rogers, and M. S. Webster. 2010. Population diversity and the portfolio effect in an exploited species. *Nature* 465:609–12.

Schmitz, O. J., J. J. Lawler, P. Beier, C. Groves, G. Knight, D. A. Boyce Jr, and J. Bulluck. 2015. Conserving biodiversity: Practical guidance about climate change adaptation approaches in support of land-use planning. *Natural Areas Journal* 35 (1): 190–203.

Stein, B. A., P. Glick, N. Edelson, and A. Staudt, eds. 2014. *Climate-Smart Conservation: Putting Adaptation Principles into Practice*. Washington, DC: National Wildlife Federation.

Todd, B., L. Read, and T. Olliff. 2013. *Climate Change Workshop Summary Report*. Lakewood, CO, February 5. IMR/DSC–Planning.

US Department of Agriculture, Forest Service. 2011. *Navigating the performance scorecard: A guide for national forests and grasslands*. Version 2. Washington, DC: USDA FS, July 15. http://www.fs.fed.us/climatechange/advisor/scorecard/scorecard-guidance-08-2011.pdf.

US Fish and Wildlife Service. 2010. *Rising to the Urgent Challenge: Strategic Plan for Responding to Accelerating Climate Change*. US Department of the Interior. http://www.fws.gov/home/climatechange/pdf/CCStrategicPlan.pdf.

US Forest Service. 2011. *National Roadmap for Responding to Climate Change*. US Department of Agriculture. http://www.fs.fed.us/climatechange/pdf/Roadmap final.pdf.

USFWS National Conservation Training Center. 2012. *Climate-Smart Conservation Course Map*. Santa Fe, NM, January 27–29.

West, J. M., S. Julius, P. Kareiva, C. Enquist, J. J. Lawler, B. Petersen, A. E. Johnson, and M. R. Shaw. 2009. U.S. natural resources and climate change: Concepts and approaches for management adaptation. *Environmental Management* 44:1001–21.

Chapter 14

Perspectives on Responding to Climate Change in Rocky Mountain National Park

Ben Bobowski, Isabel W. Ashton,
and William B. Monahan

Thousands of years before Rocky Mountain National Park was established, people came to the area to experience its beauty and its wildlife and, for many, to re-create themselves. And it remains so today. How do we protect this special place so future generations may have the same opportunities?

At first glance, it may appear that Rocky Mountain National Park (hereafter Rocky Mountain NP) in Colorado is well protected—it is a national park, a wilderness, a Class I airshed, the headwaters for the Colorado and Platte river drainages, and a biosphere reserve. Each of these facts indicates a land that has significant protection. But are these designations enough? They are important and valuable as steps toward protection, yes, but without such management activities as restoration, monitoring, research, education, and collaboration with partners on issues that extend beyond park borders, protection of this highly valued park cannot be supported, especially in light of climate change.

Climate change is the largest environmental issue in our human history, in terms of both scale and scope, and there are no new monies provided directly to parks to increase capacity for addressing this issue. In 2007, Rocky Mountain NP's Continental Divide Research Learning Center hosted a workshop to define the following: what we know, what we think we know, and what has yet to be discovered regarding climate change in Rocky Mountain NP (National Park Service 2007a). As one might suspect, there were more questions than answers, and the first glimpses of the limitations to our perspectives of the problem were exposed.

Thinking beyond our borders is still relatively novel, scientifically and practically, for various reasons: traditions of practice, the lure of place-based science, funding restrictions in time (short-term support) and space (limited resources preclude a large science footprint), complex political landscapes, and paradigms by both practitioners and scientists that limit our understanding, perspectives, and successes.

For several generations of scientists, academics, and practitioners, the "referent" for success was to manage an area within the historic range of variability. How do we go forward within the paradigm of climate change? What novel ecosystems will emerge? Indeed, what is the way forward? The climate change stories we are learning through science and history are growing, but there is so much more to learn.

The challenges that surfaced through the Continental Divide Research Learning Center's workshop were reframed within the context of numerous successes—through what we know. From diatoms to the park's largest native deer, the Rocky Mountain elk (*Cervus elaphus*), we know more about the species and the ecological systems of Rocky Mountain NP today than at any other time in history. Thanks to the hundreds of research projects in the park over the past decades, we've connected to other landscapes in many ways, including our understanding of migratory species, forest health, air quality, and snowpack and water availability for the communities that benefit from Rocky Mountain NP's protection. In the end, the workshop presented a way forward that would prove meaningful through the development of partnerships—a shared conservation approach for the future in which scarce resources, leveraged among partners, could likely provide the science and social capacity needed to adapt to change. We have come to recognize that the conservation of species within the park requires science, understanding, and working with people beyond the park borders (Bobowski 2013).

In this chapter, we begin by highlighting climate change issues facing Rocky Mountain NP and follow up with insights into some of the key management challenges and opportunities. We then elaborate on the adaptive management of climate change in Rocky Mountain NP and offer three case studies demonstrating where we have managed to make significant progress. We conclude with our ideas for a way forward, with hopes that they will encourage dialogue on management within Rocky Mountain NP and similar protected areas.

Climate Change Issues

Rocky Mountain NP turned one hundred years old in 2015, and already we are experiencing effects from extreme weather that demand a response. During the past five years, the park has experienced more large fires than in the previous ninety-five years combined. The Fern Lake Fire, a human-caused fire that began in October 2012, endured during the driest November on record in more than one hundred years (fig. 14-1). Because of large expanses of beetle-killed forests (a phenomenon related to climate change), indirect firefighting tactics were needed as a primary firefighting approach to mitigate risks to firefighters.

The Fern Lake Fire burned through winter and over snow. In the dark of early morning on December 1, 2012, firefighters evacuated a portion of the town of Estes Park when an improbable weather event pushed the

FIGURE 14-1   A helicopter helps suppress the Fern Lake Fire in Rocky Mountain National Park in 2012. The fire persisted in winter conditions after the driest November in the one-hundred-year climate record. (Photo taken by Mike Lewelling, December 6, 2012.)

FIGURE 14-2 The September 2013 flood caused damage to roads, buildings, and other infrastructure in Rocky Mountain National Park. The road shown in this photo has since been replaced, with a large culvert now added that should withstand future flooding. (Photo from the Rocky Mountain National Park digital photo library, August 27, 2014.)

fire more than three miles toward town in approximately thirty-five minutes. Firefighters were pre-positioned to thwart the improbable advance and succeeded because of the firefighters' extensive experience and despite fire models that clearly no longer could account for all the variables in play. As the days continued, frozen lakes precluded many water-drop operations, and the sounds of Christmas music struck responders as strange as more than six hundred firefighters engaged in operations above eight thousand feet in elevation. The fire went silent as heavy snows finally arrived and the year came to an end.

Less than one year later, in September 2013, the park experienced one of the largest rain events in its history, resulting in floods that erased roads and infrastructure from the park to the Front Range of Colorado (fig. 14-2). The area became part of a national disaster area, and recovery continues to the present. Geologists who were asked to evaluate the numerous landslides and erosion in preparation for rescue operations and restoration indicated that the last time the soils of this region were this saturated was during postglacial melting thousands of years ago.

These extreme events are what we can expect more of in the future as a consequence of climate change, even though the precise link to climate change is less certain. While it is impossible to predict when and where extreme weather events will occur, the park has the capacity to better prepare for these events and to work under the assumption that they will occur more regularly. Fire managers now recognize that fires can and do occur in any season in Rocky Mountain NP. As infrastructure was repaired and replaced after the 2013 flood, larger, more flood-resistant designs were chosen. For instance, culverts along the flood-damaged Old Fall River Road were replaced with ones large enough to withstand the higher flows expected with climate change.

Beyond extreme weather events, Rocky Mountain NP recognizes many other significant ecological responses to longer-term patterns of climate change, such as the continental impacts of the mountain pine beetle (*Dendroctonus ponderosae*) to forests, and the range expansion of exotic species within the park, as evidenced by the documentation of cheatgrass (*Bromus tectorum*) moving more than two thousand feet in elevation in just ten years.

To better understand these and many other impacts of climate change on the park (e.g., hydrology—earlier spring melt, earlier peak flow), we work with numerous partners, particularly universities, to cultivate research within the park, resulting in dozens of projects each year. For example, on the biodiversity front, Rocky Mountain NP has documented more than three thousand species, but we anticipate that there are thousands more yet to discover and that climate change may extirpate some of these new species before they are found. We have engaged on a pathway toward biodiversity discovery and understanding through sponsoring bioblitz events, supporting biodiversity research, and partnering with universities and nongovernmental organizations, such as the E.O. Wilson Foundation and Discover Life in America, to support graduate student engagement in the park.

Overstating the science is possible as it continues to be difficult to have causal evidence at the local level, due to the inherent complexities of ecological systems. This fact reaffirms that science is indeed a process. Therefore, integrating science into management outcomes continues to rely on a preponderance-of-evidence approach to affirming direct and indirect climate change impacts to the park. Engaging diverse audiences to build coalitions of support that can adapt to new information is the foundation for a way forward.

Management Challenges and Opportunities

Our referent for management is fundamentally shifting from our understanding of the historic range of variability to educated guesses of future scenarios and novel systems (Monahan and Fisichelli 2014). With this change comes real credibility issues in which the actions of the past that were informed by science can appear to some to be in conflict with what scientists are saying today. For example, after decades of restoration of greenback cutthroat trout (*Oncorhynchus clarkii*), recent genetic techniques are redefining the species. This realignment of what a "greenback" is confuses the general public (and scientists too, from time to time) on how to reinterpret past conservation actions. As it turns out, from the late 1800s until around 1920, more than 60 million trout were transplanted in the region for recreational purposes. These actions resulted in the conundrum today: now that we have (through genetics) identified the two remnant populations of greenback cutthroat trout, how do we proceed in conservation efforts that may affect existing populations of differing genetics, and how will these restoration efforts be affected by changing climate and the emergence of novel ecosystems? These efforts require *diplomacy* through science to bring staff and stakeholders along in this complicated scientific process in which social values are so strong.

Many resource professionals, particularly those in leadership, were educated in universities before climate change was integrated into curricula and discussions. There are and will continue to be relatively few financial support options for training or continuing education of resource professionals within the agency—online learning and webinars do not replace in kind the value of conferences or on-site field learning. The Continental Divide Research Learning Center at Rocky Mountain NP engages in a variety of information and education processes to transfer knowledge, build science literacy, and anticipate concerns or information gaps among staff and stakeholders. For example, the resource staff at Rocky Mountain NP participates in directed workshops on science topics, and in the summer they attend a weekly series focused on professional development activities. In addition, the Continental Divide Research Learning Center sponsors a scientific discourse at the park through the Rocky Mountain National Park Biennial Research Conference.

Most National Park Service leaders (e.g., management teams of parks or positions higher in the organization) do not have science or even resource management backgrounds. This is simply a numbers game in which at a given park those in resource stewardship are one division of five (20 percent), typically have less than 15 percent of the budget with smaller staff, and have even less of a presence in the leadership roles of superintendent, regional directorates, or among directors throughout National Park Service history. Most often, these leadership positions are filled by those emerging from the other park divisions: facilities management, visitor and resource protection, interpretation, or administration. As a result, the education curve can often be steep within the organization, as it is among the public.

We thus engage in numerous efforts to educate and make climate change information relevant. Opportunities to advance understanding include brown-bag lectures, workshops, and park science conferences intended to expose park staff to current science in the region. As part of the research permit process, presentations are given to the management team to explain the importance of new and large research projects in the park. The recent development of a frequently asked questions document has helped both our staff and the public better understand climate change (National Park Service 2014).

Historical accounts of knowledge accumulation suggest that knowledge of a discipline can double every ten years (Fuller 1981). The relatively recent development of computing power has, in some cases, resulted in knowledge doubling much quicker, perhaps as quickly as every one year or less (Schilling 2013). As a consequence of these advances, especially in the climate change arena, the lack of science synthesis significantly limits our success. Most science efforts support individual investigations within a limited footprint of time, space, and scope. What we find we need more of is the timely synthesis of existing science into working hypotheses to inform action. Consider that action will occur regardless, and consequently, being informed can make the difference. Efforts are evolving to engage in more timely synthesis efforts of both science and culture. We encourage partners to do the same. As stated by Thomas (1979): "The knowledge necessary to make a perfect analysis of the impacts of potential courses of . . . management action . . . does not exist. It probably never will. But more knowledge is available than has yet been brought to bear on this problem. To be useful, that knowledge must be organized so it makes sense. . . . To say we don't know enough is to take refuge behind a half-truth and ignore the fact that decisions will be made regardless of the amount of information available."

Given these challenges, finding a way forward necessarily requires a shift in practitioners' paradigms and "tool boxes." We are at a significant place in conservation history where the foundation of what science we do—and how we synthesize, interpret, and understand that science—demands new lenses in which to view it. Furthermore, how we communicate that science within the park, how we integrate it with our collective culture, and how we communicate with the public matters more than ever. It matters because we are at risk of losing credibility relatively easily by overinterpreting the science through linking cause and effect prematurely, by making decisions on the land that are shortsighted, by being "hands off" because it is perceived that there is nothing we can do and, at the same time, by "doing too much" where much of society feels compelled to do "something" in the face of this unprecedented change. Therefore, Rocky Mountain NP's biggest challenge for "responding to climate change" is the need to tell the "so what, and what can we really do" piece of the climate change story, and to tell that story in biologically and culturally meaningful ways in order to increase understanding and support from park visitors and partners.

Adaptive Management of Climate Change

Rocky Mountain NP has increasingly used the concept of adaptive management to guide decision making (e.g., National Park Service 2007b). An adaptive management framework provides for the process of science, the ability of an organization and community of partners to learn and develop future scenarios, and embraces the inherent uncertainty associated with stewardship of our resources. The concept has been around for decades and is affirmed as among the best strategies to manage for climate change (Lawler et al. 2010; National Park Service 2010).

At Rocky Mountain NP, addressing and adapting to the consequences of climate change has initially focused on three steps within the adaptive

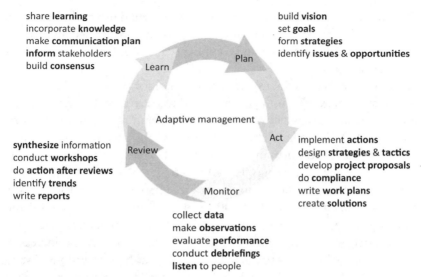

share **learning**
incorporate **knowledge**
make **communication plan**
inform stakeholders
build **consensus**

build **vision**
set **goals**
form **strategies**
identify **issues** & **opportunities**

Plan

Learn

Adaptive management

Act

synthesize information
conduct **workshops**
do **action after reviews**
identify **trends**
write **reports**

Review

implement **actions**
design **strategies** & **tactics**
develop **project proposals**
do **compliance**
write **work plans**
create **solutions**

Monitor

collect **data**
make **observations**
evaluate **performance**
conduct **debriefings**
listen to people

FIGURE 14-3   Working model of the adaptive management framework for Rocky Mountain National Park.

management framework: (1) research and monitoring, (2) synthesizing information, and (3) shared learning (fig. 14-3). First, we have worked to gather evidence and data to document that climate change is happening and is impacting ecological systems within the park. To this end, Rocky Mountain NP supported over 120 active research groups in 2014. Second, we have worked to review the evidence through synthesis reports and conducting workshops, such as the 2007 climate change workshop described earlier. Third, we have focused on transferring knowledge using a climate change story that resonates with people and works to build consensus, understanding, and support among park managers, the public, the local community, and stakeholders, as described with our frequently asked questions (National Park Service 2014).

Given the complexity of climate change phenomena, its novelty on the landscape, and the relative scarce history of specific research projects, the first priority has been to characterize climate change within the park and to record and understand the cascading effects of such change. Research and monitoring are central to making this happen, and Rocky Mountain NP has collaborated with federal agencies and research partners to establish and maintain eighteen weather stations within the park (Davey, Redmond, and Simeral 2007). Just outside of the park boundary is Grand Lake

(Northwest), one of the stations with the longest and most reliable climate records in Colorado. The park has a history of supporting such installations through the research permits and compliance process, funding, and day-to-day logistical support.

Data from these stations have been critical for providing the evidence of climate change. From these data, we have seen that Rocky Mountain NP experienced a significant increase in temperature during the last century (chap. 4). Precipitation has been more variable, and there has been no change in annual precipitation, but since the 1970s there has been a decline in April 1st snow water equivalent of 0.5 to 1.1 inches (1.2 to 2.7 centimeters) per decade at stations in and near the park (Clow 2010). The large elevation gradient in the park (7,630 to 14,259 feet, or 2,325 to 4,346 meters) and complex topography make it likely that different areas of the park are experiencing different rates of climate change. There is also increasing evidence that climate change may be amplified at high elevations (Wang, Fan, and Wang 2014). The park and researchers have recently recognized these complexities of climate monitoring and have begun the process of documenting topoclimate (i.e., climate at scales of 10 to 30 meters). In addition, efforts to expand our climate monitoring network to high elevations are ongoing.

Park planning also requires an understanding of how climates may continue to change in the future. The inherent uncertainty in emissions scenarios and climate models is embraced, but it is still useful to understand the most likely futures. For example, in Colorado, most climate models predict that warming will continue and that there will be a decline in snowpack, but there is no consensus on whether total precipitation will change (Lukas et al. 2014). There have been some efforts to derive downscaled climate models for Rocky Mountain NP, including the National Aeronautics and Space Administration Earth Exchange (NEX) Downscaled Climate Projections produced in support of the Landscape Climate Change Vulnerability Project (Thrasher et al. 2013). We have learned from these projections that subalpine limber pine (*Pinus flexilis*)—a species of management concern related to whitebark pine (chaps. 9 and 10)—is generally expected to move upslope into areas that are currently alpine and be less dominant at lower elevations in the park in response to increasing competition with shade-tolerant conifers (Monahan et al. 2013).

These insights are now being integrated into a conservation strategy (Schoettle et al., in press). In addition, the NEX Terrestrial Observation and Prediction System forecasts ecosystem processes important to park resources, such as the amount of soil water, gross primary productivity,

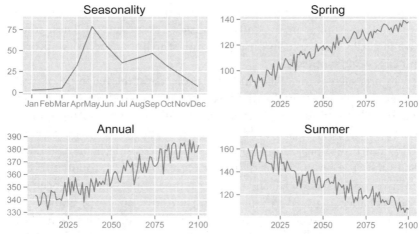

FIGURE 14-4   NASA Earth Exchange Terrestrial Observation and Prediction System forecasts of evapotranspiration for the Rocky Mountain National Park PACE. The predictions are based on an ensemble average of about thirty individual climate models, evaluated for a "business as usual" emissions scenario (RCP 8.5 W/m²). From these predictions, we see how evapotranspiration: (1) peaks in late spring (upper left plot); (2) increases dramatically in the spring (upper right); and (3) experiences a net annual increase (bottom left) despite decreasing in summer (bottom right).

runoff, and evapotranspiration (fig. 14-4). Insights from the Terrestrial Observation and Prediction System reaffirm that we are moving beyond the range of historic variability (Monahan and Fisichelli 2014), potentially into "no analog" futures (Williams, Jackson, and Kutzbach 2007). When and if this happens, we expect climatic changes and ecological responses to occur at unprecedented rates, and in unpredictable directions, potentially resulting in novel ecosystems.

Understanding the ecological consequences of climate change—and developing the climate change stories—has been less straightforward. Evidence from research studies across the western United States or other areas can provide useful hypotheses and predictions but cannot replace more local knowledge and research. Fortunately, researchers are attracted to work in Rocky Mountain NP because it provides a natural laboratory to study the effects of climate change where there is a striking elevational gradient, sensitive ecosystems, and a relative lack of confounding factors (e.g., invasive species, agriculture, development). Hundreds of researchers have been working in the park and assisting with documenting the effects of climate change on a diversity of topics, including hydrology, plant and animal species, wetlands, alpine tundra, and the spread of exotic species. These studies have generally been split into short-term research, long-term monitoring, and large-scale syntheses.

Efforts such as the "Pikas-in-Peril" project and the mapping of glacial change in the park exemplify the first category. From these short-term research studies, the park has learned that populations of the American pika (*Ochotona princeps*) appear stable in Rocky Mountain NP (Jeffress et al. 2013) but future projections to 2100 show decreasing habitat suitability (Schwalm et al., in press). Although glaciers in Rocky Mountain NP are sensitive to spring snowfall and summer temperatures, there has been only modest change in glacial area since the early 1900s (fig. 14-5), and this may be due to increasing cloud cover (Hoffman, Fountain, and Achuff 2007). While these short-term studies have increased our understanding of climate change in Rocky Mountain NP, and offer a basis for telling the climate change story, they often lack the wealth of evidence gained from long-term field studies.

Increasingly, the value of long-term studies at Rocky Mountain NP is being recognized as critical to understanding the effects of climate change, and, where possible, we are supporting these investments. Traditionally, park monies have been limited to funding one- to five-year projects, and researchers are often limited to a similar short-term funding cycle. With a mandate from the National Parks Omnibus Management Act of 1998 (P.L. 105-391), the National Park Service Inventory and Monitoring Division and the Rocky Mountain Inventory and Monitoring Network were established and funded to assist Rocky Mountain NP and other natural resource parks with this long-term monitoring need. The establishment of the Rocky Mountain Inventory and Monitoring Network is fairly recent, but they have developed operational monitoring protocols for alpine vegetation (Ashton et al. 2010), wetlands (Schweiger et al. 2015a), and streams (Schweiger et al. 2015b). They have also provided ongoing financial support to snowpack chemistry monitoring (Ingersoll et al. 2009).

Beyond the short-term research and long-term monitoring are syntheses of both that yield valuable case studies for telling the climate change and adaptive management story at Rocky Mountain NP. It is only in the exceptional cases that we have managed to move forward to planning and implementing management actions—water and air quality, forests, and exotics. Below, we highlight the exemplary case studies, with the hope that they can guide other such efforts in our own park and others.

FIGURE 14-5   Andrews Glacier in Rocky Mountain National Park, comparing 1916 (atop) and 2013 (bottom). Both photos were taken in August of their respective years. (Photos obtained from http://glaciers.us: (a) W. T. Lee, US Geological Survey Photographic Library, ID 322; (b) T. Fegel, ID 24881.)

Case Study 1: Water and Air Quality

More than three decades ago, Rocky Mountain NP had the foresight to support a study in the park to examine watershed-scale responses to atmospheric deposition and climate. In 1983, scientists instrumented the Loch Vale watershed, a 1,630-acre (660-hectare) alpine and subalpine catchment, with weather stations and stream gauges (Baron 1992). Since then, there has been a steady stream of research and monitoring of climate, snowpack, streamflow, water chemistry, atmospheric deposition, and glaciers. Loch Vale is now one of the most studied watersheds in the world, with over 150 research publications. Using such reliable climate records, it has been possible to link surface water chemistry trends with changes in summer temperature and to model the potential effects of continued warming (Baron et al. 2000a; Baron, Schmidt, and Hartman 2009; Heath and Baron 2014). These studies suggest that future warming will likely cause earlier snowmelt and peak discharge, increased forest productivity, and increased release of nitrogen from melting ice and rock glaciers (Baron et al. 2000a; Baron, Schmidt, and Hartman 2009).

Prior to documenting the climate change signal, Loch Vale data painted a vivid picture of how air pollution affected park resources, and the park was able to use this information to direct management. Results and comparisons to the larger park showed that soils, water, vegetation, and diatom communities were altered by increased nitrogen availability (Baron et al. 2000b; Wolfe, Van Gorp, and Baron 2003; Baron 2006). In 2004, the Colorado Department of Health and Environment, the National Park Service, and the US Environmental Protection Agency signed a memorandum of agreement to facilitate interagency response to the degradation of air quality and park resources. After three years, the group developed the Nitrogen Deposition Reduction Plan and adopted a target of 1.5 kilograms nitrogen per hectare per year by 2032, based on adverse effects seen at that level of deposition (National Park Service 2007b).

The Nitrogen Deposition Reduction Plan involves a range of strategies to improve air quality, including voluntary and mandatory measures to curb automobile and factory emissions and a series of agricultural best practices. The plan has largely succeeded at reducing nitrogen (until 2014, nitrogen deposition had been decreasing) and fostering a diverse regional collaboration (Morris et al. 2014). The park is currently developing ways to leverage the Loch Vale research to inform meaningful management action regarding climate change, and the research group has been instrumental in ongoing discussions, climate change workshops, and scenario-planning exercises that extend well beyond the park's boundary.

Case Study 2: Forests

Conifer forests blanket much of Rocky Mountain NP, and for decades researchers have been interested in studying patterns of forest composition and health across elevation, aspect, and level and history of disturbance (e.g., Peet 1978, 1981). At the present time, subalpine forests are generally found in the park at between 9,000 and 11,400 feet (2,743 and 3,475 meters) elevation and are dominated by Engelmann spruce (*Picea engelmannii*), subalpine fir (*Abies lasiocarpa*), and limber pine. Montane forests are found below this elevation, and ponderosa pine (*Pinus ponderosa*), quaking aspen (*Populus tremuloides*), lodgepole pine (*Pinus contorta*), and Douglas-fir (*Pseudotsuga menziesii* var. *menziesii*) dominate. Paleoecological and glacial studies have shown that the density and elevational distribution of tree species have tracked changes in climate since the last glaciation (Benedict et al. 2008; McWethy et al. 2010; Higuera, Briles, and Whitlock 2014).

Warmer temperatures have also contributed to more recent forest change in Rocky Mountain NP. While current fires are not outside the range of natural variation for these forests (Sibold, Veblen, and González 2006), they have been outside the range of experience for park staff and local communities. Most research suggests that this pattern of increased frequency and severity will continue as climate warms (e.g., Higuera, Briles, and Whitlock 2014). Climate change projections show that many forest species will likely migrate upslope to find suitable climate conditions (Stohlgren, Owen, and Lee 2000; Monahan et al. 2013). Studies within the park have already shown some indication of upslope movement (Monahan et al. 2013; Esser 2014).

Finally, like much of the western United States, Rocky Mountain NP has experienced forest insect epidemics. In the early 2000s, a mountain pine beetle outbreak started and has resulted in extensive mortality of lodgepole pine and limber pine (fig. 14-6; Raffa et al. 2008; Diskin et al. 2011). The timing and magnitude of the outbreak and subsequent tree mortality were influenced by warmer temperatures, drought-stressed trees, and a history of fire suppression (Raffa et al. 2008).

As a result of fires and mountain pine beetle–caused tree mortality, the forested landscape in Rocky Mountain NP has changed more in the past decade than in the last century. The altered landscape can be shocking to

FIGURE 14-6 In the past decade, forest landscapes have changed in Rocky Mountain National Park due to a large outbreak of mountain pine beetles. (Photo from the Rocky Mountain National Park digital photo library, date not available.)

visitors, and one of the biggest challenges has been to educate park staff and the public that change is not always equivalent to loss or destruction. Park researchers have helped broaden the discussion by documenting that this type of forest stand change is not unusual (e.g., Sibold, Veblen, and González 2006) and that, while the mortality is dramatic, there are many survivors (Rocca and Romme 2009).

To date, Rocky Mountain NP has not chosen to pursue active management of fire frequency (e.g., prescribed burns, thinning) or beetle epidemics (e.g., broadcast spraying, thinning). These decisions have been related to an already increased fire frequency in recent years and are consistent with other agencies' approaches to addressing the continental-scale beetle epidemic. Active management efforts have focused on reducing thousands of hazard trees at campsites, visitor centers, and roadways, and indirect firefighting tactics have resulted in increased acreages exposed to fire. The way forward includes continued education and interpretation of the dynamic nature of forests and how climate will continue to catalyze change, developing strategies to include fire on this fire-adapted landscape, and working with the public to adapt to new science, scenarios, and strategies.

Case Study 3: Exotics

Thanks to cool temperatures, short growing seasons, high elevations, and relatively low levels of past agricultural use and development, Rocky Mountain NP is less invaded by exotic plant species than many other national park units (Allen, Brown, and Stohlgren 2009). Unfortunately, a pattern of increasing exotic plant cover has emerged in the past decade at a time when visitation is reaching record highs and the effects of climate change are being documented.

Cheatgrass, an exotic annual grass, is of particular concern because of its seemingly increasing abundance, its spread to higher elevations, and the large impact it has on other ecosystems. Rocky Mountain NP has supported numerous research projects on exotics and was able to verify the assumption that cheatgrass cover has increased since the 1990s (Bromberg et al. 2011), but it is still mainly restricted to roads and disturbed areas (Banks and Baker 2011). Climate change is likely one of the driving factors for the increased distribution, and we anticipate this spread to continue over the next decades (West et al. 2015).

Rather than passively accepting this change, the park has begun exploring options for actively reducing the spread and—where possible—eradicating populations of cheatgrass. For instance, since cheatgrass is adept at invading after fires, Rocky Mountain NP has committed to prescribed burning only when funds and personnel are available for postfire cheatgrass control. A long-term experiment on chemical treatment showed that the agent imazapic may reduce cheatgrass cover within the park with no effect on native plants (Davis, Brown, and Esser 2015). Finally, the park is working on a revised exotic plant management plan that will strategically address this recent expansion to protect the park from increasing plant invasions.

A Way Forward

The complexity of the science, the challenges of integrating science into management, the difficulty of communicating climate change issues into stories of action—any of these could leave a person or community paralyzed. The sense of urgency seems greater than ever, and the tools in a practitioner's "tool box" as limited as ever. What success looks like often appears less clear. And yet, there is a way forward.

As in the past, the way forward is through persistent investment in science and people to build understanding and coalitions of support for action, or inaction, as the case may be. Practically speaking, the way forward at Rocky Mountain NP has been to invest in long-term relationships with a diversity of partners to assist us in finding ways to integrate science into management sooner rather than later. Consider that the time from the inception of science to the time it is applied can be decades. How can we make science timelier in order to inform and—when appropriate—to drive management considerations and outcomes? We have been exploring the following approaches over the past decade in anticipation of climate change and how to organizationally adapt to it:

1. *Opening our minds and expanding our horizons*: We often work within a footprint that is small relative to the conservation of a species or system, and experience suggests that we often consider baselines of a hundred years, or less, to understand the past or the future. These filters can limit understanding and resolution of the issue at hand, lead us to invest in a potentially less successful solution, ignore long-term trajectories as influenced by postglaciation climate or long-term environmental histories, and limit developing plausible future scenarios. Through practice, it has become clear that applying context competencies (i.e., those competencies that add relevancy to an issue, such as scales of time and space, paradigms, and cultures) to conservation issues provides for more sustainable outcomes than if we are left only to our traditions of content competencies (i.e., those competencies that increase our knowledge about a particular topic, such as wildlife biology, botany, geology, and geographic information systems). We continue to explore better ways to understand issues and to develop solutions.

2. *Communication, communication, communication*: It has long been known that communication is the key to conservation success, be it with individuals, communities, or international partners. Rocky Mountain NP has supported numerous research and synthesis efforts to make the vast amount of research done in the park more accessible to managers and other researchers. One such example is the biennial science conference, fostering the sharing of timely information among scientists, practitioners, park staff, and the public. This opportunity provides for interaction among all interested parties, free of charge, and in open forums for dialogue, debate, and inspiration. Other strategic communication examples include a climate

change workshop document (National Park Service 2007a), the State of the Alpine Report (Ashton et al. 2013), the Natural Resource Condition Assessment (Theobald et al. 2010), and the Vital Signs Summary report (Franke et al. 2015).

Some of these research projects have been in-house synthesis efforts. In other cases, Rocky Mountain NP has worked with academic and partner agencies to develop them and has asked researchers to present and train management staff at workshops where current knowledge is synthesized and discussed. We have also found it helpful to produce strategic communication products, such as short videos, seminars, and the Rocky Mountain NP climate change frequently asked questions brochure (National Park Service 2014). These efforts are not simply informational but have a purpose, a message, and anticipated audiences.

At the end of the day, it is clear that we need to be better storytellers. Those trained in the sciences aren't generally known for having that skill. Thus, we need to not only enhance our storytelling abilities but also partner with those who tell stories well: our interpreters and educators, our public information officers, and those in our communities who excel at capturing the attention of others. In doing this, we need to help as many as possible (e.g., community leaders, elected officials, and young scientists) understand climate change and enable each to be able to share their own stories.

3. *A business case*: Although likely far from the interests of what sparked a conservationist to engage in this profession, business concepts have come to dominate contemporary paradigms of conservation and land management agencies in the United States. Consider that language can matter and that reframing our actions as investments, expanding expectations of return on investment to be long term with societal relevance, and leveraging investments toward greater outcomes are all simple business concepts that can translate into financial and institutional support. Business concepts meet society halfway in what will be a long-term dialogue about conservation and financing a way forward.

4. *Understanding that the landscape we need to work in is dynamic*: The footprint of where we need to work and the communities of people we need to collaborate with change as issues emerge. The boundary of Rocky Mountain NP might be considered our core area of protection, but if we want to address conservation issues, we also need to collaborate outside our boundaries (fig. 14-7).

5. *Adaptive management and doing the best with what we have*: Engagement cannot wait for the perfect study or full funding; financial resources

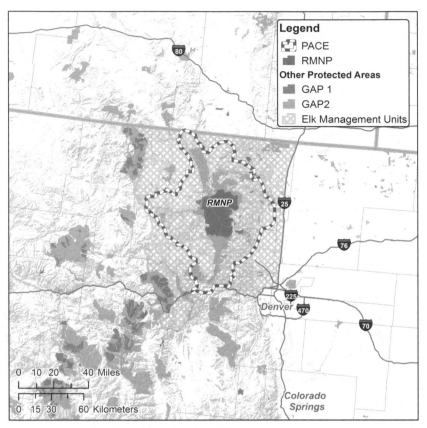

FIGURE 14-7   Map demonstrating that different ecological issues relevant to the protection of Rocky Mountain National Park (RMNP) can have varied footprints, reinforcing the need to work beyond the boundaries of a park to ensure success. The protected area centered ecosystem (PACE) encompassing RMNP is one such footprint, along with the elk management units from Colorado Parks and Wildlife. RMNP is surrounded by a number of other important and permanently protected areas (GAP 1 and 2 protected areas).

will always be scarce relative to the magnitude of the issues we address, and the perfect study is often elusive. Instead, investment in partnerships, in long-term relationships with people and institutions, will allow resource stewardship to be resilient in the face of uncertainty. Certainly, the way forward is not alone. Nor is it in a vacuum. Integrating information into an adaptive framework of management (fig. 14-3) can provide the discipline necessary to continually acquire, integrate, learn, and adapt to new information that others have described and envisioned (e.g., Holling 1978; Walters 1986; Lee 1993).

Conclusion

The opportunity to share thoughts relevant to how Rocky Mountain National Park has been responding to climate change through science is Rocky Mountain NP's story and by no means was it shared to suggest that we have all the right answers. We are continually learning from our diverse park staff, partners, and others about better ways to do business, tell our stories, and find more effective ways of conserving resources. That said, we do hope that this opportunity to share ideas, working hypotheses, and experiences stimulates discussion and debate, and helps in our collective efforts.

We have all made mistakes, and we will do so again in the future. Have you ever looked at an action on the landscape done a generation ago and wondered "What in the world were they thinking to do *that*? It makes no sense!" Of course the action made sense to someone at the time and may even have reflected the dominant paradigm of the day, as informed by science. So, too, will people ask that same question a generation from now. It is a sign of growth and the ever-evolving understanding of nature, through science and culture. However, let's work smartly together to ensure that when the question is asked a generation from now it is followed up with the statement, "Oh, that makes sense!"

What we often fail to understand as "ologists" is that land management decisions are rarely driven by science; rather, human values drive management decisions—always have, always will. This will continue until we as a society begin to value science as fundamental to how we make decisions.

References

Allen, J. A., C. S. Brown, and T. J. Stohlgren. 2009. Non-native plant invasions of United States national parks. *Biological Invasions* 11:2195–2207.

Ashton, I., E. W. Schweiger, J. Burke, D. Shorrock, D. Pillmore, and M. Britten. 2010. *Alpine Vegetation Composition Structure and Soils Monitoring Protocol: 2010 Version.* Natural Resource Report NPS/ROMN/NRR—2010/277. Fort Collins, CO: National Park Service, Natural Resource Program Center.

Ashton, I., J. Visty, E. W. Schweiger, and J. R. Janke. 2013. *State of the Alpine Report for Rocky Mountain National Park: 2010 Summary Report.* Natural Resource

Data Series. NPS/ROMN/NRDS—2013/535. Fort Collins, CO: National Park Service.

Banks, E. R., and W. L. Baker. 2011. Scale and pattern of cheatgrass (*Bromus tectorum*) invasion in Rocky Mountain National Park. *Natural Areas Journal* 31:377–90.

Baron, J. S., ed. 1992. *Biogeochemistry of a Subalpine Ecosystem: Loch Vale Watershed*. New York: Springer-Verlag.

Baron, J. S. 2006. Hindcasting nitrogen deposition to determine an ecological critical load. *Ecological Applications* 16:433–39.

Baron, J. S., M. D. Hartman, L. E. Band, and R. B. Lammers. 2000a. Sensitivity of a high-elevation Rocky Mountain watershed to altered climate and CO_2. *Water Resources Research* 36:89–99.

Baron, J. S., H. M. Rueth, A. M. Wolfe, K. R. Nydick, E. J. Allstott, J. T. Minear, and B. Moraska. 2000b. Ecosystem responses to nitrogen deposition in the Colorado Front Range. *Ecosystems* 3:352–68.

Baron, J. S., T. M. Schmidt, and M. D. Hartman. 2009. Climate-induced changes in high elevation stream nitrate dynamics. *Global Change Biology* 15:1777–89.

Benedict, J. B., R. J. Benedict, C. M. Lee, and D. M. Staley. 2008. Spruce trees from a melting ice patch: Evidence for Holocene climatic change in the Colorado Rocky Mountains, USA. *The Holocene* 18:1067–76.

Bobowski, B. 2013. Research connections . . . a story. *Rocky Mountain Nature Association Quarterly*, 10–11. Rocky Mountain Nature Association.

Bromberg, J. E., S. Kumar, C. S. Brown, and T. J. Stohlgren. 2011. Distributional changes and range predictions of downy brome (*Bromus tectorum*) in Rocky Mountain National Park. *Invasive Plant Science and Management* 4:173–82.

Clow, D. W. 2010. Changes in the timing of snowmelt and streamflow in Colorado: A response to recent warming. *Journal of Climate* 23:2293–2306.

Davey, C. A., K. T. Redmond, and D. B. Simeral. 2007. *Weather and Climate Inventory, National Park Service, Rocky Mountain Network*. Natural Resource Technical Report NPS/ROMN/NRTR—2007/036. Fort Collins, CO: National Park Service.

Davis, C., C. S. Brown, and S. M. Esser. 2015. Managing cheatgrass with imazapic in Rocky Mountain National Park: Lessons learned from a six year study. In *Rocky Mountain National Park Research Conference Proceedings*. Estes Park, CO. http://www.nps.gov/rlc/continentaldivide/upload/CDRLC_Proceedings_2015-4.pdf.

Diskin, M., M. E. Rocca, K. N. Nelson, C. F. Aoki, and W. Romme. 2011. Forest developmental trajectories in mountain pine beetle disturbed forests of Rocky Mountain National Park, Colorado. *Canadian Journal of Forest Research* 41:782–92.

Esser, S. M. 2014. *Topography, Disturbance and Climate: Subalpine Forest Change 1972–2014, Rocky Mountain National Park, USA*. Fort Collins, CO: Colorado State University.

Franke, M. A., T. L. Johnson, I. W. Ashton, and B. Bobowski. 2015. *Natural Resource Vital Signs at Rocky Mountain National Park*. Natural Resource Report. NPS/ROMO/NRR—2015/946. Fort Collins, CO: Natural Park Service.

Fuller, B. 1981. *Critical Path*. New York: St. Martin's Press.

Heath, J., and J. S. Baron. 2014. Climate, not atmospheric deposition, drives the biogeochemical mass-balance of a mountain watershed. *Aquatic Geochemistry* 20:167–81.

Higuera, P. E., C. E. Briles, and C. Whitlock. 2014. Fire-regime complacency and sensitivity to centennial-through millennial-scale climate change in Rocky Mountain subalpine forests, Colorado, USA. *Journal of Ecology* 102:1429–41.

Hoffman, M. J., A. G. Fountain, and J. M. Achuff. 2007. 20th-century variations in area of cirque glaciers and glacierets, Rocky Mountain National Park, Rocky Mountains, Colorado, USA. *Annals of Glaciology* 46:349–54.

Holling, C. S., ed. 1978. *Adaptive Environmental Assessment and Management*. New York: John Wiley & Sons.

Ingersoll, G. P., D. Campbell, A. Mast, D. W. Clow, L. Nanus, and B. Frakes. 2009. Snowpack Chemistry Monitoring Protocol for the Rocky Mountain Network: Narrative and Standard Operating Procedures. US Geological Survey Administrative Report. Fort Collins, CO.

Jeffress, M. R., T. J. Rodhouse, C. Ray, S. Wolff, and C. W. Epps. 2013. The idiosyncrasies of place: Geographic variation in the climate–distribution relationships of the American pika. *Ecological Applications* 23:864–78.

Lawler, J. J., T. H. Tear, C. Pyke, M. R. Shaw, P. Gonzalez, P. Kareiva, L. Hansen, L. Hannah, K. Klausmeyer, A. Aldous, C. Bienz, and S. Pearsall. 2010. Resource management in a changing and uncertain climate. *Frontiers in Ecology and the Environment* 8:35–43.

Lee, K. N. 1993. Compass and gyroscope: Integrating science and politcs for the environment. Washington, DC: Island Press.

Lukas, J. J., J. J. Barsugli, N. Doesken, I. Rangwala, and K. Wolter. 2014. Climate change in Colorado: A synthesis to support water resources management and adaptation. Boulder: Western Water Assessment, Cooperative Institute for Research in Environmental Sciences, University of Colorado Boulder.

McWethy, D., S. T. Gray, P. E. Higuera, J. S. Littell, G. T. Pederson, A. J. Ray, and C. Whitlock. 2010. *Climate and Terrestrial Ecosystem Change in the U.S. Rocky Mountains and Upper Columbia Basin: Historical and Future Perspectives for Natural Resource Management*. Natural Resource Report NPS/GRYN/NRR—2010/260. Fort Collins, CO: National Park Service.

Monahan, W. B., T. Cook, F. Melton, J. Connor, and B. Bobowski. 2013. Forecasting distributional responses of limber pine to climate change at management-relevant scales in Rocky Mountain National Park. *PLOS ONE* 8: e83163.

Monahan, W. B., and N. A. Fisichelli. 2014. Climate exposure of US national parks in a new era of change. *PLOS ONE* 9:e101302.

Morris, K., A. Mast, D. W. Clow, G. Wetherbee, J. S. Baron, C. Taipale, T. Blett, D. Gay, and J. Heath. 2014. *2012 Monitoring and Tracking Wet Deposition at Rocky Mountain National Park*. Natural Resource Report NPS/NRSS/ARD/NRR-2014/757. Denver, CO: National Park Service.

National Park Service. 2007a. *Climate Change in Rocky Mountain National Park: Preservation in the Face of Uncertainty*. Estes Park, CO: National Park Service.

National Park Serivce. 2007b. Rocky Mountain National Park Nitrogen Deposition Reduction Plan: Memorandum of Understanding. https://www.colorado.gov/pacific/sites/default/files/AP_PO_Nitrogen-Deposition-Reduction-Plan-NDRP.pdf.

National Park Service. 2010. *National Park Service Climate Change Response Strategy*. Fort Collins, CO: National Park Service, Climate Change Response Program.

National Park Service. 2014. *Climate Change in Rocky Mountain National Park: Frequently Asked Questions*. Estes Park, CO: Rocky Mountain National Park. http://www.nps.gov/romo/learn/nature/upload/Climate_Change_RMNP_FAQ.pdf.

Peet, R. K. 1978. Forest vegetation of the Colorado Front Range: Patterns of species diversity. *Vegetatio* 37:65–78.

Peet, R. 1981. Forest vegetation of the Colorado Front Range. *Vegetatio* 45:3–75.

Raffa, K. F., B. H. Aukema, B. J. Bentz, A. L. Carroll, J. A. Hicke, M. G. Turner, and W. H. Romme. 2008. Cross-scale drivers of natural disturbances prone to anthropogenic amplification: Dynamics of biome-wide bark beetle eruptions. *BioScience* 58:501–17.

Rocca, M. E., and W. H. Romme. 2009. Beetle-infested forests are not "destroyed." *Frontiers in Ecology and the Environment* 7:71–72.

Schilling, D. R. 2013. Knowledge doubling every 12 months, soon to be every 12 hours. Industry Tap, April 19. http://www.industrytap.com/knowledge-doubling-every-12-months-soon-to-be-every-12-hours/3950.

Schoettle, A. W., C. M. Cleaver, K. S. Burns, and J. J. Connor. In press. *Limber Pine Conservation Strategy for the Greater Rocky Mountain National Park Area*. Gen. Tech. Rep. RMRS-GTR-XXX. Fort Collins, CO: US Department of Agriculture, Forest Service, Rocky Mountain Research Station.

Schwalm, D., C. W. Epps, T. J. Rodhouse, W. B. Monahan, J. A. Castillo, C. Ray, and M. R. Jeffress. In press. Habitat availability and gene flow influence diverging local population trajectories under scenarios of climate change: A place-based approach. *Global Change Biology*.

Schweiger, E. W., E. Gage, K. M. Haynes, D. Cooper, L. O'Gan, and M. Britten. 2015a. *Rocky Mountain Network Wetland Ecological Integrity Monitoring Protocol: Narrative, Version 1.0*. Natural Resource Report NPS/ROMN/NRR—2015/991. Fort Collins, CO: National Park Service.

Schweiger, E. W., L. O'Gan, E. Borgman, and M. Britten. 2015b. *Rocky Mountain Network Stream Ecological Integrity Monitoring Protocol: Narrative, Version 1.0*. Natural Resource Report NPS/ROMN/NRR—2015/1011. Fort Collins, CO: National Park Service.

Sibold, J. S., T. T. Veblen, and M. E. González. 2006. Spatial and temporal variation in historic fire regimes in subalpine forests across the Colorado Front Range in Rocky Mountain National Park, Colorado, USA. *Journal of Biogeography* 33:631–47.

Stohlgren, T. J., A. J. Owen, and M. Lee. 2000. Monitoring shifts in plant diversity in response to climate change: A method for landscapes. *Biodiversity and Conservation* 9:65–86.

Theobald, D. M., J. S. Baron, P. Newman, B. Noon, J. B. Norman III, I. Leinwand, S. E. Linn, R. Sherer, K. E. Williams, and M. Hartman. 2010. *A Natural Resource Condition Assessment for Rocky Mountain National Park*. Natural Resource Report. NPS/NRPC/WRD/NRR—2010/228. Fort Collins, CO: National Forest Service, Natural Resource Program Center.

Thomas, J. W. 1979. Preface. In *Wildlife Habitats in Managed Forests—the Blue Mountains of Oregon*, ed. J. W. Thomas, 6–7. Department of Agriculture Handbook No. 553. Washington, DC: US Government Printing Office.

Thrasher, B., J. Xiong, W. Wang, F. Melton, A. Michaelis, and R. Nemani. 2013. Downscaled climate projections suitable for resource management. *Eos, Transactions American Geophysical Union* 94:321–23.

Walters, C. 1986. *Adaptive Management of Renewable Resources*. New York: Macmillan.

Wang, Q., X. Fan, and M. Wang. 2014. Recent warming amplification over high elevation regions across the globe. *Climate Dynamics* 43:87–101.

West, A. M., S. Kumar, T. Wakie, C. S. Brown, T. J. Stohlgren, M. Laituri, and J. Bromberg. 2015. Using high-resolution future climate scenarios to forecast *Bromus tectorum* invasion in Rocky Mountain National Park. *PLOS ONE* 10: e0117893.

Williams, J. W., S. T. Jackson, and J. E. Kutzbach. 2007. Projected distributions of novel and disappearing climates by 2100 AD. *Proceedings of the National Academy of Sciences of the United States of America* 104:5738–42.

Wolfe, A. P., A. C. Van Gorp, and J. S. Baron. 2003. Recent ecological and biogeochemical changes in alpine lakes of Rocky Mountain National Park (Colorado, USA): A response to anthropogenic nitrogen deposition. *Geobiology* 1:153–68.

Chapter 15

Case Study: Whitebark Pine in the Greater Yellowstone Ecosystem

Karl Buermeyer, Daniel Reinhart, and Kristin Legg

Whitebark pine (*Pinus albicaulis*), an iconic tree species generally associated with upper subalpine ecosystems, provides an excellent case study for studying the potential impacts of climate change on a species at the landscape level and how it affects the conservation of that species. Whitebark pine is considered a keystone species in that it dominates areas where other tree species grow poorly or not at all and has broad effects on ecosystem processes. Whitebark pine canopies help regulate snowmelt, extending the length of spring runoff and reducing erosion (Tomback et al. 2001; Farnes 1990). Its large, calorie-rich seeds are a valuable food source for a variety of wildlife species, including grizzly bears (*Ursus arctos horribilis*), which obtain the seeds almost exclusively by raiding red squirrel (*Tamiasciurus hudsonicus*) middens (Reinhart and Mattson 1990; Mattson, Tomback, and Reinhart 2001). Upon establishment on high-elevation slopes and other harsh sites, whitebark pine provides favorable microsites for the growth of other plant species, thus increasing ecosystem biodiversity (Keane et al. 2012).

Whitebark pine abundance has declined throughout its range in recent years due to multiple biotic and abiotic factors (Tomback, Arno, and Keane 2001; Schwandt 2006). Potential impacts to whitebark pine communities associated with climate change include changes in fire dynamics, changes in suitability of habitat for whitebark pine and competing tree species, and changes in temperature and moisture regimes as they affect insect and disease pests, particularly the native mountain pine beetle (*Dendroctonus ponderosae*) and the occurrence and distribution of white pine blister rust (Koteen 1999; Schwandt 2006; US Fish and Wildlife Service 2011). This chapter will focus on the ecology of whitebark pine in the Greater Yellowstone Ecosystem (GYE), current and potential future climate-related impacts, and how federal land managers are addressing these challenges.

Ecology of Whitebark Pine in the Greater Yellowstone Ecosystem

The Greater Yellowstone Ecosystem is a relatively intact ecosystem with Yellowstone National Park at its center (fig. 15-1). The GYE encompasses approximately 23 million acres, with vegetative communities ranging from sagebrush-steppe and riparian communities at lower elevations around 4,000 feet (1,200 meters) to alpine and treeline communities at elevations approaching 14,000 feet (4,300 meters). In the GYE, whitebark pine is represented within approximately 10 percent of the ecosystem (fig.15-1), based on where whitebark pine is dominant in the overstory. There are stands where individual trees, small clumps, and understory components occupy a much wider range, including those within open sagebrush and aspen communities below 7,000 feet (2,100 meters) in elevation that are not represented in currently available maps.

Whitebark pine is a relatively long-lived (from four hundred to more than one thousand years), slow-growing conifer species (Keane and Parsons 2010). Its dominance in mixed-species forest stands is typically limited to higher-elevation, harsher sites primarily by competition from other, faster-growing conifer species (Mattson and Reinhart 1990; Tomback 2001). Whitebark pine cone crops vary by year; heavy cone crops are produced approximately every three to five years (McCaughey and Tomback 2001). The mode of seed dispersal and regeneration is almost entirely through a mutualistic relationship with Clark's nutcracker (*Nucifraga columbiana*), a bird that removes the wingless seeds from indehiscent cones that do not open to release seeds upon maturity (Tomback 2001). Whitebark pine trees produce cones at the end of branches, which makes it easy for nutcrackers to detect and collect seeds. Nutcrackers cache the seeds in the ground across the landscape up to 6 to 39 miles (10 to 62 kilometers) away from the source for future retrieval (Lorenz et al. 2011; Tomback 2001). Seeds that are not retrieved may germinate and grow if conditions are suitable. Because of this adaption, and for maximum cone production for regenera-

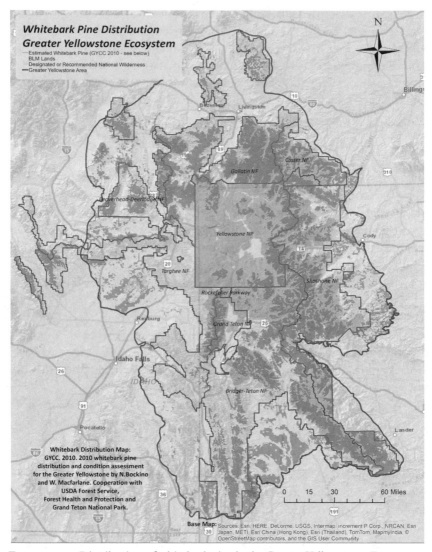

FIGURE 15-1 Distribution of whitebark pine in the Greater Yellowstone Ecosystem.

tion and wildlife food, it is ideal for whitebark pines to grow in open conditions, having large crowns with light on all sides.

Whitebark pine has an adaptive capacity to survive in many different habitats and has survived on the landscape through wide climatic variations going back to the last ice age (Iglesias, Krause, and Whitlock 2015). It is considered moderately shade tolerant; seedlings and saplings can persist in the forest understory from one to three hundred years, but their growth and reproductive maturity can be severely restricted by lack of sunlight (Arno and Hoff 1989). Information is limited on their ability to respond to release and to mature subsequent to the removal of the overstory.

On moister, wind-sheltered sites, whitebark pine may become established in the wake of fire or some other stand-replacing disturbance. However, its long-term growth and dominance on these sites is limited by competition from other conifer species, including lodgepole pine (*Pinus contorta*), Engelmann spruce (*Picea engelmannii*), and particularly subalpine fir (*Abies lasiocarpa*) (Keane and Parsons 2010). With advance forest succession, whitebark pine cover and abundance tend to diminish over time, and whitebark pine reduces its cone production capacity under dense subalpine fir canopies (Mattson and Reinhart 1990; Keane 2001). At higher elevations and at more exposed locations, whitebark pine will outcompete its competitors as a result of its wind and drought tolerance (Arno and Hoff 1989; Mattson and Reinhart 1990). Open sites also favor the development of whitebark pine at lower elevations, from 6,400 to 6,900 feet (1,950 to 2,103 meters), such as in northeastern Nevada (Critchfield and Allenbaugh 1969) and other locations where whitebark pine occurs (Weaver 2001), possibly due to lack of competition from other tree species.

Threats to Whitebark Pine—Past, Present, and Future

Three primary factors have caused the numbers and stand dominance of whitebark pine to decline across its range over the past fifty to sixty years: fire management and forest succession, white pine blister rust, and mountain pine beetle (Kendall and Keane 2001; USFWS 2011). Because whitebark pine is a seral species on more favorable subalpine environments, historical fire exclusion by humans of wildfire has removed an important disturbance agent required by whitebark pine to become established and gain competitive advantage over faster-growing but less fire tolerant species, such as subalpine fir (Keane 2001; Keane et al., in press). The introduction of the fungus *Cronartium ribicola*, the causal agent for white pine blister rust, to North America around 1910 has resulted in up to 90 percent mortality of whitebark pine in many stands throughout its range, particularly in more mesic areas, such as the Cascade Mountains, northern Idaho, and western Montana (McDonald and Hoff 2011). In the GYE, recent monitoring found blister rust infection rates in whitebark pine to be between 20 and 30 percent (Shanahan et al. 2014). Finally, mountain pine beetle, a native species that can kill stands of pine trees, including whitebark pine, has undergone past outbreaks, including a particularly severe outbreak beginning in the early 2000s.

These causal agents of decline have occurred in past decades in the GYE and throughout the range of whitebark pine (Keane et al. 2012). However, recent climate conditions, in varying degrees and spatial occurrences, have influenced whitebark pine mortality and survivorship. This was evidenced recently with record drought conditions in the GYE causing high mortality as a result of epidemic mountain pine beetle infestations (Macfarlane, Logan, and Kern 2010; Buotte 2015).

Future climate scenarios have the potential to further affect whitebark pine distribution as conditions for these agents change (Chang, Hansen, and Piekielek 2014; Keane et al., in press). In the past, severe cold events (successive days of temperatures below −30 degrees F) have generally prevented mountain pine beetle populations from building to outbreak levels in high-elevation environments. However, warmer temperatures since 2000 have allowed beetle populations to soar, causing significant mortality in the larger trees, with a 90 to 100 percent loss of the whitebark pine overstory in some stands (Macfarlane, Logan, and Kern 2010). Altered fire regimes, increased competition from other tree species, white pine blister rust, and increased frequency of more severe mountain pine beetle outbreaks will separately or in combination negatively affect whitebark pine in the GYE. Primary climate-related impacts to whitebark pine in the GYE would likely involve a wide and dynamic myriad of interacting factors as described above and would be evidenced within varying microsites across the ecosystem.

Current climate change predictions demonstrate a decline of whitebark pine (Chang, Hansen, and Piekielek 2014). Two overarching factors make impacts to whitebark pine from long-term changes in climatic conditions difficult to predict. First, while temperatures are predicted to increase, resulting precipitation patterns are more uncertain. Precipitation in the GYE is predicted to increase over time, but with a preponderance of this increase coming in the winter, followed by more arid summers (Chang, Hansen, and Piekielek 2014). Past and projected future climatic conditions in the GYE are described in detail in chapters 4 and 7. While warmer, drier climate conditions might increase whitebark pine mortality from fire occurrence and mountain pine beetle attacks, it is also possible that drier conditions could benefit some habitats of whitebark pine by increasing its competitive advantage over such competitors as subalpine fir as a result of its high drought tolerance (Arno and Hoff 1989; Keane et al., in press).

In many cases, seed caching sites selected by Clark's nutcrackers give the establishment of whitebark pine an advantage over species with wind-dispersed seeds, in that whitebark pine seeds tend to be placed near stumps and other landmarks, which can provide an ideal microsite, as opposed to seeds that are randomly distributed over the surface. This dispersal mechanism also gives whitebark pine an advantage over wind-dispersed species because of the distances that nutcrackers can fly to caching sites (Tomback, Arno, and Keane 2001). However, another disadvantage of warmer, drier climate would be a decline in regeneration (seed germination and survival of seedlings) for all tree species. Seedlings are much more susceptible to heat and drought since their root systems are limited to upper soil layers, which dry more quickly, making them more sensitive to environmental changes than mature trees (Monleon and Lintz 2015).

Understanding the relationship between temperature and precipitation is important in understanding how whitebark pine trees and other vegetation may be affected by climate change. In temperature-limited environments, such as higher elevations in the GYE, as temperatures increase so does evapotranspiration, the amount of water taken up by plants. This can lead to soil moisture deficits that lead to less water available for plants if there is not enough precipitation during the year, especially during the growing season, to maintain soil moisture (Thoma, Rodman, and Tercek 2015).

The second factor that complicates the prediction of climate change effects is that changes in conditions may have opposing effects on causal agents. For example, warmer temperatures may exacerbate mountain pine beetle pressure by making higher elevations more habitable to beetle infestations, favoring more frequent outbreaks, but alternatively may increase drought stress on stand tree competitors, thereby reducing competition stresses. Impacts on fire regimes can have varying effects on whitebark pine; low- and mixed-intensity fires tend to favor whitebark pine by removing competitors that are more fire susceptible and creating openings that are favorable to regeneration (Campbell et al. 2011; Keane et al. 2012). However, severe fires will also kill mature, seed-producing stands as well as young trees.

The long-term predictions for wildfire in the GYE is for more frequent large, severe fires, which could inhibit some forest species from regenerating (Westerling et al. 2011). An advantage that whitebark pine has over other conifer species following large stand-replacing fires is seed establishment by Clark's nutcrackers dispersing seeds over a far greater distance than the wind-dispersed seeds of competitors (Tomback 2001). This assumes

TABLE 15-1. Matrix of potential climate effects.

Scenario	Fire	Competition	Mountain Pine Beetle	White Pine Blister Rust (WPBR)
Warmer/ drier	↑reduced interspecies competition ↓ stand-replacing fire kills whitebark pine trees	↑ drier conditions could favor whitebark pine trees over some competitors ↔ regeneration of all species reduced	↓increase in outbreaks	↑could reduce spread of WPBR ↔ could increase WPBR hosts and potential infection of trees
Warmer/ wetter	↔depends on temporal distribution of precipitation	↓ competing tree species favored	↓ increase in outbreaks	↓could increase spread of WPBR

Note: Up arrows indicate a potential benefit to whitebark pine; down arrows indicate a potential detriment.

that adequate seed sources exist and that the numbers of seeds cached exceed those that will be retrieved and consumed (Tomback 2001; Keane et al., in press). Because of whitebark pine's adaptive capacity to a wide range of environmental conditions, understanding the interaction of whitebark pine ecology, regeneration mechanisms, and environmental variables is critical to predicting the future of whitebark pine in the GYE and how managers will respond to this future.

Table 15-1 summarizes potential impacts of climate change on various factors affecting whitebark pine survival for two general climate scenarios: warmer and drier, and warmer and wetter. Up arrows indicate a potential benefit to whitebark pine; down arrows indicate a potential detriment.

Climate envelope models predict that suitable habitat for whitebark pine could be reduced by up to 95 percent in the GYE over the next century (Chang, Hansen, and Piekielek 2014). Another modeling effort found that whitebark pine is more likely to die with increases in winter and average fall temperatures that promote mountain pine beetle survival (Buotte et al., in review). While each model has limitations, such as not accounting for other factors that may affect whitebark pine, including fire, competition, blister rust, and ongoing management, they do provide a foundation

FIGURE 15-2   Whitebark pine (left side of photo) in association with Douglas-fir, aspen, and sagebrush at 6,900 feet elevation, Buffalo Valley, northwest Wyoming. (Photo from US Forest Service.)

to address whitebark pine conservation and management into the future (Pearson and Dawson 2003). Recognizing the importance of conserving whitebark pine in the GYE, the 2011 whitebark pine management strategy (Greater Yellowstone Coordinating Committee 2011) identified the need to understand the impacts of climate change on whitebark pine. The above studies have helped managers to understand and discuss strategies to adapt management efforts now and into the future.

Within the GYE, local managers and technicians have noted the presence of whitebark pine beyond the dominant overstory stands depicted in formal forest mapping methods. We observed white bark pine at lower elevations in association with such stands as quaking aspen (*Populus tremuloides*), Rocky Mountain juniper (*Juniperus scopulorum*), and Douglas-fir (*Pseudotsuga menziesisii* var. *glauca*). We have also observed whitebark pine as a prevalent understory component of mid-elevation lodgepole pine stands in the GYE. These observations of whitebark pine within other habitats demonstrate the adaptive capacity of whitebark pine to disperse and compete beyond what has been depicted in existing whitebark pine distribution maps. Localized site conditions, stand structures relatively free of competition (fig. 15-2), and potential genetic tree characteristics on low-elevation and mid-elevation sites that are currently mapped as "unsuitable" for whitebark pine in species distribution models (chap. 10) could provide valuable information on where and how to manage and restore whitebark pine as suitable habitat changes in light of potential climate scenarios.

Whitebark Pine Management in the Greater Yellowstone Ecosystem

Four main components of managing whitebark pine in the GYE (and elsewhere throughout its range) have been established (Greater Yellowstone Coordination Committee 2011; Keane et al. 2012):

- **Monitoring.** Tracking the current and trending condition of whitebark pine throughout its range and conducting effective monitoring on restoration activities. Monitoring results are used to guide protection and restoration activities.
- **Protection.** Preventing or minimizing damage to existing trees and stands from insects, disease, fire, and so forth.
- **Restoration.** Replanting or creating conditions that favor natural regeneration and dominance of whitebark pine on the landscape.
- **Tree Improvement.** Identifying and propagating genotypes that have resistance or tolerance to such adverse factors as drought and white pine blister rust.

Whitebark pine management in the GYE is coordinated largely by the Greater Yellowstone Coordinating Committee (GYCC) Whitebark Pine Subcommittee, which consists of representatives from five national forests, two national parks, the US Bureau of Land Management, and the Greater Yellowstone Inventory and Monitoring Network of the National Park Service. The subcommittee prepared a management strategy for whitebark pine in the GYE (Greater Yellowstone Coordinating Committee 2011). The strategy outlines objectives (similar to those described earlier and below) and methods for protecting and restoring whitebark pine on the landscape based on their past, current, and projected conditions.

This umbrella document acknowledged that climate science is evolving and that research is continually providing new information to consider in the management of whitebark pine. As a result, the subcommittee developed a shorter-term Adaptive Management Action Plan, which details the site-specific protection and restoration actions that take into account results from research and monitoring. Managers in the GYE recognize that not all whitebark pine stands can be, or should be, protected from all forms of risks and mortality. The subcommittee works together to prioritize trees and stand locations where protection and restoration actions can be implemented in order to provide regionwide whitebark pine conservation and meet agency policy.

Monitoring

Recognizing the importance of monitoring a species over time, the National Park Service Greater Yellowstone Network, in collaboration with the US Forest Service, the US Geological Survey, Yellowstone and Grand Teton national parks, and Montana State University, initiated a long-term monitoring program to track and document the health and status of whitebark pine across the GYE. This subgroup of the GYCC Whitebark Pine Subcommittee, known as the Greater Yellowstone Whitebark Pine Monitoring Working Group, developed the Interagency Whitebark Pine Monitoring Protocol for the Greater Yellowstone Ecosystem (Greater Yellowstone Whitebark Pine MWG 2011) and initiated monitoring in 2004.

After 2011, the group analyzed data collected from two sampling time periods to describe changes observed since 2004 (the next comparison will be made after 2015). This ground-based monitoring program found that there had been no significant change in the blister rust infection rates of 20 to 30 percent during the two time periods and that approximately 27 percent of the whitebark pine (greater than 4.6 feet [1.4 meters] in height) in the GYE died during the recent mountain pine beetle outbreak. Of the tracked trees that had died, more than 70 percent were in the largest-diameter size class, which is indicative of mountain pine beetle preference for larger-diameter trees. Fire also played a role in tree loss, as especially noted in the 2013 Millie Fire on the Gallatin National Forest, Montana, when more than 170 tagged and all understory whitebark pine trees died (Shanahan et al. 2014).

In addition, the subcommittee monitors the effectiveness of restoration actions such as plantings. As results from these monitoring activities become available, managers can adapt management actions to improve the effectiveness of restoration and protection actions.

Protection

With the exception of a failed historic attempt to control *Ribes* spp., an alternate host for white pine blister rust, in Yellowstone National Park from

1945 to 1978 (Kendell and Asebrook 1998), whitebark pine protection efforts have largely been directed toward protecting trees and stands from mountain pine beetle attack. The most recent outbreak began in 2000, peaked in 2009/2010, and has currently returned to endemic levels in most of the GYE (Greater Yellowstone Whitebark Pine MWG 2014). Two methods are used to protect trees from mountain pine beetles: pheromones and insecticides. Artificial pheromones mimic a natural anti-aggregating pheromone emitted by the beetles indicating that a tree or stand is fully occupied—effectively a "no vacancy" sign. More effective in protecting trees in extremely high beetle populations is the direct spray application of an insecticide to the boles of individual trees that will kill the beetles when they attack the tree and attempt to chew through the bark.

Because the protection of individual trees over large areas is labor intensive and logistically problematic, trees prioritized for protection include genetically superior trees (discussed later in this chapter), areas with high genetic diversity, and stands that are relatively intact and produce large amounts of seed. In addition to protection against bark beetles, selected individual or small stands of whitebark pines, particularly "plus" trees (described below), are protected from wildland fire by clearing a radius of 15 to 30 feet (5 to 9 meters) around them using a treatment called "daylighting" (Keane et al. 2012).

Restoration Treatments

Restoration is reestablishing the dominance of whitebark pine on the landscape following catastrophic or anthropogenic losses. A number of methods are used, including planting seedlings, creating openings conducive to the natural regeneration of whitebark pine, and removing trees and vegetation that directly compete with existing trees, saplings, and seedlings. Planting generally occurs in burned areas, since they are devoid of overstory and have reduced vegetative competition (fig. 15-3). Stands where the overstory has been lost to bark beetles or blister rust generally maintain well-developed understories of trees, including whitebark pine, precluding the need to plant. However, there may be a need to remove seedlings and saplings of other species to reduce competition. Other restoration efforts include creating openings in mature stands to promote the caching of seeds by nutcrackers and to release existing understory seedlings and saplings, and clearing around individual saplings or smaller trees to minimize competition and promote the development of large, broad-crowned trees that maximize tree growth and cone production.

FIGURE 15-3   Planting whitebark pine in the Salt Lick Burn, western Wyoming, September 2014. (Photo from US Forest Service.)

Tree Improvement

An important component of whitebark pine conservation in light of current and future threats such as climate change is tree improvement, or the detection and harnessing of genetic traits that will favor whitebark pine's survival and growth. The primary objective of this improvement program is to provide a source of whitebark pine seedlings that are resistant to white pine blister rust infections; in addition, the program seeks to target traits that are drought and cold tolerant.

As described earlier in this chapter, drought resistance may prove to be an important adaptation to climate change. Beginning in 2000, around 240 trees throughout the GYE were identified as having potential resistance to white pine blister rust, based on the lack of observable symptoms of blister rust when it was present in the surrounding stand. These trees are referred to as "plus" trees (Mahalovich and Hipkins 2011). Seeds are collected from these trees and propagated in a nursery setting to be tested for blister rust resistance; those that show resistance are used to populate a seed orchard that will produce resistant seed for future restoration efforts. In addition, seedlings grown from these plus trees or stands showing little blister rust infection are used directly in restoration planting activities. Attention is given to where the seeds are collected so that seedlings are planted in the same geographic location to improve chances of survival.

Implementation Considerations and Challenges

Given the range of management tools available to protect and restore whitebark pine in the GYE, an important consideration and limiting factor is the spatial extent to which these tools may be employed. Overall management direction for the various land management agencies, as well as specific activities currently accepted and practiced within land management designations, is presented in table 15-2, along with the corresponding percentage of currently mapped whitebark pine habitat. Lack of road access to transport personnel, supplies, and equipment into remote backcountry such as wilderness and inventoried roadless areas often limits the practicality of conducting management activities in these areas. Priority stands identified by the GYCC Whitebark Pine Subcommittee for restoration activities are generally within two miles of an access road.

As presented in table 15-1, one of the most likely impacts of a warming climate on whitebark pine is the potential for more frequent and severe outbreaks of mountain pine beetle at higher elevations (Logan, Macfarlane, and Willcox 2010; Macfarlane, Logan, and Kern 2010). The most recent outbreak is one of the most severe and widespread in recorded history in relation to its effects on whitebark pine (Macfarlane, Logan, and Kern 2010). Endemic populations and additional outbreaks of mountain pine beetle are predicted to continue as temperatures are projected to increase and host trees grow into a desirable size for beetles (Buotte et al., in review). To the degree that surviving whitebark pine stands are accessible, protection of the best cone-bearing trees or stands, especially ones that have demonstrated resistance to white pine blister rust, to provide a seed source for natural regeneration is an acceptable strategy, given current land management policy and how much managers are willing to spend. The strategy of protecting select seed-bearing trees from mountain pine beetles has been used by land managers with some success during the most recent outbreak, but regionwide effects of losing mature trees and other ecological functions over large expanses that are either inaccessible or where management direction restricts its implementation, may limit this protection strategy on a regional scale.

TABLE 15-2. Current management direction on mapped whitebark pine habitat.

Greater Yellowstone Ecosystem

Legal Direction/ Agency/ Allocation	WBP Restoration Management Philosophy	Tools Allowed and Likely to Be Used	% Whitebark Pine
National Forests • Custer Gallatin National Forest (NF) • Caribou-Targhee NF • Bridger-Teton NF • Beaverhead-Deerlodge NF • Shoshone NF	Multiple use while maintaining ecological integrity	• Planting seedlings sowing seeds • Pruning • Seed collection • Wildland and prescribed fire use • Targeted fire suppression • Mechanical thinning • Research/monitoring • Protection from mountain pine beetle	5%
National Forests – Wilderness Area (Designated Wilderness, Wilderness Study Areas, etc.)	Maintain natural and untrammeled conditions	• Wildland fire use • Seed collection • Research/monitoring	54%
National Forests – Inventoried Roadless Areas	Actions less restricted but accessibility is an issue	• Planting seedlings/ sowing seeds • Wildland and prescribed fire use • Seed collection • Research/monitoring • Mechanical thinning (requires USDA secretarial approval) • Protection from mountain pine beetle	27%
National Parks • Yellowstone National Park • Grand Teton National Park and John D Rockefeller Jr. Mem. Parkway	National Park Service policy: "Take no action that would diminish the wilderness eligibility of an area" AND/BUT	• Seed collection • Wildland fire use • Research/monitoring • Limited protection from mountain pine beetle	13%

TABLE 15-2. *(Continued)*

Greater Yellowstone Ecosystem

Legal Direction/ Agency/ Allocation	*WBP Restoration Management Philosophy*	*Tools Allowed and Likely to Be Used*	*% Whitebark Pine*
• Occurs mostly in recommended wilderness	"Management actions…should be attempted because of anthropogenic past impacts and only when knowledge and tools exist to accomplish clearly articulated goals."		
Bureau of Land Management	Multiple use while maintaining ecological integrity	• Planting seedlings/ sowing seeds • Pruning • Seed collection • Wildland and prescribed fire use • Targeted fire suppression • Mechanical thinning • Research/monitoring • Protection from mountain pine beetle	1%
Bureau of Land Management – Wilderness Study Areas	Manage so as not to impair the suitability of such areas for preservation as a wilderness; limited management; valid existing rights recognized	• Planting seedlings/ sowing seeds • Pruning • Seed collection • Wildland and prescribed fire use	1%

While current monitoring suggests that white pine blister rust infection appears relatively unchanged in whitebark pine in the GYE (Shanahan et al. 2014), future trends are uncertain (table15-1) and managers remain committed to addressing the current and future threats of this nonnative pathogen (Greater Yellowstone Coordinating Committee 2011). With no feasible methods available to prevent the spread of this disease through stands or to protect individual trees, the protection of individuals resistant to blister rust and especially the planting of resistant seedlings remain paramount to restoration efforts throughout the range of whitebark pine (Keane et al. 2012; Keane et al., in press). GYE land managers have initiated the development of blister rust–resistant planting stock for whitebark pine. However, as with protection, planting resistant trees at this time will be limited to areas that can logistically be planted based on such factors as access, rockiness of the soil, competing vegetation, and land designations that are administratively authorized, such as nonwilderness areas.

As managers incorporate knowledge gained through effectiveness monitoring of plantings and understanding where natural regeneration is successful, they can refine where to plant trees in the future. Moreover, managers throughout the GYE will continue discussions about native species restoration on designated and recommended wilderness and on national park lands to determine the appropriate strategy for conserving whitebark pine on the landscape. The information gained through monitoring and research will be included in future climate modeling efforts that may be able to better predict where to plant trees. Current work to incorporate climate change into the Range-Wide Strategy for Whitebark Pine (Keane et al. 2012; Keane et al., in press) demonstrates that whitebark pine will diminish substantially under climate change scenarios, similar to the other modeling efforts (Chang, Hansen, and Piekielek 2014; Buotte et al., in review), if no management and restoration is applied, but implementation of restoration activities will allow for whitebark pine to be present on the landscape long into the future even under the predicted climate scenarios (Keane et al., in press).

Recognizing that climate change science is constantly evolving, the GYCC Whitebark Pine Subcommittee will hold periodic science reviews, as new information becomes available, to help inform updates to the adaptive management strategy. As mentioned earlier, the GYCC Whitebark Pine Subcommittee has developed an Adaptive Management Action Plan to implement identified and prioritized strategies for whitebark pine conservation across the GYE landscape. As part of this planning process, they have developed a checklist that takes into account current as well as future suitability given projected climate change scenarios (table 15-3).

There are a number of potential impacts of climate change that need further exploration, through cooperation between scientists and managers, to predict how they may affect whitebark pine and conservation efforts to maintain it on the landscape. These include the following:

TABLE 15-3. Planting site assessment checklist, developed by the Greater Yellowstone Coordinating Committee Whitebark Pine Subcommittee. It identifies variables to consider prior to initiating proposed actions and takes into account key research findings important for successful restoration of whitebark pine.

Planting Site Assessment Checklist	
Variables/Components	
Landscape-level characteristics	Climate projection models results
	Biorefugia/habitat suitability identified by field or modeling data using biological and abiotic variables
Site-level characteristics	Edaphic characteristics: slope, aspect, soil type, elevation
	Climate variables at the site level
	Available microsites—type, quantity, quality
	For burned sites, time since fire
	Overstory rust infection rate or potential (low-medium-high)
	Competing species—density, presence/absence
	Suitability of site for competitor
	Follow USFS seed planting guidelines
Planting effectiveness monitoring framework	Develop and implement sampling and analysis plan for long-term effectiveness monitoring of planting sites.
	Monitor natural regeneration within planting sites and compare to natural regeneration in nonplanted sites.

1. *Effects of increased wildland fire on whitebark pine and its competitors.* Will low- to mixed-severity fires benefit whitebark pine by removing competing tree species, or will stand-replacing fires dominate, killing all trees, including whitebark pine seed-producing trees? From a management perspective, fire has the most potential for wide-scale use throughout the GYE. For instance, lightning-caused wildland fires can be managed to burn in wilderness, parks, and other areas, within constraints of risk to life and

property, to increase natural regeneration as well as to promote stand resiliency of a more mosaic landscape in these areas (Keane et al., in press).

2. *Level of investment, in labor and other costs, that will be required to protect mature, cone-bearing whitebark pines from increased levels of mountain pine
beetle outbreak.* What level of protection and restoration will be required to
keep ecologically functional populations across the GYE, and how will this
be affected spatially by areas where protection is impractical or administratively restricted? If it will be possible only to keep representative populations on the landscape, what will be the ecological implications?

3. *Future spread and severity of white pine blister rust.* Since areas that
can logistically and administratively be replanted with rust-resistant seedlings are currently limited, how will large areas lose significant seed-producing capacity and other ecosystem functions as a result of blister rust
infection?

4. *Changes in microsite distributions.* What microclimates (those sites
with greater resolution than those of current climate models) could be
more or less conducive to the growth and survival of whitebark pine, ir-
respective of larger-scale climatic conditions?

5. *Trend and magnitude of drought conditions.* To what degree will
drought and attendant changes in soil water balance help or hurt the ability
of whitebark pine to compete with other tree species?

6. *Changes in ecosystem resiliency.* If site conditions become less favorable to whitebark pine regeneration, how long do we have to establish
stands under current favorable conditions so that they can become more
tolerant of these conditions with age, and how might this be affected by
microclimates? To what degree will resiliency increase with tree size and
age?

7. *Keystone impacts on associated species.* What impacts will climate have
on other species that are critical to whitebark pine, such as Clark's nutcracker, and what could be done to keep these functionally on the landscape, if needed?

8. *Establishment and spread of new pests or diseases under future climates.*
Could we predict and prepare for such eventualities?

9. *Potential for growth in response to overstory changes.* Although not directly related to climate change but important to the resiliency of current
and future stands, will potentially suppressed understory whitebark pine be
able to respond and grow to reproductive maturity following changes in
the overstory?

10. *Changes in technology.* Will improved modeling techniques include
how the factors presented in table 15-1 will be affected by climate change

and, in turn, impact whitebark pine? How can these and other modeling techniques help managers better predict where and how whitebark pine can be managed and restored most effectively?

Recognizing the importance of conserving whitebark pine on the GYE landscape in the face of predicted changes in climatic conditions, managers should continue to develop, update, and implement conservation strategies to protect and conserve important whitebark pine habitats and should collaborate with scientists to continue research that informs the implementation of the strategy. This would include allowing natural ecological processes, such as wildland fire, to occur while identifying and protecting targeted mature, cone-producing, and genetically important trees across the GYE, managing stands to lessen widespread competition from other tree species, and protecting key whitebark pine trees and stands from mountain pine beetle attacks where appropriate. Additionally, collecting and storing whitebark pine seed from key representative areas, as well as genetic strains resistant to white pine blister rust, will hedge against future loss of these trees across varying GYE landscapes. This is already occurring to some degree as part of the GYE-wide genetic "plus" tree improvement program, as well as through general gene conservation seed collections and the identification of genetically diverse populations.

Of particular interest would be to protect and study whitebark pine in atypical areas, such as at lower elevations, until it is understood why these sites, or the genetic makeup of these populations, are conducive to their survival. Sagebrush and lower-treeline species are the only habitats predicted to increase with climate change (chap. 10), so understanding the current adaptive capacity of whitebark pine to survive in and around these types of habitats could be informative for long-term conservation.

Land managers must remain receptive to advancing research, updated climate change models, and adaptive strategies and techniques to maintain and restore whitebark pine. As climate models are refined in precision and spatial resolution, specific areas where whitebark pine is predicted to be more successful in competing and regenerating can be identified and prioritized for protection and treatment. Committing scarce resources to areas with higher probability of success could maximize the economic efficiency of restoration efforts. Given the longevity of whitebark pine and the relative sensitivity of tree seedlings, establishing trees while site conditions are still conducive to regeneration could ensure their presence on these sites well into the future. Maintaining the distribution of these stands over the landscape can preserve ecosystem functions, such as long-distance seed dispersal by nutcrackers, and resilience to climate change effects.

Conclusion

Throughout discussions of whitebark pine management, maintaining long-term ecosystem-wide monitoring programs, management of effectiveness monitoring, and research studies are critical to understanding how whitebark pine is responding to climate change and disturbance factors, including fire, white pine blister rust, mountain pine beetle, and other factors that may not be known at this time. The information garnered from these monitoring and research activities will be valuable as managers determine the best strategies for managing and conserving whitebark pine. As we learn more about the effects of climate change, we will be able to understand changes to ecosystem dynamics, adapt management strategies in order to maximize their effectiveness and success, and inform managers how to most effectively maintain this important ecological species across the GYE landscape into the future.

Acknowledgments

The authors would like to thank Dr. Robert Keane and S. Thomas Olliff for a scientific review and administrative review, respectively. We would also like to express our heartfelt gratitude to the GYCC Whitebark Pine Subcommittee for the development of the Whitebark Pine Strategy for the Greater Yellowstone Area and the committee's continued efforts conserving whitebark pine.

References

Arno, S. F., and R. J. Hoff. 1989. Silvics of Whitebark Pine (*Pinus albicaulis*). GTR INT-253. Ogden, UT: US Forest Service, Intermountain Research Station.

Buotte, P. C. 2015. Understanding and capturing geographic variability in vulnerability of natural resources to climate change through modeling and workshops. Dissertation. University of Idaho, Moscow.

Campbell, E. M., R. E. Keane, E. Larson, M. P. Murray, A. W. Schoettle, and C. Wong. 2011. Disturbance ecology of high-elevation five-needle pine ecosystems in Western North America. In *The Future of High-Elevation, Five-Needle White Pines in Western North America: Proceedings of the High Five Symposium, 28–30 June 2010, Missoula, MT*, edited by R. E. Keane, D. F. Tomback, M. P. Murray, and C. M. Smith, 154–63. Proceedings RMRS-P-63. Fort Collins, CO: US Forest Service, Rocky Mountain Research Station.

Chang, T., A. J. Hansen, and N. Piekielek. 2014. Patterns and variability of projected bioclimatic habitat for *Pinus albicaulis* in the Greater Yellowstone Area. *PLOS ONE* 9 (11): e111669. doi: 10.1371/journal.pone.0111669.

Critchfield, W. B., and G. L. Allenbaugh. 1969. The distribution of *Pinaceae* in and near northern Nevada. *Madrono* 20 (1): 12–26.

Farnes, P. E. 1990. SNOTEL and snow course data describing the hydrology of whitebark pine ecosystems. In *Proceedings—Symposium on Whitebark Pine Ecosystems: Ecology and Management of a High-Mountain Resource*, compiled by W. C. Schmidt and K. J. McDonald, 302–4. General Technical Report INT-270. Ogden, UT: US Forest Service Intermountain Research Station.

Greater Yellowstone Coordinating Committee, Whitebark Pine Subcommittee. 2011. Whitebark Pine Strategy for the Greater Yellowstone Area. Report.

Greater Yellowstone Whitebark Pine Monitoring Working Group (MWG). 2011. *Interagency Whitebark Pine Monitoring Protocol for the Greater Yellowstone Ecosystem, Version 1.1.* Bozeman, MT: Greater Yellowstone Coordinating Committee. https://irma.nps.gov/App/Reference/ Profile/660369.

Greater Yellowstone Whitebark Pine Monitoring Working Group (MWG). 2014. *Summary of Preliminary Step-Trend Analysis from the Interagency Whitebark Pine Long-Term Monitoring Program—2004–2013.* Prepared for the Interagency Grizzly Bear Study Team. Natural Resource Data Series NPS/GRYN/NRDS-2014/600. Fort Collins, CO: National Park Service.

Iglesias, V., T. R. Krause, and C. Whitlock. 2015. Complex response of white pines to past environmental variability increases understanding of future vulnerability. *PLOS ONE* 10 (4): e0124439. doi: 10.1371/journal.pone.0124439.

Keane, R. E. 2001. Successional dynamics: Modeling an anthropogenic threat. In *Whitebark Pine Communities, Ecology and Restoration*, edited by D. F. Tomback, S.F. Arno, and R. E. Keane, 159–92. Washington, DC: Island Press.

Keane, R. E., L. M. Holsinger, M. F. Mahalovich, and D. F. Tomback. In press. *Restoring Whitebark Pine Ecosystems in the Face of Climate Change.* General Technical Report RMRS-GTR-xxx. Fort Collins, CO: US Forest Service, Rocky Mountain Research Station.

Keane, R. E., and R. A. Parsons. 2010. *Management Guide to Ecosystem Restoration Treatments: Whitebark Pine Forests of the Northern Rocky Mountains.* General Technical Report RMRS-GTR-232. Fort Collins, CO: US Forest Service, Rocky Mountain Research Station.

Keane, R. E., D. F. Tomback, C. A. Aubry, A. D. Bower, E. M. Campbell, C. L. Cripps, M. B. Jenkins, M. F. Mahalovich, M. Manning, S. T. McKinney, et al. 2012. *A Range-Wide Restoration Strategy for Whitebark Pine* (Pinus albicaulis). Gen. Tech. Rep. RMRS-GTR-279. Fort Collins, CO: US Forest Service, Rocky Mountain Research Station.

Kendall, K. C., and J. M. Asebrook. 1998. The war against blister rust in Yellowstone National Park, 1945–1978. *Historical Perspectives on Science and Management in Yellowstone National Park* 15 (4).

Kendall, K. C., and R. E. Keane. 2001. Whitebark pine decline: Infection, mortality, and population trends. In *Whitebark Pine Communities, Ecology and*

Restoration, edited by D. F. Tomback, S. F. Arno, and R. E. Keane, 221–42. Washington, DC: Island Press.

Koteen, L. 1999. Climate change, whitebark pine, and grizzly bears in the Greater Yellowstone Ecosystem. In *Wildlife Responses to Climate Change*, edited by S. H. Schneider and T. L. Root, 343–64. Washington, DC: Island Press.

Logan, J. A., W. W. Macfarlane, and L. Willcox. 2010. Whitebark pine vulnerability to climate-driven mountain pine beetle disturbance in the Greater Yellowstone Ecosystem. *Ecological Applications* 20 (4): 895–902.

Lorenz, T. J., K. A. Sullivan, A. V. Bakian, and C. A. Aubry. 2011. Cache-site selection in Clark's nutcracker (*Nucifraga columbiana*). *The Auk* 128 (2): 237–47.

Macfarlane, W. W., J. A. Logan, and W. R. Kern. 2010. Using the Landscape Assessment System (LAS) to Assess Mountain Pine Beetle-Caused Mortality of White Bark Pine, Greater Yellowstone Ecosystem, 2009. Project report prepared for the Greater Yellowstone Coordinating Committee, Whitebark Pine Subcommittee, Jackson, WY.

Mahalovich, M. F., and V. D. Hipkins. 2011. Molecular genetic variation in whitebark pine (*Pinus albicaulis Engelm.*) in the Inland West. In *The Future of High-Elevation, Five-Needle White Pines in Western North America: Proceedings of the High Five Symposium*, edited by R. E. Keane, D. F. Tomback, M. P. Murray, and C. M. Smith. Proceedings RMRS-P-63. Fort Collins, CO: US Department of Agriculture, Forest Service, Rocky Mountain Research Station. http://www.fs.fed.us/rm/pubs/rmrs_p063.html.

Mattson, D. J., and D. P Reinhart. 1990. Whitebark pine on the Mount Washburn massif. In *Proceedings—Symposium on Whitebark Pine Ecosystems: Ecology and Management of a High Elevation Resource*, compiled by W. C. Schmidt and K. J. McDonald, 106–17. GTR INT-270. Ogden, UT: US Forest Service, Intermountain Research Station.

Mattson, D. J., K. C. Kendall, and D. P. Reinhart. 2001. Whitebark pine, grizzly bears and red squirrels. In *Whitebark Pine Communities: Ecology and Restoration*, edited by D. F. Tomback, S. F. Arno, and R. E. Keane, 12–136. Washington, DC: Island Press.

McCaughey, W. W., and D. F. Tomback. 2001. The natural reproduction process. In *Whitebark Pine Communities, Ecology and Restoration*, edited by D. F. Tomback, S. F. Arno, and R. E. Keane, 105–20. Washington, DC: Island Press.

McDonald, G. I., and R. J. Hoff. 2001. Blister rust: An introduced plague. In *Whitebark Pine Communities, Ecology and Restoration*, edited by D. F. Tomback, S. F. Arno, and R. E. Keane, 193–220. Washington, DC: Island Press.

Monleon, V. J., and H. E. Lintz. 2015. Evidence of tree species' range shifts in a complex landscape. *PLOS ONE* 10 (1): e0118069. doi: 10.1371/journal.pone.0118069.

Pearson, R. G., and T. P. Dawson. 2003. Predicting the impacts of climate change on the distribution of species: Are bioclimate envelope models useful? *Global Ecology and Biogeography* 12:361–71.

Reinhart, D. P., and D. J. Mattson. 1990. Red squirrels in the whitebark zone. In *Proceedings—Symposium on Whitebark Pine Ecosystems: Ecology and Management of a High Elevation Resource*, compiled by W. C. Schmidt and K. J. McDonald, 256–63. GTR INT-270. Ogden, UT: US Forest Service, Intermountain Research Station.

Shanahan, E., K. M. Irvine, D. Roberts, A. Litt, K. Legg, and R. Daley. 2014. *Status of Whitebark Pine in the Greater Yellowstone Ecosystem: A Step-Trend Analysis Comparing 2004–2007 to 2008–2011*. Natural Resource Technical Report NPS/GRYN/NRTR—2014/917. Fort Collins, CO: National Park Service.

Schwandt, J. W. 2006. *Whitebark Pine in Peril: A Case for Restoration*. Report R1-06-28. Missoula, MT: US Department of Agriculture Forest Service.

Thoma, D., A. Ray, A. Rodman, and M. Tercek. 2015. Interpreting climate change impacts using a water balance model. *Yellowstone Science* 23 (1): 29–35.

Tomback, D. F. 2001. Clark's nutcracker: Agent of regeneration. In *Whitebark Pine Communities, Ecology and Restoration*, edited by D. F. Tomback, S. F. Arno, and R. E. Keane, 89–104. Washington, DC: Island Press.

Tomback, D. F., S. F. Arno, and R. E. Keane. 2001. The compelling case for management intervention. In *Whitebark Pine Communities, Ecology and Restoration*, edited by D. F. Tomback, S. F. Arno, and R. E. Keane, 3–28. Washington, DC: Island Press.

US Fish and Wildlife Service. 2011. Endangered and threatened wildlife and plants; 12-month finding on a petition to list *Pinus albicaulis* as endangered or threatened with critical habitat. *Federal Register* 76 (138): 42631–54.

Weaver, T. 2001. Whitebark pine and its environment. In *Whitebark Pine Communities, Ecology and Restoration*, edited by D. F. Tomback, S. F. Arno, and R. E. Keane, 41–88. Washington, DC: Island Press.

Westerling, A. L., M. G. Turner, E. A. H. Smithwick, W. H. Romme, and M. G. Ryan. 2011. Continued warming could transform Greater Yellowstone fire regimes by mid-21st century. *Proceedings of the National Academy of Sciences of the United States of America* 108 (32): 13165–70. doi: 10.1073/pnas.1110199108.

Chapter 16

Insights from the Greater Yellowstone Ecosystem on Assessing Success in Sustaining Wildlands

Andrew J. Hansen and Linda B. Phillips

An overarching question in natural resource management is, "How well are we sustaining entire ecosystems under climate and land use change?" The chapters of this book have dealt with understanding and managing landforms, tree species, and fish and vegetation communities in the face of changing climate. Application of the Climate-Smart Conservation framework is typically in the context of species, communities, or ecological processes deemed to be the most vulnerable to climate change. The effectiveness of adaptation options for vulnerable elements can be evaluated through adaptive management in which multiple treatments are implemented, monitored, and compared (chap. 13).

Ultimately, however, it is at the scale of entire ecosystems that management effectiveness needs to be evaluated. It is the interactive effects of stressors (such as climate change and land use change), ecosystem processes, and biodiversity and the multiple management strategies employed across all natural resources that determine the ecological condition of an ecosystem (Parks Canada Agency 2008). Unfortunately, monitoring the ecological status and trends across entire ecosystems is seldom done, and methods for doing so and for communicating results to stakeholders are underdeveloped. Consequently, we often do not know whether ecosystems are approaching tipping points at which additional small changes in stressors can result in large reductions in ecosystem services (Scheffer et al. 2015). Yet, if stakeholders had such information, they might factor it into their personal-, business-, and governmental-level decision making in ways that further sustain ecological condition and services.

Insights into the challenges and benefits of monitoring entire wildland ecosystems can be obtained from a look at the Greater Yellowstone Ecosystem (GYE) in the US northern Rockies. The GYE is widely considered to be one of the iconic wildland ecosystems in the world's temperate zones. The designation of Yellowstone as the first national park in 1872 was instrumental in the subsequent development of the global protected area network (Schullery 1997). Today, the GYE remains one of the largest areas of relatively natural lands in the contiguous United States. Federal lands make up most of the ecosystem, and this portion of the GYE is managed with the benefit of the considerable resources and expertise of the US government. Moreover, conservation on private lands has been promoted by numerous conservation organizations and land trusts. Thus, the GYE represents a large and relatively well managed wildland ecosystem.

The natural characteristics for which the GYE is so well known, however, are now attracting many new residents and businesses entrepreneurs who are seeking natural amenities. Consequently, the GYE is best described as a rapidly humanizing wildland ecosystem. Human population size, rural housing, and backcountry recreation are growing rapidly within the ecosystem (Hansen et al. 2002; Gude, Hansen, and Jones 2007). The potential impacts of human development on the ecosystem are exacerbated by current and projected future climate change. These trends lead to the fundamental question: "How well are we sustaining GYE as a wildland ecosystem?"

We address this question by evaluating trends in ecological integrity across the public and private lands that comprise the GYE (fig. 16-1; Hansen et al. 2002). This delineation of the GYE was based on both ecological and socioeconomic factors and includes a larger private land component than the Yellowstone Protected Area Centered Ecosystem defined in chapter 1. We use the GYE as the area of analysis for this chapter because local stakeholders and policy makers typically consider both socioeconomic and ecological factors when making decisions relating to resource management, and the GYE boundaries include the areas of strong socioecological interactions.

An ecosystem is said to have integrity when "the structure, composition and function of the ecosystem are unimpaired by stresses from human activity; natural ecological processes are intact and self-sustaining, the ecosystem evolves naturally and its capacity for self-renewal is maintained; and the ecosystem's biodiversity is ensured" (Parks Canada Agency 2008;

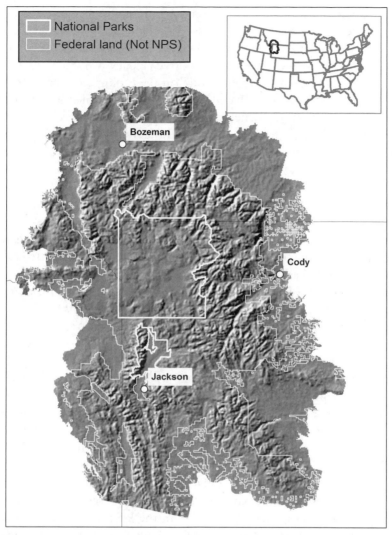

FIGURE 16-1   Shaded relief map of the Greater Yellowstone Ecosystem as delineated by Hansen et al. (2002). Yellowstone National Park (center) and other public
lands are outlined in white.

see also Karr 1987). Ecological integrity has been recommended as a basis
for monitoring the status of protected areas (Parrish, Braun, and Unnasch
2003; Chape et al. 2005), and this approach is in use in Canada (Parks
Canada Agency 2008) and in the Great Northern Landscape Conservation
Cooperative (Greater Northern LCC; Finn et al. 2015).

We first review the history of land use and climate change in the GYE. The influences of these human forces on ecological processes and biodiversity are summarized. We then evaluate how well measures of ecological integrity are being monitored and reported. Finally, we draw on current information to present a "report card" on the ecological integrity of the GYE. This review is largely based on previously published literature and data. We do, however, add a new analysis of the extent of "developed" lands across the GYE and the extent to which this development has fragmented habitats of various types.

A Brief History of the Greater Yellowstone Ecosystem

The Greater Yellowstone Ecosystem is centered on the Yellowstone Plateau and the surrounding mountain ranges. The region has been uplifted as the continental plate has moved over the Yellowstone volcanic hot spot over the past 2 million years (Mogk et al. 2012). The resulting high elevations and midcontinental setting causes climate to be relatively harsh across the GYE. Winters are long, cold, and with heavy snowpack at higher elevations, and frost can occur any time of year (Despain 1990). The volcanic soils are relatively infertile. Plant productivity is relatively low over most of the area, but higher in valley bottoms, where climate and soils are more favorable (Hansen et al. 2000). The extensive snowpack is a source of water for the eight major rivers that flow from the GYE. These rivers feed the three major watersheds of the western United States—the Missouri, Colorado, and Columbia—and thus the GYE is called the "water tower" of the West.

The remoteness of the GYE and its harsh climate influenced the history of human settlement of the ecosystem. Prior to the presence of EuroAmericans, the GYE was at the intersections of the lands occupied by several Native American groups (Janetski 1987). While these tribes hunted in the GYE, only small groups of Shoshones lived in what is now Yellowstone National Park, presumably due to the harshness of the climate. Starting with Lewis and Clark in 1806, EuroAmerican explorers and trappers passed through the lower elevations of the GYE, but the Yellowstone Plateau and the higher mountains remained relatively unknown until the Folsom Expedition of 1869 and the Washburn Expedition of 1870 (Haines 1977).

Westward expansion from the densely settled eastern United States was well under way at this time. However, the main route to the northwestern United States, the Oregon Trail, bypassed the GYE to the south. Direct access to the GYE from the population centers in the East was largely blocked by Native Americans until after the Battle of the Little Bighorn in 1876 (Garcia 1967). Approaching by the long southwestern route, gold miners arrived in southwestern Montana in early 1860s (Black 2012). Ranchers, farmers, and townspeople began moving into the area soon afterward. They settled largely in the more fertile river valleys, and these were claimed as private lands.

In recognition of its scenic and geothermal wonders, the Yellowstone Plateau was designated as the world's first national park in 1872 with repeated park expansions until the 1920s (Haines 1977). Further protection followed when the surrounding mountainous lands were designated as the first national timber reserves in the 1890s and then became national forests in 1905. Today, Yellowstone National Park and the surrounding federal and state lands comprise 64 percent of the GYE, with private lands being largely in the valley bottoms and lower elevations surrounding the public lands. Despite the creation of the national park, EuroAmerican migration into the GYE was hindered by the remoteness of the area until the 1970s. The GYE population increased slowly during the 1900s relative to many places in the western United States and remained below about 220,000 until the late 1960s.

Starting in the 1970s, the mountain wilderness of the GYE that had dissuaded settlement became an attractant. Particularly by the 1990s, many people and businesses moved into the GYE because of the scenery, access to public lands, and outdoor recreation and other "natural amenities" (Rasker and Hansen 2000). The population of the twenty counties of the GYE has more than doubled since 1970 to the current level of 470,000 (fig. 16-2), with some counties in the ecosystem being among the fastest increasing in the nation. Many of the natural amenity migrants choose to live "out of town" on ranchettes and in exurban subdivisions. Consequently, exurban home density has increased dramatically since 1970 (Gude et al. 2006). The proportion of private lands with low densities of homes (rural, 0.012–0.025 homes per acre, or 0.031–0.063 homes per hectare) has steadily declined, while the proportion in exurban (0.025–0.05 homes per acre, or 0.063–0.145 homes per hectare) and suburban/urban densities (>0.58 homes per acre, or > 1.45 homes per hectare) and commercial /industrial densities has increased (fig. 16-3). Today, exurban housing extends in a radius of commuting distance around most of the towns and cities of the GYE and fringes the public lands boundaries. Population growth is projected to continue in future decades, with the US Census forecasting a population of 750,000 by 2040 (Davis and Hansen 2011).

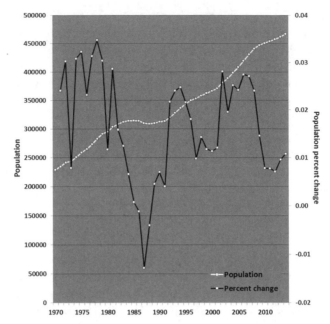

FIGURE 16-2   Change in human population (left Y axis) and in annual percent change (right Y axis) for the twenty counties of the Greater Yellowstone Ecosystem.

One of the natural amenities sought by GYE residents and visitors is outdoor recreation. Fishing, hunting, hiking, backcountry skiing, mountain biking, and off-road vehicle use are all means by which people access wildlands. Growth in population would suggest increases in these backcountry activities. Unfortunately, data on change in backcountry use across GYE are sparse. Total visitation to Yellowstone and Grand Teton national parks increased rapidly until the early 1990s, plateaued until about 2008, and has been rising since then (fig. 16-4; Yellowstone Center for Resources 2011). The number of backcountry overnight visitations in Yellowstone National Park increased until about 1995 and has been relatively stable since then. Skier days in area ski resorts have risen rapidly over the last few decades (fig. 16-5). Backcountry skiing rates have likely increased also, but data are not available.

In addition to changes in land use, the GYE has also been undergoing changes in climate (chap. 5; Chang and Hansen 2015). Average annual temperature has increased 1.1 degrees F (0.6 degrees C) per century since 1900, primarily due to rises in minimum temperatures in spring and summer (1.5 degrees F [0.8 degrees C] per century) and reduced winter

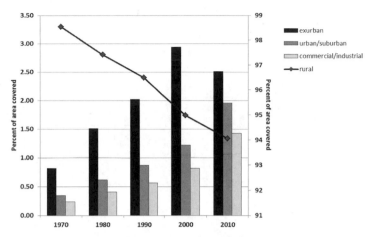

FIGURE 16-3   Proportion of the Greater Yellowstone Ecosystem in housing density classes in decades since 1970. (Data from Theobald 2014.)

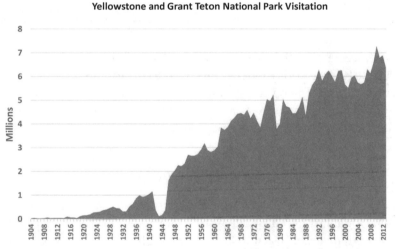

FIGURE 16-4   Visitation rates to Yellowstone and Grand Teton national parks since 1970. (Data from https://irma.nps.gov/Stats/Reports/Park.)

extreme cold. Very cold winter monthly temperatures (< 0 degrees F, or < –18 degrees C) occurred about every six years prior to 1993 and have not occurred since then. Most of the warming has occurred since 1980. Precipitation has increased slightly in the past three decades, but not enough to offset evaporation due to increased temperatures; thus, aridity has increased slightly. By 2100, temperature is projected to increase from 3.6 to

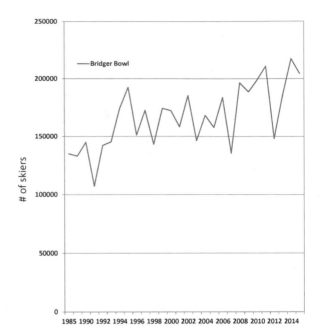

FIGURE 16-5
Trends in annual
skier days at Bridger
Bowl ski resort in
the Greater Yellow-
stone Ecosystem.

16.2 degrees F (2 to 9 degrees C) above the average for the reference period
1900–2010. The aridity index is projected to increase moderately or substantially, depending on climate scenario. Forecasts project that the current
climate changing pattern we have experienced for the past thirty years will
continue and become more severe.

Changes in Ecosystem Processes

Although the Greater Yellowstone Ecosystem has relatively high levels of
natural variability in climate and disturbance, trends in ecosystem processes
related to human land use and climate warming have emerged. The observed
warming has reduced snowpack (chap. 7). April 1 snow water equivalent
(SWE) is currently 20 percent lower than the average for the period 1200–
2000 (Pederson et al. 2011). Moreover, 1900–2000 represents the longest
period of below-average snowpack in the 800-year record, and the decade
of the 1990s was among the lowest in the century.

　　Climate change and land use practices over the past century, such as
irrigated agriculture, have also influenced streamflow and temperature.
Stream discharge declined during 1950–2010 in 89 percent of the streams
analyzed in the central Rocky Mountains, including those in the GYE
(Leppi et al. 2012). Reduced flows were most pronounced during the sum-

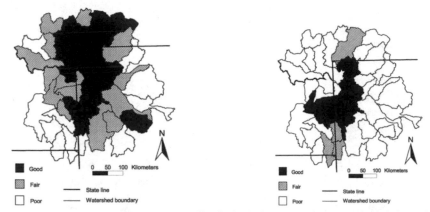

FIGURE 16-6   Spatial distribution of hydrological integrity by watershed (left) and status of native trout and grayling by watershed (right) across the Greater Yellowstone Ecosystem. (Redrawn from Van Kirk and Benjamin 2001.)

mer months, especially in the Yellowstone River. Stream temperatures have warmed across the region by about 1.8 degrees F (1 degree C) over the past century (Al-Chokhachy et al. 2013). Stream warming during the decade of the 2000s exceeded that of the Great Dustbowl of the 1930s and represents the greatest rate of change over the past century.

Most rivers in the GYE have impoundments or water withdraws for human use. These human alterations influence peak flows, water temperature, stream sediment, channel morphology, and habitat structure. An assessment of hydrologic integrity based on reservoir surface area, total surface water withdrawals, and total consumptive water use was done for forty-one watersheds across the GYE (Van Kirk and Benjamin 2001). The assessment found that watersheds at lower elevations around the perimeter of the GYE had substantially lower hydrological integrity scores than those in headwater areas (fig. 16-6).

Water quality remains high across most of the public lands of the GYE (Yellowstone Center for Resources 2011). Streams and rivers flowing through developed private lands, however, incur inputs of nitrogen, phosphorous, sediment, bacteria, herbicides, pesticides, and chemical hormones. Sampling of water quality is currently inadequate to quantify trends in water quality on private lands or impacts on aquatic ecosystems (Van Kirk and Benjamin 2001). Ad hoc sampling of individual streams, such as Sourdough Creek downstream of Bozeman, Montana, in the northern portion of the GYE, reveals elevated levels of nitrate, phosphate, and *E. coli* (W. F. Cross, personal communication, Montana State University

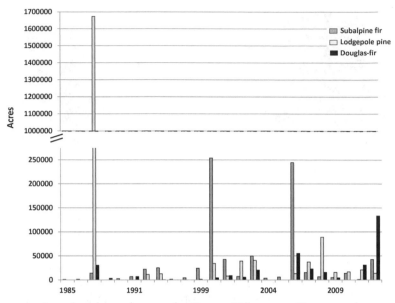

FIGURE 16-7 Area burned across the Greater Yellowstone Ecosystem by vegetation type since 1984. The inset shows the large area burned in the extreme 1988 fires. (Data from the National Monitoring Trends in Burn Severity database.)

Water Quality Extension). This level of pollution may influence aquatic ecosystems by altering the structure of communities and the rates of ecosystem processes, such as primary production and decomposition, that underpin stream ecosystem health (Allan and Castillo 2007). The pervasiveness of such degradation of water quality and ecological effects across the private land portion of the GYE is currently unknown but should be of high concern.

Land use and climate change have also influenced natural disturbance regimes in uplands. Although human activities have not yet had a discernible impact on fire regimes in subalpine forests in the GYE (Romme and Despain 1989), fire regimes at the lower-forest treeline appear to have undergone two regime shifts since EuroAmerican settlement. The relatively frequent and low-intensity natural fires in this zone were largely excluded between 1880 and 2000 by cattle grazing, land development, and human fire suppression (Littell 2002). As a consequence of this fire exclusion, forests expanded into sage/grasslands in this zone and became denser, increasing fuel loads (Powell and Hansen 2007). Since 2000, several large and intense fires have occurred in the lower forest zone due to drought and high fuel loads (fig. 16-7). This may be an early formulation of the fundamental shift in fire toward the increased fire frequency projected for the coming century under climate change (Westerling et al. 2011).

Another consequence of warming has been an outbreak of forest pests and forest die-off (chaps. 10 and 15). Mild winter temperatures have been found to directly relate to the survivorship of overwintering mountain pine beetle (*Dendroctonus ponderosae*), the major pest of the whitebark pine (*Pinus albicaulis*) (Logan, Macfarlane, and Willcox 2010; Logan and Powell 2009). Arid summers likely provide a compounding effect of increasing pine beetle development rates and increasing resource stress on whitebark pine. Since 1999, an eruption of mountain pine beetle events has been observed that exceeds the frequencies, impacts, and ranges documented during the past 125 years (Raffa et al. 2008; Macfarlane, Logan, and Kern 2013). Aerial assessment of whitebark pine species populations within the GYE has indicated a 79 percent mortality rate of mature trees. At lower elevations, elevated rates of Douglas-fir (*Pseudotsuga menziesii*) and lodgepole pine (*Pinus contorta*) mortality have also been recorded.

Trends in Biodiversity

Changes in climate and land use also influence plant and animal species and communities. For example, land use development fragments natural habitats, and the proportion of original habitat remaining today is a measure of human impact.

In previous studies, we drew on the concept of human disturbance zone to quantify the cumulative effects of human land uses across the GYE (Gude, Hansen, and Jones 2007; Jones et al. 2009; Davis and Hansen 2011; Piekielek and Hansen 2012). The disturbance zone approach (Theobald et al. 1997) is based on a functional relationship between effect on habitat and distance from development. Many studies have found that roads, rural homes, and other types of human development degrade habitat quality some distance beyond the actual location of the human infrastructure. Mapping the cumulative "disturbance zone" among land use types is an index of the proportion of a wildland ecosystem that has been degraded by human development. In addition to using this approach to quantify development effects around protected areas, as in the studies cited above, Theobald (2014) used the approach to develop an index of landscape integrity, which is defined as the inverse of cumulative development and analyzed patterns across the United States (see chapter 6 for an application with the Great Northern Landscape Conservation Cooperative).

Because our previous analyses of disturbance zone across the GYE were based on data through 2007, we redid the analyses with newer data on land use (including through 2013). Exurban residential housing was mapped based on groundwater well data and county assessor tax data. For seven counties in eastern Idaho, tax assessor records were used to identify the number of homes in each quarter section. For all other counties, domestic wells data were used to represent rural homes. In Montana, this data was provided by the Montana Bureau of Mines and Geology, and in Wyoming the data was provided by the State Engineer's office. Data for urban, suburban, and agricultural lands were derived from the US Geological Survey's National Land Cover 2011 edition data set (NLCD) (Jin et al. 2013). Road data from US Census Bureau Tiger/Line files used and included primary and secondary roads (feature class codes A11 to A41; US Census Bureau 2014).

Because NLCD labels areas that are close to all roads as "developed," we masked this NLCD developed class within public lands if there was no other evidence of human modification. The distance of the disturbance zone around each land use type was estimated based on previous studies (see the references cited earlier). We buffered all rural homes and wells by 3,280 feet (1 kilometer), NLCD classes by 1,640 feet (500 meters), and roads by 328 feet (100 meters), consistent with Gude, Hansen, and Jones (2007). The resulting map of developed lands covers about 30 percent of the area of the GYE (fig. 16-8).

We obtained habitat data for several vegetation types, wildlife species, and irreplaceable biophysical settings (table 16-1). We overlaid developed lands on these habitat types to estimate reduction in habitat area due to land use. We found that loss of area of habitat types that are centered on higher elevations and on public lands has been minor (10 to 13 percent for subalpine coniferous forests and grizzly bear, *Ursus arctos horribilis*) (fig. 16-9). Habitat loss was intermediate (25 to 32 percent) for vegetation types at mid-elevations (Douglas-fir, aspen, upland deciduous) and for elk habitats. Habitat types most reduced (39 to 57 percent) were those overlapping lower elevations and private lands, including sagebrush/grasslands, bird hot spots, moose habitat, pronghorn habitat, and large river riparian zones. Within private lands, habitat loss was 50 percent or more for all of the habitat types and 89 percent for large river riparian zones.

Although habitat area has been lost for many elements of biodiversity, large mammals of high conservation concern have expanded dramatically in population size and range. The GYE is the heart of the major large mammal restoration in the northern Rockies (Picton and Lonner 2008).

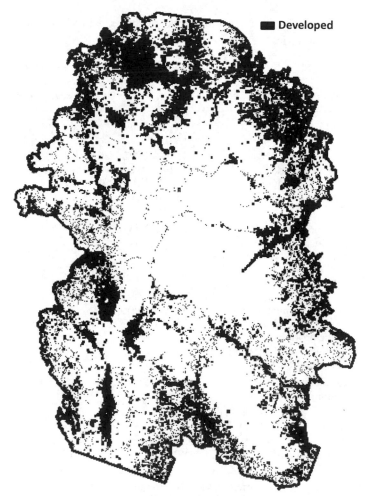

■ Developed

FIGURE 16-8   Locations within the Greater Yellowstone Ecosystem that are considered "developed" as defined as locations in or near urban, suburban, rural residential, agricultural lands, or roads (see chapter text for details).

Large mammals were decimated in the region in the late 1800s as a result of market hunting. The grey wolf was entirely extinct, and small populations of bison, grizzly, elk, bighorn sheep and pronghorn antelope persisted in remote locations, such as the area that is now Yellowstone National Park. Focused conservation efforts have allowed these populations to grow and flourish to the point that the GYE is now a major source area for wildland species that are currently repopulating other parts of the western United States.

TABLE 16-1.   Habitat data for several vegetation types, wildlife species, and irreplaceable biophysical settings.

Habitat Type	Definition	Source	Proportion Overlaying Land Use Development	
			Private Lands	All Lands
Aspen	Stands dominated by aspen	Brown et al. (2006)	56	27
Riparian habitat	Rivers buffered by 256 meters and adjacent deciduous habitat	USGS 1992 National Land Cover; USGS/EPA 1999 Hydrography; USFS 1990–2001 stand map	89	57
Sage/ grassland	Nonforest vegetation dominated by sagebrush and grassland communities	USGS NLCD	68	39
Upland woody deciduous			63	32
Douglas-fir forest	Forests dominated by Douglas-fir, which occurs in the productive lower-treeline to mid-elevation portion of the GYE	FIA, Forest Type data	50	25
Subalpine coniferous forest	Coniferous forests dominated by lodgepole pine, subalpine fir, Engelmann spruce, or whitebark pine	FIA, Forest Type data	50	10
Bird hot spots	Areas of > 70% of maximum bird diversity and abundance	Hansen et al. (2002)	65	41
Irreplaceable areas	Multicriteria assessment based on habitat and population data for terrestrial and aquatic GYE species	Noss et al. (2002)		
Pronghorn	Habitat suitability; expert opinion	Montana Fish, Wildlife, and Parks; Wyoming Game and Fish	66	51
Moose	Habitat suitability; expert opinion	Montana Fish, Wildlife, and Parks Wyoming Game and Fish	64	44
Grizzly bear	Edge of composite polygon of fixed-kernel ranges from all grizzly locations (1990–2000)	Schwartz et al. (2002)	61	13
Elk winter	Habitat suitability; expert opinion	Rocky Mountain Elk Foundation	56	30

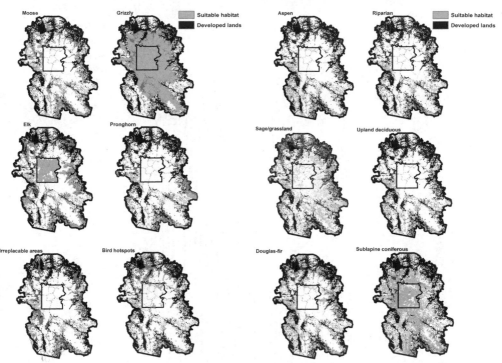

FIGURE 16-9   Spatial distribution of suitable habitats for various elements of biodiversity overlaid on lands classified as developed across the Greater Yellowstone Ecosystem. Habitat types are described in the chapter text.

The grizzly bear was declared a threatened species in 1975 when the population was about 136 individuals. The current population size exceeds 600 (fig. 16-10), and the area occupied by the population has expanded by more than 50 percent from a core in Yellowstone National Park to most of the GYE (Yellowstone Center for Resources 2011). The bison (*Bison bison*) population grew from about 50 individuals in 1900 to more than 4,500 today and has expanded its migratory range to lower-elevation grasslands outside of Yellowstone National Park. In the twenty years since the grey wolf (*Canis lupus*) was reintroduced into GYE, the population has grown to more than 500 individuals and the population range has expanded dramatically, with individuals wandering as far as Oregon, Colorado, and Arizona.

This wildlife success story has created a new set of management challenges relating to humans learning to live with dangerous wildlife. As these large mammals expand from core public lands to private lands, there is considerable concern about conflicts with people relating to risk to humans,

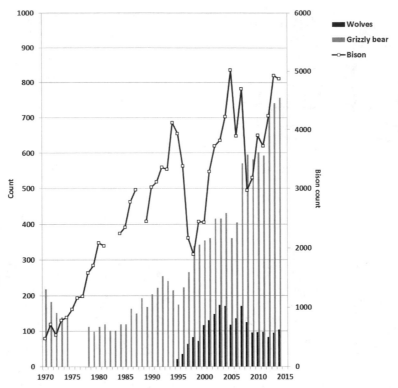

FIGURE 16-10   Population sizes of select large mammals in the Greater Yellowstone Ecosystem since 1970. (Data from Center for Natural Resources, Yellowstone National Park.)

livestock and pet depredation, spread of disease, and damage of fences and property. In fact, the largest source of mortality for adult grizzly bears results from conflict with people due to accidental death (e.g., highway fatalities) or euthanasia of "problem" bears (Schwartz, Haroldson, and White 2010). Expansion of bison into winter habitats in the lower-elevation valley bottoms surrounding the national parks has been constrained by livestock managers concerned about the spread of brucellosis. A great deal has been learned about minimizing wildlife-human conflict involving both modifying human behavior and controlling wildlife in locations of high negative impact (e.g., Mattson et al. 1996; Bangs et al. 2005). As the human population continues to expand in the GYE, means and methods of living with dangerous wildlife will be essential for maintaining these wildlife populations.

Most bird species in the GYE are migratory, and their population dynamics reflect population dynamics over subcontinental- to continental-sized areas. Birds of prey, such as bald eagle (*Haliaeetus leucocephalus*) and peregrine falcon (*Falco peregrinus*), have largely been increasing in abundance across the GYE as they have been continentally since the use of pesticides such as DDT was curtailed in the 1960s. Habitat for songbirds associated with willow, aspen, and riparian habitats have expanded in the past two decades in the Yellowstone Northern Range following the reintroduction of wolves, reduced elk herbivory, and expansion of woody deciduous plant species (Baril et al. 2011). Bird diversity and abundance, however, are concentrated in the GYE in valley bottoms and toe slopes with more favorable climate, better soils, and faster-growing plants (Hansen et al. 2002). These bird hot spots are disproportionately on private lands, where they overlap with rural home development and agriculture (Gude, Hansen, and Jones 2007). These land uses favor midsized native and exotic predators (Hansen et al. 2005). Consequently, songbird reproduction in hot spots near human development is depressed by predation, and these habitats are now population "sinks," where reproduction is below mortality and the populations would go extinct if not supplemented by immigrants from other locations (Hansen and Rotella 2002).

In contrast to large mammals, native fish in the GYE have been declining in recent decades (chap. 12). All four subspecies of cutthroat trout (*Oncorhynchus clarkii*) native to the GYE, as well as the native Montana grayling (*Thymallus arcticus montanus*), were suggested by conservation organizations for endangered species protection in the 1990s. The primary threat to these native fish are exotic fishes. The nonnative brown trout (*Salmo trutta*), rainbow trout (*O. mykiss*), and brook trout (*Salvelinus fontinalis*) are now widespread throughout the GYE, and lake trout are found in many GYE lakes and reservoirs, including Yellowstone and Jackson lakes. More recently, eastern warm-water species have moved into the GYE, including smallmouth bass (*Micropterus dolomieu*) and northern pike (*Esox lucius*).

These exotic species reduce native fish distributions and populations through hybridization, predation, and competition. The impacts of the exotic fish species are exacerbated by reduced river flows, increased pollution, impoundments, loss of connectivity between headwater populations, and the exotic whirling disease. Van Kirk and Benjamin (2001) assessed the status of native fish across the GYE based on current distribution relative to historic distribution. They ranked the status of native salmonids as "good" in eight of forty-one watersheds (20 percent), "fair" in four (10 percent), and "poor" in the remaining twenty-nine. All watersheds in which native salmonid status was either good or fair occurred in the Upper Yellowstone, Upper Snake, and Bear River basins (fig. 16-7). Watersheds with poor status were in lower watersheds on both private and public lands.

Except for the high levels of mortality in whitebark pine described earlier, vegetation in the GYE appears to be resilient to the level of climate change that has occurred to date. Climate change in the coming decades, however, is projected to exceed the climate tolerances of most tree species in portions of their current range in the GYE (chaps. 9 and 10). In general, climate suitability for tree species is projected to shift upslope. Climate suitability for whitebark pine, now at upper treeline, is projected to be largely lost in the GYE. The suitability of subalpine species shifts to the current alpine zone, mid-elevation montane shifts to the current subalpine zone, and low- to mid-elevation forests become suitable for the nonforest sagebrush/grassland and juniper communities now in the valley bottoms.

Ecological Integrity Report Card

Given the changes described above, how well has the ecological integrity of GYE been sustained? To what extent are the criteria from the Parks Canada Agency (2008) quoted near the beginning of this chapter being met, including "the structure, composition and function of the ecosystem unimpaired by stresses from human activity; natural ecological processes intact and self-sustaining, the ecosystem evolves naturally and its capacity for self-renewal is maintained; and the ecosystem's biodiversity is ensured"?

The information required to answer this question, largely reviewed above, is far from complete (table 16-2). The highest level of monitoring, analysis, and reporting of results is within the national parks (e.g., Yellowstone Center for Resources 2011). Virtually no monitoring is done by the other federal land management agencies in the GYE, with the exception of those mentioned below. Monitoring on private lands is very limited. The only measures of ecological integrity that are regularly monitored across the private and public lands are those monitored by national programs (climate, river discharge, snow water equivalent, tree species, breeding and wintering birds), and none of these is summarized and reported at the level of the GYE. Monitoring and reporting is done across the entire ecosystem for high-profile mammal species that are largely restricted to public lands (e.g., grizzly bear and gray wolf). Consequently, environmental decision makers, such as county commissioners, do not have the benefit of including

TABLE 16-2. Status of monitoring and reporting of indicators of ecological integrity within the Greater Yellowstone Ecosystem.

Class of Vital Sign	Vital Sign	Quantified By	Frequency	Extent	Summarized and Reported by Media GYE-wide Periodically
Stressors					
Population	Human density	US Census	Decadal since 1900	US	No
Developed lands	Rural homes, suburban, urban, commercial, agriculture, wells, roads	NPS I&M; Gude, Hansen, and Jones 2007; Piekielek and Hansen 2012	Once	GYE	No
Backcountry recreation	Public lands visitation	NPS	Annually	YNP, GTNP	National Parks only
	Backcountry visitor nights	NPS	Annually	YNP	YNP only
	Hunting licenses	States	Annually	Within states	Within states only
	Fishing licenses	States	Annually	Within states	Within states only
	Resort skiing	Ski areas	Annually	Individual resorts	No
	Mountain biking, backcountry skiing, wildlife viewing	Not quantified			No
Air	Air quality	NPS	Annually	YNP, GTNP	National Parks only
Climate	Temperature	NPS I&M	Annually	GYE	No
	Precipitation	NPS I&M	Annually	GYE	No
Exotics	Aquatic	No			No
	Terrestrial plants	No			No
Ecological Processes Response					
Snowpack	Apr 1 SWE	USGS	Annually since 1950	US	Within Watersheds only
Water	River discharge	USGS	Annually since 1950	US	Within Watersheds only
	Quality	NPS		YNP, GTNP	National parks only
	Temperature	No			No
	Hydrologic integrity	Van Kirk and Benjamin 2001	Once	GYE	No

TABLE 16-2. (*Continued*)

Class of Vital Sign	Vital Sign	Quantified By	Frequency	Extent	Summarized and Reported by Media GYE-wide Periodically
Forest mortality	Whitebark pine mortality	NPS I&M;	Annually since 2004	GYE	Yes
Fire	Acres burned	USFS	Annually	US	No
Biodiversity Response					
Habitat fragmentation	See fig. 16-9	This chapter	Once	GYE	No
Trees	Basal area by species	USFS FIA	Five-year intervals since 1990s	US	No
Mammals	Wolves	USFWS	Annually since 1996	GYE	Yes
	Grizzly bear	NPS	Annually since 1975	GYE	Yes
	Elk	NPS, States	Annually	Within hunting districts or ranges	Individual herds only
	Moose	No			No
	Pronghorn	NPS, States	Annually	Within hunting districts or ranges	No
	Mule deer	NPS, States	Annually	Within hunting districts or ranges	No
	Bighorn sheep	NPS, States	Annually	Within herds	No
Birds	Breeding birds	BBS	Annually since 1980s	US	No
	Wintering birds	CBC	Annually since 1900s	US	No
Amphibians	Four species	NPS I&M	Once	YNP	No
Fish	Cutthroat throat	NPS	Annually	Yellowstone Lake	No
	Arctic grayling	No			No
Biodiversity indices	Elk migration	Wyoming Coop Unit	Current	GYE	No
	Connectivity	Theobald 2010, 2014	Once	US	No
	Bird hot spots	Hansen et al. 2002	Once	GYE	No
	Irreplaceable areas	Noss et al.	Once	GYE	No

Note: US = United States; GYE = Greater Yellowstone Ecosystem; GTNP = Grand Teton National Park.

TABLE 16-3. Greater Yellowstone Ecosystem ecological integrity report card.

| Vital Sign | Public Lands | | Private Lands | | All Lands |
	Grade	Certainty	Grade	Certainty	Grade
Snowpack and river discharge	C	High	D	High	C
Air quality	A	High	B	High	B
Hydrologic integrity and water quality	A	High	D	Low	C
Forest mortality	C	High	B	High	C
Wildfire deviation from natural range	A	Medium	C	Medium	B
Habitat fragmentation	A	High	C	High	B
Large mammals	A	High	A	High	A
Breeding birds	A	Medium	C	Medium	B
Fish and amphibians	C	Medium	E	Medium	D
Connectivity to other wildlands	B	Medium	B	Medium	B

impacts on GYE ecological integrity in the information they consider when evaluating development projects such as subdivision approvals.

Based on the information that is available, ecological integrity for the GYE can be summarized in the form of a report card (table 16-3) that grades public lands, private lands, and the GYE as a whole. This exercise is subjective and based on inadequate data. Our goal is to make a case for more rigorous monitoring and reporting and for development of criteria to more objectively assess trends in ecological integrity. Examples of other attempts to quantify the ecological condition of wildland eco-systems include Lookingbill et al. (2014), Hansen et al. (2014), and Scheffer et al. (2015).

Public lands in the GYE continue to have all native species present, and their structure, function, and composition are largely well within the natural range of variation. The restoration of degraded or extirpated populations of large mammals (wolves) is a major conservation success story. Thus, high grades are assigned for most elements of ecosystem function and biodiversity on public lands. The initial responses to climate warming are most evident at higher elevations, however. Snowpack and runoff have declined substantially in recent decades. Mountain pine beetles are benefiting from the warming, and whitebark pine are undergoing mass mortality due to beetle infestation. Thus, grades for these elements are lower. This is also the case with fish and amphibians as a result of both climate change effects and, to a higher degree, the effects of exotic fish moving upstream from private lands.

The private lands in GYE, in contrast, have substantially declined in ecological integrity. Several habitat types have been largely converted to rural home plots, farms, suburbs, or cities. Rivers have been impounded, diverted, constrained, dewatered, and possibly polluted. Natural disturbance regimes have been altered. Riparian zone diversity has been reduced below impoundments by lack of flooding. Lower-elevation forests, long adapted to frequent low-intensity fire, are now experiencing severe fire, with unknown consequences for erosion, sedimentation into streams, and invasion of nonnative species. Exotic species, noxious weeds, and over-abundant mesocarnivore native species dominate large river and valley bottom communities and are reducing the viability of native species. Bird hot spots that were likely source areas for subpopulations at higher elevations are now population sinks. Native fish have been largely extirpated from the most impaired rivers.

The differences in ecological integrity across the GYE largely coincide, of course, with land allocation and management effectiveness. The public lands in GYE arguably are well managed due to the considerable expertise and resources of federal and state agencies. This is a considerable achievement because these public land managers have been challenged by the large size of the GYE and the complex mix of management jurisdictions.

It is, of course, the places where people live, work, and grow food that nature conservation is most challenged. In mountainous systems like the GYE, private lands and human endeavors are concentrated in the small portion of the system that is most important for native species and key ecosystem processes. The locations with more favorable climate, better soils, surface water, and groundwater that attract people are also locations of high ecological productivity, native species diversity, key seasonal habitats, and higher demographic performance. Consequently, the condition of these lower elevation private lands is vital to the ecological integrity of the broader ecosystem. Considerable progress has been made on conservation of some private lands in the GYE. Private lands protected by conservation easements have increased exponentially since 1970 and include some of the most ecologically important lands. Moreover, people are increasingly learning how to live with dangerous wildlife, such as grizzly bears and wolves. Much more progress will be needed, however, to retain or restore ecological integrity on the private lands of the GYE in the face of growing human populations and climate change.

Conclusion

Like many other remaining wildland ecosystems across the world, the GYE is at a crossroad. The natural factors that inhibited human expansion here in the past are now major attractants for people and businesses that value access to high-quality nature. The resulting land use pressures on private lands and climate change stresses on public lands have reduced the ecological integrity of the GYE. These forces are projected to increase in the coming decades, raising questions about the future prospects for sustaining the GYE as a wildland ecosystem. If Yellowstone, as the first national park, inspired the creation of the global system of protected areas, can the GYE inspire progress toward sustainable human and wildland systems?

Application of the Climate-Smart Conservation framework is one important element of a strategy for sustaining the GYE. By identifying the most vulnerable natural resources and focusing management attention on them, there is hope of adapting to the most pervasive impacts of climate and land use change. Other elements include developing across the public and private lands of the ecosystem a vision and benchmarks for ecological integrity, establishing a system for monitoring and evaluating key ecological metrics, and communicating to stakeholders in an annual report card or other form the status and trends of the ecosystem. It is critically important that communities of the GYE come together to discuss future prospects and develop means of working toward a desirable future.

Acknowledgments

Patty Gude provided data on rural homes. S. Thomas Olliff, Virginia Kelly, John E. Gross, Woody Turner, and Gary Tabor shared insights on the topics in this chapter. David M. Theobald, S. Thomas Olliff, and Virginia Kelly provided review comments on earlier drafts of the chapter.

References

Al-Chokhachy, R., J. Alder, S. Hostetler, R. Gresswell, and B. Shepard. 2013. Thermal controls of Yellowstone cutthroat trout and invasive fishes under climate change. *Global Change Biology* 19:3069–81. doi: 10.1111.

Allan, J. D., and M. M. Castillo. 2007. *Stream Ecology. Structure and Function of Running Waters*. Dordrecht, The Netherlands: Springer.

Bangs, E. E., J. A. Fontaine, M. D. Jimenez, F. J. Meier, E. Bradley, C. C. Niemeyer, D. W. Smith, C. M. Mack, V. Asher, and J. K. Oakleaf. 2005. Managing wolf-human conflict in the northwestern United States. Chap. 21 in *People and Wildlife: Conflict or Coexistence*, edited by R. Woodroffe, S. Thirgood, and A. Rabinowitz. London: Cambridge University Press.

Baril, L. M., A. J. Hansen, R. Renkin, and R. Lawrence. 2011. Songbird response to increased willow (*Salix* spp.) growth in Yellowstone's northern range. *Ecological Applications* 21 (6): 2283–96.

Black, G. 2012. *Empire of Shadows: The Epic Story of Yellowstone*. New York: St. Martin's Press.

Brown, K., A. J. Hansen, R. E. Keane, and L. J. Graumlich. 2006. Complex interactions shaping aspen dynamics in the Greater Yellowstone Ecosystem. *Landscape Ecology* 21:933–51.

Chang, T., and A. J. Hansen. 2015. Historic and projected climate change in the Greater Yellowstone Ecosystem. *Yellowstone Science* 23 (1): 14–19.

Chape, S., J. Harrison, M. Spalding, and I. Lysenko. 2005. Measuring the extent and effectiveness of protected areas as an indicator for meeting global biodiversity targets. *Philosophical Transactions of the Royal Society B* 360:443–55. doi: 10.1098/rstb.2004.1592.

Davis, C. R., and A. J. Hansen. 2011. Trajectories in land-use change around US national parks and their challenges and opportunities for management. *Ecological Applications* 21 (8): 3299–3316.

Despain, D. 1990. *Yellowstone Vegetation*. Boulder, Co: Roberts Rinehart.

Finn, S., Y. Converse, T. Olliff, M. Heller, R. Sojda, E. Beever, S. Pierluissi, J. Watkins, N. Chambers, and S. Bischke. 2015. Great Northern Landscape Cooperative Science Plan 2015–2019. http://greatnorthernlcc.org/sites/default/files/documents/gnlcc_science_plan.pdf.

Garcia, A. 1967. *Tough Trip through Paradise*. Boston: Houghton Mifflin.

Gude, P. H., A. J. Hansen, and D. A. Jones. 2007. Biodiversity consequences of alternative future land use scenarios in Greater Yellowstone. *Ecological Applications* 17 (4): 1004–18.

Gude, P. H., A. J. Hansen, R. Rasker, and B. Maxwell. 2006. Rate and drivers of rural residential development in the Greater Yellowstone. *Landscape and Urban Planning* 77:131–51.

Haines, A. L. 1977. *The Yellowstone Story: A History of Our First National Park*. Yellowstone National Park, WY: Yellowstone Library and Museum Association.

Hansen, A. J., R. Knight, J. Marzluff, S. Powell, K. Brown, P. Hernandez, and K. Jones. 2005. Effects of exurban development on biodiversity: Patterns, mechanisms, research needs. *Ecological Applications* 15 (6): 1893–1905.

Hansen, A. J., N. Piekielek, C. Davis, J. Haas, D. Theobald, J. Gross, W. Monahan, T. Olliff, and S. Running. 2014. Exposure of U.S. national parks to land use and climate change 1900–2100. *Ecological Applications* 24 (3): 484–502.

Hansen, A. J., R. Rasker, B. Maxwell, J. J. Rotella, J. Johnson, A. Wright Parmenter, U. Langner, W. Cohen, R. Lawrence, and M. V. Kraska. 2002. Ecological causes and consequences of demographic change in the New West. *BioScience* 52 (2): 151–68.

Hansen, A. J., and J. J. Rotella. 2002. Biophysical factors, land use, and species viability in and around nature reserves. *Conservation Biology* 16 (4): 1–12.

Hansen, A. J., J. J. Rotella, M. L. Kraska, and D. Brown. 2000. Spatial patterns of primary productivity in the Greater Yellowstone Ecosystem. *Landscape Ecology* 15:505–22.

Janetski, J. C. 1987. *Indians of Yellowstone National Park*. Salt Lake City: University of Utah Press.

Jin, S., L. Yang, P. Danielson, C. Homer, J. Fry, and G. Xian. 2013. A comprehensive change detection method for updating the National Land Cover Database to circa 2011. *Remote Sensing of Environment* 132:159–75.

Jones, D. A., A. J. Hansen, K. Bly, K. Doherty, J. P. Verschuyl, J. I. Paugh, R. Carle, and S. J. Story. 2009. Monitoring land use and cover around parks: A conceptual approach. *Remote Sensing of Environment* 113:1346–56.

Karr, J. R. 1987. Biological monitoring and environmental assessment: A conceptual framework. *Environmental Management* 11:249–56.

Leppi, J. C., T. H. DeLuca, S. W. Harrar, and S. W. Running. 2012. Impacts of climate change on August stream discharge in the Central-Rocky Mountains. *Climatic Change* 112:997–1014.

Littell, J. S. 2002. Determinants of Fire Regime Variability in Lower Elevation Forests of the Northern Greater Yellowstone Ecosystem. Thesis. Montana State University, Bozeman, MT.

Logan, J. A., W. W. Macfarlane, and L. Willcox. 2010. Whitebark pine vulnerability to climate-driven mountain pine beetle disturbance in the Greater Yellowstone Ecosystem. *Ecological Applications* 20 (4): 895–902.

Logan, J. A., and J. A. Powell. 2009. Ecological consequences of climate change altered forest insect disturbance regimes. In *Climate Change in Western North America: Evidence and Environmental Effects*, edited by F. H. Wagner. Salt Lake City: University of Utah Press.

Lookingbill, T. R., J. P. Schmit, S. M. Tessel, M. Suarez-Rubio, and R. H. Hilderbrande. 2014. Assessing national park resource condition along an urban–rural gradient in and around Washington, DC, USA. *Ecological Indicators* 42:147–59.

Macfarlane, W. W., J. A. Logan, and W. Kern. 2013. An innovative aerial assessment of Greater Yellowstone Ecosystem mountain pine beetle-caused whitebark pine mortality. *Ecological Applications* 23:421–37.

Mattson, D. J., S. Herrero, R. G. Wright, and C. M. Pease. 1996. Science and management of Rocky Mountain grizzly bears. *Conservation Biology* 10:1013–25. doi: 10.1046/j.1523-1739.1996.10041013.x.

Mogk, D., D. Henry, P. Mueller, and D. Foster. 2012. Origins of a continent. *Yellowstone Science* 20 (2): 22–32.

Noss, R. F., C. Carroll, K. Vance-Borland, and G. Wuerthner. 2002. A multicriteria assessment of the irreplaceability and vulnerability of sites in the Greater Yellowstone Ecosystem. *Conservation Biology* 16 (4): 895–908.

Parks Canada Agency. 2008. *Parks Canada Guide to Management Planning*. Gatineau, Quebec: Parks Canada Agency.

Parrish, J. D., D. P. Braun, and R. S. Unnasch. 2003. Are we conserving what we say we are? Measuring ecological integrity within protected areas. *BioScience* 53 (9): 851–60. doi: 10.1641/0006-3568(2003)053[0851:AWCWWS]2.0.CO;2.

Pederson, G. T., S. T. Gray, C. A. Woodhouse, J. L. Betancourt, D. B. Fagre, J. S. Littell, E. Watson, B. H. Luckman, and L. J. Graumlich. 2011. The unusual nature of recent snowpack declines in the North American cordillera. *Science* 333 (6040): 332–35.

Picton, H. D., and T. N. Lonner. 2008. *Montana's Wildlife Legacy: Decimation to Restoration*. Bozeman, MT: Media Works.

Piekielek, N. B., and A. J. Hansen. 2012. Extent of fragmentation of coarse-scale habitats in and around US national parks. *Biological Conservation* 155:13–22.

Powell, S. L., and A. J. Hansen. 2007. Conifer cover increase in the Greater Yellowstone Ecosystem: Frequency, rates, and spatial variation. *Ecosystems* 10:204–16.

Raffa, K. F., B. H. Aukema, B. J. Bentz, A. L. Carroll, J. A. Hicke, M. G. Turner, and W. H. Romme. 2008. Cross-scale drivers of natural disturbances prone to anthropogenic amplification: The dynamics of bark beetle eruptions. *BioScience* 58 (6): 501–17.

Rasker, R., and A. J. Hansen. 2000. Natural amenities and population growth in the Greater Yellowstone region. *Human Ecology Review* 7 (2): 30–40.

Romme, W. H., and D. G. Despain. 1989. Historical perspective on the Yellowstone fires of 1988. *BioScience* 39 (10): 695–99.

Scheffer, M., S. Barrett, S. R. Carpenter, C. Folke, A. J. Green, M. Holmgren, T. P. Hughes, S. Kosten, I. A. van de Leemput, D. C. Nepstad, et al. 2015. Creating a safe operating space for iconic ecosystems. *Science* 347:1317–19.

Schullery, P. 1997. *Searching for Yellowstone*. Boston: Houghton Mifflin.

Schwartz, C. C., M. A. Haroldson, and G. C. White. 2010. Hazards affecting grizzly bear survival in the Greater Yellowstone Ecosystem. *Journal of Wildlife Management* 74:654–67.

Theobald, D. M. 2014. Development and applications of a comprehensive land use classification and map for the US. *PLOS ONE* 9 (4): e94628. doi: 10.1371/journal.pone.0094628.

Theobald, D. M., J. R. Miller, and N. T. Hobbs. 1997. Estimating the cumulative effects of development on wildlife habitat. *Landscape and Urban Planning* 39:25–36.

US Census Bureau. 2014. 2014 TIGER/Line Shapefiles (machine-readable data files). Prepared by the US Census Bureau.

Van Kirk, R. W., and L. Benjamin. 2001. Status and conservation of salmonids in relation to hydrologic integrity in the Greater Yellowstone Ecosystem. *Western North American Naturalist* 61 (3): 359–74.

Westerling, A. L., M. G. Turner, E. A. Smithwick, W. H. Romme, and M. G. Ryan. 2011. Continued warming could transform greater Yellowstone fire regimes by mid-21st century. *Proceedings of the National Academy of Sciences of the United States of America* 108 (32): 13165–70.

Yellowstone Center for Resources. 2011. *Yellowstone National Park: Natural Resource Vital Signs*. YCR-2011-07. Mammoth Hot Springs, WY: National Park Service.

Chapter 17

Synthesis of Climate Adaptation Planning in Wildland Ecosystems

Andrew J. Hansen, David M. Theobald,
S. Thomas Olliff, and William B. Monahan

The chapters of this book have delved into the timely and important topic of science and management of wildland ecosystems in the face of climate and land use change. The period of the book's development (2011–2015) was one of rapid advancement in science, policy, agency infrastructure, and understanding of climate change adaptation (chaps. 2, 3, and 13). During this period, evidence of climate change and its consequences was ever more apparent. This was the warmest five-year period on record (http://www .ncdc.noaa.gov). Extreme climate events, such as droughts in California, Amazonia, and Australia, caused fundamental changes in allocating water to people and managing human risk from fire. Evidence of the ecological impacts of climate change became pervasive, including forest die-offs in many parts of the world and massive bleaching of coral ecosystems.

During this time of rapid climate change, our ability to understand and develop ways to manage under climate change has evolved substantially. In essence, this book tells the story of the progress made in climate adaptation on federal lands within the Great Northern and Appalachian landscape conservation cooperatives (LCCs) during this time period. This concluding chapter synthesizes what we have learned about climate change and its consequences in these regions, evaluates progress in linking climate science with management to develop and implement adaptation options, identifies "lessons learned," and explores challenges and opportunities for applications elsewhere in the United States and internationally.

Global Change in the Rocky and Appalachian Mountains

Our reconstruction and projection of climate for 1900–2100 revealed corresponding reductions in snowpack and runoff and increases in severe fires, forest mortality, and exotic species. In the coming century, changes in temperatures and aridity are projected to meet or exceed those that have occurred in the fourteen thousand years since the last glacial advance. Paleoecological studies have revealed that vegetation has been resilient to climate fluctuations in past millennia. The extremely rapid rate of change projected for the coming decades, however, is likely to dramatically alter wildland ecosystems, pushing portions of them into novel states with no current or recent analogs.

Both the Rockies and the Appalachians have undergone variation in climate over the past century, likely associated with long-term natural cycles operating over periods of ten to thirty years (chaps. 4 and 5). Although the southern Appalachians do not show strong trends in temperature over the past century, the Rockies have warmed substantially and aridity has increased, especially since 1980, suggesting a signal of human-induced climate change. Ecological responses to this climate change have emerged in the past few decades (chap. 7). Snowpack and runoff have declined, fire frequency has increased, forest pests have expanded leading to high levels of tree mortality (chaps. 14 and 15), and nonnative fish—better able to tolerate warm water—have increasingly impacted native species (chap. 12).

Projections for the coming century indicate dramatic warming in both the Rockies and the Appalachians (chaps. 4 and 5). Among the many tree species in the Appalachian region, some are projected to cope well with climate change and even expand in distribution. Others, such as those in the subalpine and in moist cove forests, are projected to decline dramatically. The Rockies in the coming century are projected to undergo major transformations from snow-fed to rain-fed watersheds, from cold rivers with thriving trout populations to warm and flashy rivers supporting some eastern warm-water fish, from short summer growing seasons to complex spring and fall growing seasons, and from late-seral forest–dominated mountain landscapes to desert scrub–dominated landscapes.

In both of these mountainous regions, human land use has intensified. This is most pronounced in the Appalachians, where wildlands on private parcels have been converted to farms, rural home plots, and urban areas since the 1940s (Hansen et al. 2014). These trends were slower to emerge in the US northern Rockies, but now private lands there also have high rates of conversion (chap. 16). Climate and land use change are often synergistic in influencing ecosystems (chap. 6). Both destroy habitat for some species, typically with climate change felt most strongly at higher elevations and land use at lower elevations. Land use constrains options for managing under climate change—for example, by limiting the use of natural and pre-scribed fire and not allowing flooding to accomplish ecological objectives. Land use also constrains the abilities of native species to relocate under climate change and facilitates the expansion of harmful nonnative species. Climate warming is likely a positive feedback for land use in the Rock-ies, attracting more residents as the harsh climate becomes more equitable, leading to further land use intensification.

Thus, wildland ecosystems in our study regions are likely to undergo dramatic reductions in ecological integrity with major shifts in structure, composition, and function. The consequences of this for people living in and around these ecosystems will likely be substantial. Infrastructure, such as roads, culverts, and buildings, may require retrofitting, as has already occurred following the extreme fire and flooding events in Rocky Moun-tain National Park (chap. 14). The costs of managing these ecosystems will likely increase substantially. Fire protection, weed management, research, and adaptive management needs are expanding under climate change, and budgets will need to increase to allow effective stewardship. The communi-ties surrounding wildlands are particularly tied to ecosystem services. Re-ductions in water yield, agricultural production, and forest and forage pro-ductivity will have strong negative economic effects. Rural homes will be increasingly at risk of wildfire. Natural amenities, including snow-related and coldwater fishery–related tourism, and the revenue they generate, will decline.

In essence, wildland ecosystems are in a similar situation as small island nations around the world. These places have contributed little to the green-house gas emissions that are driving global climate change but dispropor-tionately bear the risks and costs of the climate change.

Progress in Climate Adaptation

This book clearly demonstrates the importance of climate change in wild-land ecosystems. Fortunately, it also provides evidence of the rapid evolu-tion in our ability to understand and manage wildland ecosystems in the face of this threat (chaps. 2, 3, and 13). Scientific data, tools, models, and knowledge on climate change and its consequences have advanced rapidly. Policy mandating adaptation to climate change has recently emerged from the federal and state governments. Conceptual approaches for climate adaptation, such as the Climate-Smart Conservation framework, have been developed and refined, and institutional infrastructure, such as LCCs, has radically expanded in the past five years.

Since 2010, the major national and international programs dealing with climate adaptation have made considerable progress in staffing, infrastructure, research capacity, policy, and education (chaps 2, 3, and 13; fig. 17-1; Halofsky et al. 2015). Ken Salazar's Secretarial Order 3289 (US Department of the Interior 2009) might be considered the dawn of climate change science and adaptation on public lands in the United States. This order established an approach throughout the Department of the Interior (DOI) for "applying scientific tools to increase understanding of climate change and to coordinate an effective response to its impacts."

The DOI programs put in place by this order have hired staff and developed infrastructure, built partnerships with stakeholders, funded research, and provided guidance on climate adaptation. For example, the North Central Climate Science Center, which includes the US northern Rockies, has developed and is implementing a five-year plan to deliver climate, data, tools, and scientific knowledge to managers. Activities include a foundational science team and plan, a facility for training scientists and managers in quantitative climate change vulnerability assessment, projects with funding and aimed at demonstrating links between climate science and management, and symposia to communicate major findings. Similarly, the Great Northern LCC has developed a community of partners, crafted a science plan, funded several applied projects, developed a data repository, and conducted numerous webinars and training activities.

The National Park Service initiated a Climate Change Response Program, commissioned a major science and policy report to help the National Park System cope with climate change (Colwell et al. 2012), and is now developing policy guidelines based on the recommendations in the report. The US Forest Service has required that climate change be considered during forest plan revision (Code of Federal Regulations 2012). The interagency working group that wrote the original *Scanning the Horizons* framework for climate adaptation (Glick, Stein, and Edelson 2011) has now refined and expanded the approach into the Climate-Smart Conservation framework (Stein et al. 2014). Groups of states across the United States have developed alliances on coping with land use and climate change, as illustrated by the Western Governors' Association Crucial Habitats and Corridors Initiative. Individual national parks and national forests have begun

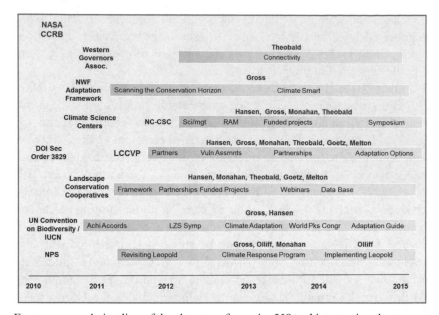

FIGURE 17-1 A timeline of development for major US and international programs on climate change adaptation, emphasizing projects and initiatives in the Rocky Mountains and Appalachian Mountains, and a listing of participation by Landscape Climate Change Vulnerability Project team members. Key: NASA CCRB = NASA Climate Change and Biological Response request for proposals; NWF = National Wildlife Federation; NC-CSC = North Central Climate Science Center; RAM = Resource for Advanced Modeling; LCCVP = Landscape Climate Change

to allocate staff and resources to planning, monitoring, and managing in the context of climate change (e.g., chap. 14).

Internationally, the United Nations Convention on Biodiversity has set goals for expanding protected areas and maintaining ecological integrity (http://www.cbd.int), and the International Union for the Conservation of Nature has been writing guides for implementation and monitoring compliance (http://www.iucn.org/about/work/programmes/gpap_home). Participation by our research team in many of these activities (fig. 17-1) has resulted in synergies and mutual benefits among the Landscape Climate Change Vulnerability Project (LCCVP) and these other efforts. In short, these US federal and international programs have developed a community of practice around land use and climate change impacts assessment.

The component of climate adaptation that has been slower to emerge is the actual implementation of on-the-ground adaptation actions (chap. 13). Many of the examples that do exist were in response to extreme events, such as fires, floods, or pest outbreaks (chaps. 14 and 15), that necessitated immediate action. The lack of management action in nonemergency cases should not be disheartening, however, because on-the-ground management is the culmination of a complex intersection of science, policy, agency infrastructure, education, stakeholder engagement, and funding. Bringing these together can take years to decades within federal agencies. As evidenced in this book, rapid progress at the intersection described above has happened in just the past five years. The rapid evolution of the elements of this intersection suggests that adaptation actions are in the pipeline and will increasingly become "shovel ready." Thus, we conclude that the wheels are in motion for a new era of natural resource management that is increasingly equipped to cope with the challenges of climate and land use change.

The contents of this book illustrate the considerable progress that has been made within the Great Northern and Appalachian LCCs. The success of the LCCVP and the progress within the two LCCs and by the North Central Climate Science Center illustrate how sustained funding and collaboration among resource managers and scientists over a five-year period can yield rapid results.

While our work has been under way, many other groups around the world have also tackled issues embedded within climate change adaptation. Examples of studies that synthesized current science and identified adaptation options include Brandt et al. (2014) for the central US hardwood forests, Gauthier et al. (2014) and Parka et al. (2014) for boreal forests, Lindner et al. (2014) and Brang et al. (2014) for European forests, and Shoo et al. (2014) for Australian rainforests and wetlands. Along with chapter 15 of this book, one of the first studies to illustrate implementation of climate adaptation options in forested ecosystems is Janowiak et al. (2014).

Lessons Learned

Collectively, the efforts described in this book and in these other studies reveal some "lessons" on how to hasten progress on climate adaptation planning. These lessons are described in more detail in chapters 3 and 13 and in Gross et al. (2011).

Science–Management Partnerships

Tackling climate adaptation planning requires diverse, interdisciplinary teams of scientists and managers. Within the science component, expertise in climatology, hydrology, population and community ecology, remote sensing, spatial analysis, and ecological modeling is needed. The management component benefits from inclusion of specialists in specific resources (e.g., vegetation, wildlife, fish, disturbance), higher-level program managers, and interpretative and communication experts. Large science–management teams are built over a number of years; familiarity, trust, and respect lead to successful outcomes and climate change adaptation efforts. There are no substitutes for repeated personal interactions for building and maintaining effective teams. Such integrated teams of scientists and managers allow for the iterative learning that is essential for success. Scientists can educate managers on the climate science, and managers can educate scientists on what science is "actionable" by federal agencies.

In the case of the LCCVP, the core science team has worked together for more than a decade. We maintain close connections through monthly team calls and biannual team meetings. Although each scientist also works on many other projects, we have found synergies whereby concepts and tools developed from our other studies are evaluated for relevance to this project, often with productive outcomes. For example, David M. Theobald's United States–wide mapping of landforms and connectivity was co-opted in our project as a basis for evaluating vulnerability of landscapes across the Great Northern LCC and the western United States for the Western Governors' Association Crucial Habitat Tool. Similarly, the core management teams in the project have collaborated for many years and started this project with the strong social ties necessary for successful outcomes.

Linking the science and management teams within the LCCVP was, and remains, a challenge. The linkages have worked best when individual scientists and individual managers invested in personal relationships that provide the glue to unite the larger partnership. Ben Bobowski and William B. Monahan played that role in the case of the Rocky Mountain Protected Area Centered Ecosystem, as did Virginia Kelly, S. Thomas Olliff, and Andrew J. Hansen in Yellowstone, Scott Goetz and Rich Evans in the case of the Delaware Water Gap National Recreation Area, and Patrick Jantz and Jim Renfro in the context of Great Smoky Mountains National Park.

Management-Relevant Issues

Although scientists often select projects out of interest, managers select them out of necessity. The most pressing management problems get moved to the front of the line for attention. In the resource- and time-limited federal agencies, climate change is often considered an issue for the distant future and thus secondary to more pressing issues. The challenge to successful science and management teams is to identify the climate change issues that are "management relevant"—or, in other words, high priorities for immediate management attention. Projected reductions in ecosystem productivity in future decades do not make the cut on this. Reductions in productivity as manifest by thousands of acres of dead trees and raging fires do make the cut. An example comes from Rocky Mountain National Park—a severe fire followed by intense rain and flooding led to active management to install larger culverts under roads to better withstand land sliding effects.

Time Required for Planning

Like other types of management actions on federal lands, climate adaptation options will typically be enacted only after several years of undertaking science and working through the federal planning process. Most climate adaptation actions are large in scope and so are likely to be subject to the National Environmental Policy Act. Large, complex projects under the act can take decades for planning. For example, the current elk and vegetation management plan in Rocky Mountain National Park was enacted about ten years after initiation. Consequently, a substantial form of "success" in applying the Climate-Smart Conservation framework is bringing a management-relevant climate project to the point of being the subject of an active planning dialogue. One of us (William B. Monahan) summed up the situation as follows: "I would thus argue that all of our work is educational, either to inform current plans being developed, or more distant ones yet to be conceived."

Smart Science

Because many scientists work on topics they find interesting, it is easy for collaborative efforts to be too broad to be highly effective. The science that is done needs to be "smart science." By this we mean science that is highly relevant to a well-defined problem at hand, designed at the scales and level of specificity appropriate for the problem, and analyzed and interpreted to be robust to uncertainty. In this project, we limited the response variables to those that management partners suggested had high potential for management relevance and that we knew how to model and analyze well. In analyzing and interpreting results, we framed the outcomes of multiple studies, scenarios, or models to identify high-level agreement or disagreement among them. Higher levels of agreement inspired more confidence in the results; lower levels indicated that more research was needed.

The whitebark pine (*Pinus albicaulis*) issue in Yellowstone National Park provides an example. The five or six key interacting factors influencing whitebark pine response to climate change reflect a level of complexity too high to be reliably tackled in this project. Thus, we focused on one factor (climate and habitat suitability) that we thought was the most important first filter for organizing management (chaps. 9 and 15). A high level of agreement among climate models and scenarios in the large projected loss of habitat suitability reinforced that this is a primary filter for designing adaptation options. Climate suitability zones are now being used as a basis for tailoring management treatments to the landscape in order to be most effective under climate change (chap. 15).

"Place" Is Critical

Resource management is about important resources in particular places. Most resource managers know their unit and resources, and thus effective resource management efforts are place based. Climate change projects necessarily require consideration of landscapes and regions much larger than typical management units. Consequently, these projects are best tackled with careful attention to scaling. A good approach is to design the research to allow data, analyses, and interpretation to be generated at multiple spatial scales, from small watersheds to management jurisdictions (e.g., national parks) to functionally important areas (e.g., protected area centered ecosystems, or PACEs) up to entire landscape conservation cooperatives.

This design allows the results to be useful for adaptation planning for the places that most matter to resource managers. For example, collaborators in Great Smoky Mountains National Park were interested in broad regional patterns of climate change and vegetation response but were more particularly interested in community types of high management concern, such as subalpine forest and cove forests. Thus, the managers indicated that the results of the LCCVP would be most useful if they could be applied to capture these fine-scale vegetation units (chap. 8).

Communication

Climate adaptation projects can ultimately be successful only if communication of key findings happens with stakeholders and decision makers.

Communication vehicles differ among scientists, resource managers, policy makers, and citizen stakeholders, as described in chapter 2. All these vehicles should be considered in these large collaborative projects. The need to communicate to resource managers, however, does not reduce the need to communicate with scientists. The credibility conveyed by the scientific peer-review process is highly valued by the management community.

Defining Impact

The Climate-Smart Conservation framework suggests that the ultimate step in climate adaptation is active implementation of an adaptation option and monitoring the consequences. We have learned, however, that working through the Climate-Smart Conservation framework typically takes several years and is contingent on critical milestones along the way. These milestones include consideration of climate change in decisions about policy, budgets, education programs, passive management and, finally, active management. The impact of a project is best judged relative to each of these milestones because they each contribute to overall success.

Developments in Rocky Mountain National Park are a case in point (chap. 14). Since their initial workshop on climate change in 2007, park resource managers have focused on monitoring, research, education, and collaboration with partners, all of which are considered critical precursors to effective active management, a stage that they have largely not yet implemented. Similarly, managers of whitebark pine in the Greater Yellowstone Ecosystem are now incorporating climate science into active management only after years of assessment, planning, strategy development, and strategy evaluation (chap. 15). The Climate-Smart Conservation framework is indeed a cycle without a fixed end point. Progress on each step is an incremental contribution to sustaining ecosystems under climate change. In this context, the federal agencies in the Rocky and Appalachian Mountain regions have made substantial advancement in climate adaptation.

Challenges and Opportunities for the Future

Throughout the book, we have emphasized that although progress has been rapid over the past five years, climate adaptation will need to continue to evolve rapidly to allow for natural resource managers to cope with climate change. The challenges that we have highlighted also represent opportunities for greater progress in the future.

- *Engage private landowners and diverse local stakeholders.* The fate of federal land management is heavily influenced by local people. Better information on how wildland ecosystems benefit local communities may enhance the engagement of local people in climate adaptation and lead to more informed decision making and better maintenance of ecological integrity. Such information should include evaluation of what attributes of wildland ecosystems are highly valued by local people. This information could provide the basis for a mission statement, a vision, and benchmarks for sustaining the attributes that are vital to both ecological integrity and healthy vibrant human communities. It is imperative that the status and trends of these ecological and human communities are communicated to people in ways that are understandable and compelling.
- *Moving from individual protected areas to protected area centered ecosystems.* Because wildland ecosystems are so linked to surrounding private lands—economically, socially, and ecologically—there is an opportunity to expand research, monitoring, and collaborative management from only federal lands to the larger PACEs. Doing so both facilitates management on the spatial scales most relevant to climate adaptation and helps to achieve the stakeholder engagement described above.
- *Revisiting constraining legislation.* We have traditionally thought of wilderness areas and other places where human impacts are prohibited by law as the lands most likely to achieve conservation goals. Under climate change, however, many species are expected to be increasingly pushed into these more restricted federal land types because they largely lie at higher elevations. Active management may be the best hope for maintaining viable populations of these species, and for consideration of current conditions in neighboring parks and ecosystems to anticipate future local issues. Thus, dialogue and debate are warranted on the acceptability in wilderness areas, roadless areas, and national parks of active human intervention for the purpose of conservation.
- *Scaling up to national and international applications.* Within the United States, the establishment of climate science centers and LCCs offers the promise of advancing climate adaptation planning in all of the nation's wildland ecosystems. Projects like the LCCVP can help break new ground and thus hasten progress in other locations. To this end, we are providing the models, data, and decision support tools developed in the projects to the climate science centers and LCCs so that they are available after the end of the project. International application is also essential. The challenge is greater in countries that have less science

and management infrastructure, financial resources, and legal commitment. Nonetheless, the ratification of the Convention on Biodiversity and progress on the implementation plan of the International Union for Conservation of Nature and Natural Resources offers an opportunity to rapidly expand climate adaptation planning to the wildland ecosystems around the world.

Conclusion

Many of the remaining wildland ecosystems across the world are at a crossroad. The natural factors that inhibited human expansion in these ecosystems in the past are now major attractants for people and business that value access to nature. The resulting land use pressures on private lands and climate change pressures on public lands have reduced the ecological integrity of many of these wildland ecosystems. These forces are projected to increase in the coming decades, raising challenges for sustaining wildland ecosystems. Application of the Climate-Smart Conservation framework is an important foundation of a strategy for sustaining these iconic places. Other elements include developing across the public and private lands of the ecosystem a vision and benchmarks for ecological integrity, a system for monitoring and evaluating key ecological metrics, and a means of communicating to stakeholders in an annual report card or other form the status and trends of the ecosystem (chap. 16). It is critically important that communities in and around wildland ecosystems come together to discuss future prospects and develop means of working toward a desirable future.

References

Brandt, L., H. He, L. Iverson, F. R. Thompson III, P. Butler, S. Handler, M. Janowiak, P. Danielle Shannon, C. Swanston, M. Albrecht, R. Blume-Weaver, et al. 2014. *Central Hardwoods Ecosystem Vulnerability Assessment and Synthesis: A Report from the Central Hardwoods Climate Change Response Framework Project.* Gen. Tech. Rep. NRS-124. Newtown Square, PA: US Department of Agriculture, Forest Service, Northern Research Station.

Brang, P., P. Spathelf, J. B. Larsen, J. Bauhus, A. Bončìna, C. Chauvin, L. Drössler, C. García-Güemes, C. Heiri, G. Kerr, et al. 2014. Suitability of close-to-nature silviculture for adapting temperate European forests to climate change. *Forestry* 87 (4): 492–503.

Code of Federal Regulations. 2012. 36 CFR Part 219. USDA Forest Service National Forest System Land Management Planning Rules and Regulations 77 (68), April 9.

Colwell, R., S. Avery, J. Berger, G. E. Davis, H. Hamilton, T. Lovejoy, S. Malcom, A. McMullen, M. Novacek, R. J. Roberts, et al. 2012. *Revisiting Leopold: Resource Stewardship in the National Parks*. Washington, DC: National Park System Advisory Board Science Committee.

Gauthier, S., P. Bernier, P. J. Burton, J. Edwards, K. Isaac, N. Isabel, K. Jayen, H. Le Goff, and E. A. Nelson. 2014. Climate change vulnerability and adaptation in the managed Canadian boreal forest. *Environmental Reviews* 22:256–85. http://dx.doi.org/10.1139/er-2013-0064.

Glick, P., B. A. Stein, and N. Edelson, eds. 2011. *Scanning the Conservation Horizon: A Guide to Climate Change Vulnerability Assessment*. Washington, DC: National Wildlife Federation.

Gross, J. E., A. J. Hansen, S. J. Goetz, D. M. Theobald, F. M. Melton, N. B. Piekielek, and R. R. Nemani. 2011. Remote sensing for inventory and monitoring of U.S. national parks. In *Remote Sensing of Protected Lands*, edited by Y. Q. Yang, 29–56. Boca Raton, FL: Taylor & Francis.

Halofsky, J. E., D. Peterson, and K. W. Marcinkowski. 2015. Climate Change Adaptation in United States Federal Natural Resource Science and Management Agencies: A Synthesis. Report. USGCRP Climate Change Adaptation Interagency Working Group.

Hansen, A. J., N. Piekielek, C. Davis, J. Haas, D. Theobald, J. Gross, W. Monahan, T. Olliff, and S. Running. 2014. Exposure of U.S. National Parks to land use and climate change 1900–2100. *Ecological Applications* 24 (3): 484–502.

Janowiak, M. K., C. W. Swanston, L. M. Nagel, L. A. Brandt, P. R. Butler, S. D. Handler, P. Danielle Shannon, L. R. Iverson, S. N. Matthews, A. Prasad, et al. 2014. A practical approach for translating climate change adaptation principles into forest management actions. *Journal of Forestry* 112 (5): 424–33. http://dx.doi.org/10.5849/jof.13-094.

Lindner, M., J. B. Fitzgerald, N. E. Zimmermann, C. Reyer, S. Delzon, E. van der Maaten, M. Schelhaas, P. Lasch, J. Eggers, M. van der Maaten-Theunissen, et al. 2014. Climate change and European forests: What do we know, what are the uncertainties, and what are the implications for forest management? *Environmental Management* 146:69–83. http://dx.doi.org/10.1016/j.jenvman.2014.07.030.

Parka, A., K. Puettmann, E. Wilson, C. Messierde, S. Kamesa, and A. Dharf. 2014. Can boreal and temperate forest management be adapted to the uncertainties of 21st century climate change? *Critical Reviews in Plant Sciences* 33 (4): 251–85. doi: 10.1080/07352689.2014.858956.

Shoo, L. P., J. O'Mara, K. Perhans, J. R. Rhodes, R. K. Runting, S. Schmidt, L. W. Traill, L. C. Weber, K. A. Wilson, and C. E. Lovelock. 2014. Moving beyond the conceptual: Specificity in regional climate change adaptation actions for biodiversity in South East Queensland, Australia. *Regional Environmental Change* 14:435–47. doi: 10.1007/s10113-012-0385-3.

Stein, B. A., P. Glick, N. Edelson, and A. Staudt, eds. 2014. *Climate-Smart Conservation: Putting Adaptation Principles into Practice*. Washington, DC: National Wildlife Federation.

US Department of the Interior. 2009. *Addressing the Impacts of Climate Change on America's Water, Land, and Other Natural and Cultural Resources*. Secretarial Order 3289. Washington, DC: US Department of the Interior.

ROBERT AL-CHOKHACHY is a research fisheries biologist at the US Geological Survey's Northern Rocky Mountain Science Center in Bozeman, Montana. He conducts applied research to improve the understanding of the impacts of land use, nonnative species, and climate change on the persistence, management, and conservation of native fishes.

ISABEL W. ASHTON is an ecologist with the National Park Service Northern Great Plains Inventory and Monitoring Network in Rapid City, South Dakota. She formerly served as director of the Continental Divide Research Learning Center at Rocky Mountain National Park.

BEN BOBOWSKI is chief of the Division of Resource Stewardship at Rocky Mountain National Park. He has also served as chief of resource management for Grant-Kohrs Ranch National Historic Site and as range ecologist for Glen Canyon National Recreation Area.

KARL BUERMEYER is a silviculturist with the US Forest Service, currently located on the Helena and Lewis and Clark national forests in Helena, Montana. He was chair of the Greater Yellowstone Coordinating Committee Whitebark Pine Subcommittee from 2012 to 2015 and is a member of the Whitebark Pine Ecosystem Foundation.

TONY CHANG is a PhD candidate in ecology at Montana State University and a NASA Earth and Space Science Fellow. His research focuses on dynamics of forest trees and their pests, using statistical and process-based ecosystem models to evaluate climate change impacts to the Greater Yellowstone Area.

TINA CORMIER is a remote sensing analyst and research associate at the Woods Hole Research Center in Falmouth, Massachusetts. She formerly worked on the remote sensing team at the Southern Nevada Water Authority in Las Vegas, Nevada.

SCOTT J. GOETZ is deputy director and senior scientist at the Woods Hole Research Center in Falmouth, Massachusetts. He has been conducting remote sensing science for thirty years, with an emphasis on ecosystem vulnerability to climate and land use change. He is currently science lead of NASA's Arctic Boreal Vulnerability Experiment and is deputy principal investigator of NASA's Global Ecosystem Dynamics Investigation.

JOHN E. GROSS is an ecologist with the National Park Service Climate Change Response Program. He conducts and applies ecological and climate science research to help better manage protected areas in the United States and worldwide.

KEVIN GUAY is a research assistant at the Woods Hole Research Center in Falmouth, Massachusetts, where he uses remote sensing to study northern forest ecosystems.

ANDREW J. HANSEN is a professor in the Ecology Department at Montana State University. He studies how land use and climate change influence plants and animals and implications for ecosystem management, especially in the context of protected areas. He currently is on the science leadership teams for the US Department of the Interior North Central Climate Science Center in Fort Collins, Colorado, and the Montana Institute of Ecosystems.

DYLAN HARRISON-ATLAS is a PhD candidate in the graduate degree program in ecology at Colorado State University. He is an NSF IGERT Fellow studying the vulnerability of freshwater fishes to climate change in the western United States.

NATHANIEL HITT is a research fish biologist in the Aquatic Ecology Branch of the US Geological Survey Leetown Science Center in Kearneysville, West Virginia. Dr. Hitt's research investigates freshwater fish ecology and community ecotoxicology from a landscape perspective, focusing on stream ecosystems in the Appalachian highlands.

PATRICK JANTZ is a research associate II at the Woods Hole Research Center in Falmouth, Massachusetts. His research focuses on forest connectivity and forest vulnerability to climate change in temperate and tropical regions.

VIRGINIA KELLY until recently was the executive coordinator for the Greater Yellowstone Coordinating Committee, a partnership of four federal land managing agencies in the Greater Yellowstone Ecosystem. She is now the forest plan revision team leader for the Custer Gallatin National Forest and is based in Bozeman, Montana.

TODD KOEL is leader of the Native Fish Conservation Program for Yellowstone National Park, holding affiliate faculty appointments at Montana State University and the University of Wyoming. His work focuses on recovery of the Yellowstone Lake ecosystem and the preservation and restoration of cutthroat trout and Arctic grayling in large remote watersheds throughout Yellowstone.

MATTHEW A. KULP is the supervisory fishery biologist for Great Smoky Mountains National Park in Gatlinburg, Tennessee. Matt is a certified fisheries professional who specializes in native fish restoration, fish population monitoring, and conservation genetics of native fish populations.

KRISTIN LEGG is an ecologist and manages the Greater Yellowstone Inventory and Monitoring Network with the National Park Service. She has been working in the field of natural resource science and management since the late 1980s and has an interest in cultivating the integration of science into land management decisions.

SHUANG LI is a research scientist in the Cooperative for Research in Earth Science and Technology at NASA Ames Research Center in Moffett Field, California. He works on remote sensing of terrestrial ecosystems, with an emphasis on the detection of drought impacts and long-term trends driven by climate change. His specific expertise is in applications of optical and radar remote sensing for ecosystem monitoring.

FORREST MELTON is a senior research scientist with the Cooperative for Research in Earth Science and Technology at NASA Ames Research Center in Moffett Field, California. His research focuses on applications of satellite data, ecosystem and agricultural models, and high-performance computing to investigate the impacts of climate change on ecosystems, water resources, and agriculture.

CRISTINA MILESI is scientific director of the Institute for Public Health and Environment, Palo Alto, California. Her research focuses on the effects of environmental change on the health and function of human-dominated ecosystems. Prior to joining the Institute for Public Health and Environment, she was a senior research scientist with NASA Ames Research Center and California State University Monterey Bay.

WILLIAM B. MONAHAN started working on this book as an ecologist with the U.S. National Park Service Inventory and Monitoring Program, which documents the current status and recent trends of park ecosystems to allow managers to make better-informed decisions and to work more effectively with other agencies and individuals for the benefit of park resources. Bill now works for Forest Health Protection, part of the State and Private Forestry Deputy Area of the USDA Forest Service, overseeing the Quantitative Analysis Program for the Forest Health Technology Enterprise Team. Bill's current work focuses on how forests across the United States are impacted by and respond to environmental changes and insect and disease disturbances operating across multiple spatiotemporal scales.

JEFFREY T. MORISETTE is director of the US Department of the Interior North Central Climate Science Center in Fort Collins, Colorado, whose goal is to provide the best available climate science to natural and cultural resource managers in the region.

RAMAKRISHNA NEMANI is a senior earth scientist with NASA Ames Research Center in Moffett Field, California. His interests are in ecological forecasting (Terrestrial Observation and Prediction System) and collaborative computing (NASA Earth Exchange, NEX).

S. THOMAS OLLIFF is a co-coordinator of the Great Northern Landscape Conservation Cooperative and division chief of landscape conservation and climate change for the National Park Service Intermountain Region. He is the natural resources representative on the NPS Revisiting Leopold Implementation Team.

LINDA B. PHILLIPS is a research scientist in the Landscape Biodiversity Lab at Montana State University. She specializes in building and analyzing large spatial databases for biodiversity and climate change studies.

NATHAN B. PIEKIELEK is an interdisciplinary spatial scientist and currently holds the position of geospatial services librarian for the Pennsylvania State University Libraries in University Park. He has a long history working for and with the National Park Service on landscape-scale protected area conservation issues.

ASHLEY QUACKENBUSH is a research scientist with the Cooperative for Research in Earth Science and Technology at NASA Ames Research Center in Moffett Field, California, where her research focuses on utilizing geographic information systems to analyze the impacts of climate change on ecosystems and water resources.

DANIEL REINHART is a supervisory vegetation ecologist at Grand Teton National Park. Dan began studying wildlife–habitat relationships while working for the Interagency Grizzly Bear Study Team, including grizzly bear use of cutthroat trout spawning streams and the relationships among bears, red squirrels, and whitebark pine. Dan worked as resource management coordinator for Yellowstone National Park and currently oversees vegetation resources in Grand Teton National Park.

ANN RODMAN is acting branch chief for physical resources and climate science at Yellowstone National Park.

BRENDAN M. ROGERS is a postdoctoral fellow at the Woods Hole Research Center in Falmouth, Massachusetts. Using a combination of remote sensing and models, he investigates how forests will be impacted by climate change in terms of their ranges, species composition, and fluxes of water, carbon, and energy.

BRADLEY B. SHEPARD is a senior fish scientist for B.B. Shepard & Associates in Livingston, Montana. He holds adjunct faculty appointments at Montana State University and the University of Idaho. He focuses on conservation of native inland trout of western North America for a diverse group of clients throughout the region.

MARIAN TALBERT is a statistician for the US Department of the Interior North Central Climate Science Center in Fort Collins, Colorado. She works to develop software and tools for various applications, including the exploration of climate data.

MICHAEL TERCEK, PhD, is proprietor of Walking Shadow Ecology in Gardiner, Montana (http://www.YellowstoneEcology.com). He specializes in data analysis, computer programming, and climate change ecology.

DAVID M. THEOBALD is a senior scientist at Conservation Science Partners in Fort Collins, Colorado, and adjunct professor at Colorado State University. He applies concepts from geography and landscape ecology and methods from spatial analysis to understand patterns of landscape change and their effects on watersheds, fish and wildlife habitat, and biodiversity.

DAVID THOMA is an ecologist with the National Park Service Inventory and Monitoring Program in Bozeman, Montana.

WEILE WANG is a research scientist with the Cooperative for Research in Earth Science and Technology at NASA Ames Research Center in Moffett Field, California. His research focuses on studying the coupled carbon-climate system with numerical models and remote sensing data sets.

JUN XIONG is a research scientist with US Geological Survey and Northern Arizona University and was recently a postdoctoral fellow at NASA

Ames Research Center in Moffett Field, California. His research focuses on global cropland change, water resources limitations, and impacts to ecosystem processes and food security under a changing climate.

SCOTT ZOLKOS is a PhD student in the Department of Biological Sciences at the University of Alberta in Edmonton, Canada. His graduate specialization in ecology stems from his fascination with the connections between earth systems, how human activities influence the resilience of these systems, and the implications of environmental change for the future of our biosphere.

Page numbers followed by "f" and "t" indicate figures and tables.

ACT framework. *See* Adaptation for
    Conservation framework
Adaptation, 17–18, 35
Adaptation approaches. *See*
    Management options
Adaptation for Conservation (ACT)
    framework, 262
*Adapting to Climate change in the
    Olympic National Forest and
    Olympic National Park* (Halofsky
    et al. 2011), 261
Adaptive capacity
    Appalachian LCC and, 213–217,
        214f, 219–224
    coarse-filter approach in Great
        Northern LCC and, 97, 99–101,
        103–106, 107f, 108f
    Delaware Water Gap PACE and,
        225–227, 226f
    impact and, 117
    vulnerability and, 20–21, 25–26,
        26t, 212
    of whitebark pine, 309–310
Adaptive management
    in GYE, 310–320
    in Rocky Mountain National Park,
        284–289, 285f, 293–297
Adaptive Management Action Plan for
    Greater Yellowstone Ecosystem,
    310–311
Adelgid. *See* Hemlock wooly adelgid
Air quality, 290–291

AmeriFlux observation network, 127
Ames Research Center, 23, 25
Appalachian Highlands, workshops
    and, 44t
Appalachian Landscape Conservation
    Cooperative. *See also Specific parks
    and regions*
    adaptive capacity estimates for,
        213–217, 219–224
    Climate-Smart Conservation
        framework and, 22
    data to support management in,
        88–914
    defining vulnerability in, 212–213,
        214f, 217–219, 218f
    Delaware Water Gap case study and,
        225–228
    ecosystem models and, 120–121
    ecosystem vulnerabilities of,
        140–144, 141t
    geography and climate of, 79–81,
        79f
    key climate variable trends in, 81–83
    landscape vulnerability assessment
        for, 219
    management implications of
        projected vegetation trends in,
        224–225
    overview of, 6, 8f
    overview of vegetation trends in,
        149–154, 160–167, 161f, 162t,
        163t, 167

overview of vulnerability assessment for, 78–79, 91
potential impact and, 213
precipitation trends in, 83t, 84f, 85f, 86t, 87–88, 87f
predictors of vegetation trends in, 157
Scanning the Conservation Horizons framework and, 10t
soil properties, vegetation trends and, 158
southeastern warming hole and, 80, 84
species vulnerability assessment for, 220–224, 221–222t, 223f
study area and approach, 154
temperature, precipitation, and land use trends in (projected), 127–129, 129t
temperature trends in, 83–87, 83t, 84f, 85f, 86t, 87f
topographic metric, vegetation trends and, 158
tree species distribution models and, 158–159
tree species modeled, 155–157, 155f, 156t
vegetation trends in, 210–212
Appalachian Mountains, overview of impacts on, 4–5, 353–354
Area under the receiver operating characteristic curve (AUC), 159, 194
Arid Lands Initiative, 110
Awareness of science, raising of, 42–43

Bald eagles, 341
Bandelier National Monument, 261
Bears, 3, 339–340, 340f
Beartooth Front (Montana), 173f
Beetle, mountain pine. *See* Mountain pine beetle
Bermuda High pressure system, 80

Bighorn Canyon National Recreation Area, 110, 113
Bioclimate envelope modeling, 174–176, 189–190
Bioclimate suitability models, 175–182
Biodiversity. *See also* Physiography
coarse-filter approach in Great Northern LCC and, 97
loss of in fish species, 233–234
monitoring and reporting of in GYE, 344t
trends in GYE, 335–342
Biogeochemical cycle models, 120–121, 290
Biome velocity, 97
BIOME-BGC model, 122–123
Bird hot spots, 338t, 339f, 341
Birds, 341, 344t. *See also Specific birds*
Bison, 339–340, 340f
Blister rust, 203, 305, 308t, 313–314, 316–317, 319
Blue Mountains, 97
Bobowski, Ben, 358
Boundaries, working across, 36–37
Bridger-Teton National Forest, 264
Business case, management strategies and, 295

Canadian Rocky Mountains, 97, 110
CanESM2 model, 60
CCSM4 model, 60
Census data, 100
Cheatgrass, 293
Chestnut blight, 149
Clark's nutcrackers, 303–304, 307–308
Climate Action Plan, 33
Climate adaption, progress in, 354–357, 356f
Climate Analyzer, 70–71
Climate Change Action Plan, 264
Climate Change Performance Scorecard for national forests, 264

Climate Change Response Program, 7, 355

Climate niche modeling, 36

Climate refugia, 57, 90, 111, 202–203

Climate science centers, 267–268

Climate suitability models, 175–182, 181–182, 183, 360

Climate velocity, 97, 99, 103–105, 104t, 108f

Climate-Smart Conservation framework
  approach of LCCVP and, 23–24
  as collaborative approach to climate adaptation, 34–43, 41f
  conservation goals review, management option evaluation and, 26–27, 189
  data and information transfer and, 28–29, 29t
  ecological models and forecasting in, 24–25, 24f
  ecosystem processes and, 119
  overview of, 9, 10t, 15, 18–19, 19f, 30, 258
  partnerships and, 27–28
  Rocky Mountain ecoregion study area and approach, 22–23
  species distribution models and, 151–152
  steps of, 20–22, 20f
  vulnerability assessment and, 25–26, 26t

Climate-Smart Conservation (Stein et al. 2014), 7, 257, 272

CMIP5. *See* Coupled Model Intercomparison Project Phase 5

CMP. *See* Crown Managers Partnership

CNRM-CM5 model, 60

Coarse-filter approach
  for Great Northern LCC, 95–97, 106–113
  methods used for Great Northern LCC, 97–101

results of habitat types and landforms of, 101–103, 102t

results of impact, adaptive capacity, and vulnerability, 103–106

Coarse-filter indicators, 23–24

Collaboration. *See also* Cooperation; Partnerships
  Climate-Smart Conservation framework and, 27–28, 34–43, 41f
  lessons learned and, 357–358
  progress in, 266–268, 355–356
  Rocky Mountain National Park and, 285–286, 294–295

Columbia Plateau (Washington), 97, 106

Communication
  importance of, 294–295, 360–361
  integrating science with federal land management and, 34–36, 40, 43

Computer simulations, 8

Connectivity, improving, 225

Conservation Science Partners, 8

Continental Divide Research Learning Center, 277–278, 282–283

CONUS-SOIL data set, 193

Cooperation, 46–47, 266–267. *See also* Collaboration

Cordilleran ice sheet, 58

Core habitats, 204–205, 205f, 269

Coupled Model Intercomparison Project Phase 5 (CMIP5)
  Appalachian Landscape Conservation Cooperative and, 81–83, 90
  GYE modeling and, 193–194
  Landscape Climate Change Vulnerability Project and, 46
  TOPS-Biogeochemical Cycle model and, 124
  vegetation trends in eastern U.S. and, 157

Crater Lake, 110

Crossing Boundaries conference,
45t
Crown Managers Partnership (CMP),
46, 266
Crown of the Continent, 97, 105,
105f, 263, 266
Crucial Habitat Tool, 358
Crucial Habitats and Corridors
Initiative, 355, 356f

Data information and transfer,
Landscape Climate Change
Vulnerability Project and, 29t
Decision-making, difficulty in, 36
Deforestation, impacts of, 149
Degree of human modification. *See*
Human modification estimates
Delaware Water Gap (DEWA)
National Recreation Area. *See also*
Appalachian LCC
land use trends in projected by
TOPS-BCG model, 128, 129t
workshops and, 44t
Delaware Water Gap PACE. *See also*
Appalachian LCC
case study, 225–228
temperature, precipitation, and land
use trends in projected by TOPS-
BCG model, 139–140
vegetation trends in, 154, 160–164,
161f, 165
Department of Agriculture (DOA), 6
Department of Interior (DOI)
IRMA and, 29, 29t
new programs and frameworks of, 6,
355, 356f
Order 3289 of, 33, 47, 355, 356f
Deteriorating habitats, 204–205, 205f,
269
DEWA. *See* Delaware Water Gap
Dinosaur National Monument, 106,
110, 113
Discover Life in America, 282

Diseases. *See* Pests and pathogens;
Specific diseases
Dispersal potential, 216–217
Disturbance events, TOPS-BGC model
and, 143
Disturbance zone approach, 335–336
Diversity. *See* Biodiversity;
Physiography
DOA. *See* Department of Agriculture
DOI. *See* Department of Interior
DOI Secretarial Order 3289, 33, 47,
355, 356f
Downscaled Climate Projections,
60, 81. *See also* Coupled Model
Intercomparison Project Phase 5
Downscaling, 90, 112f
Drought events
Appalachian Landscape
Conservation Cooperative and,
89–90
eastern forests and, 149–150
vegetation trends in eastern U.S.
and, 164–165
whitebark pine in GYE and, 306,
319

Earth Exchange. *See* NASA Earth
Exchange
Earth Observation Satellite, 122. *See
also* Terrestrial Observation and
Prediction System
Earth Science Program, 6, 7
EcoAdapt, 42
Ecological levels, framing adaptation
options to protect, 259, 260–261t
Ecological processes, monitoring and
reporting of in GYE, 343–344t
Ecosystem indicators, Terrestrial
Observation and Prediction
System and, 120–121, 121t
Ecosystem integrity
coarse-filter approach and, 95–98
defined, 326

Ecosystem integrity (*cont.*)
    effectiveness of monitoring and
        reporting in, 342–345, 343–344t
    in GYE, 326–327, 345–347, 345t
    trends in, 354
Ecosystem models. *See also Specific*
    models
    approach used in Terrestrial
        Observation and Prediction
        System, 122–123, 123f
    Climate-Smart Conservation
        framework and, 25
    overview of use of, 120–121
Ecosystem processes, 119–121,
    332–335
Electrofishing, 243
Elevation
    climate refugia and, 57
    habitat loss in GYE and, 336
    impacts in Rocky Mountain National
        Park and, 286
    overview of impacts on vegetation
        in GYE, 188–191, 194–198,
        195–196t, 197f, 198f, 206–207
    TOPS-Biogeochemical Cycle model
        and, 124t
    trout and, 236, 243–244
    vegetation trends in eastern U.S.
        and, 158
Elk, 4
Engagement, as challenge, 362
E.O. Wilson Foundation, 282
ET. *See* Evapotranspiration
Evans, Rich, 358
Evapotranspiration (ET)
    in Appalachian LCC (projected),
        138, 140, 141t
    in Great Northern LCC (projected),
        133, 136, 141t, 174
    overview of, 121, 121t
    in Rocky Mountain National Park
        PACE, 286–287, 287f
Executive Order 13514, 33

Executive Order 13635, 33
Exotic species. *See also* Salmonids,
        nonnative
    in GYE, 341–342
    impacts of, 149
    monitoring and reporting of in GYE,
        343t
    in Rocky Mountain National Park,
        281, 293
    vegetation trends in eastern U.S.
        and, 165, 166
Exposure
    coarse-filter approach in Great
        Northern LCC and, 97, 98–99,
        103, 104t, 105f, 106–113, 108f,
        109t
    elements of, 53, 78
    sensitivity and, 117
    vulnerability and, 20–21, 25–26,
        26t, 212
Exurban housing, 329

Farquhar model, 122
Feasibility matrices, 262–263, 271
Federal land management
    challenges of integrating science
        with, 34–39
    creating and promoting framework
        for, 47–48
    LCCVP's collaborative approach
        and, 39–43, 41f
    overview of, 33–34, 48
    products for, 43–47
Fern Lake Fire, 279–281, 279f
Fine-filter indicators, 24, 118
Fine-scale climate projections, 90
Fintrol, 245
Fires
    in Appalachian LCC, 90
    eastern forests and, 149
    in GYE, 334–335, 334f, 344t
    in Rocky Mountain National Park,
        4, 69, 279–281, 279f, 291–292

TOPS-BGC model and, 143
vegetation trends in eastern U.S.
and, 165
whitebark pine in GYE and, 302,
305, 307–308, 308t, 317–318
Fish. *See also* Salmonids, nonnative;
Trout
change in aquatic environments and,
235–236
in GYE, 341–342
management and conservation of in
Great Smoky Mountains National
Park, 244–246
management and conservation of
in Yellowstone National Park,
242–244
Fish barriers, 241, 242
Fisheries, coldwater, 238–239
Flooding, 89, 280f, 281
Folsom Expedition, 328
Forecasting, *Climate-Smart
Conservation* framework and,
24–25
Forest Inventory and Analysis plots,
158–159, 166–167
Forests. *See* Gross primary production;
Vegetation trends
Fragmentation
Appalachian LCC and, 210–211,
215
eastern forests and, 150
monitoring and reporting of in GYE,
344t
native trout and, 236–237
vegetation trends in eastern U.S.
and, 166
Frost days, 86, 87f, 89
Frost-free period, 89
Future habitats, 204–205, 205f,
269

Gallatin National Forest, 264
Gametic wastage, 238

Gap Analysis Program (GAP) Land
Cover data set, 126
General circulation models (GCM),
35–36
Genetic rescue programs, 245
Glacier National Park, 4, 46, 266. *See
also* Great Northern LCC
Glacier PACE, 135–136, 180–181
Glaciers
Great Northern LCC and, 59
GYE and, 198–199
Little Ice Age and, 59
melting of, 4
Rocky Mountain National Park and,
288, 289f
trout species distributions and,
234–235
Goetz, Scott, 8, 358
GPP. *See* Gross primary production
Grand Lake station, 286
Grand Teton National Park, 44t,
47–48, 330, 331f
Graphics Catalog of USGS NCCSC,
71
Grayling Creek (Yellowstone National
Park), 242
Great Northern Landscape
Conservation Cooperative
adaptive capacity estimates for,
99–101, 103–106
climate suitability models, vegetation
trends and, 175–176, 181–182
Climate-Smart Conservation
framework and, 22
coarse-filter approach for, 95–97,
106–113
consequences of projected shifts in
climate suitability in, 181–182
Crown Managers Partnership and,
266
data to support management in,
69–71
ecological integrity and, 327

Great Northern Landscape (*cont.*)
  ecosystem models and, 120–121
  ecosystem responses to changes
    projected by TOPS-BCG model,
    129–137, 130f, 131t, 132t,
    135–136t
  ecosystem vulnerabilities of predicted
    using TOPS-BGC model,
    140–144, 141t
  exposure and impact variables of,
    98–99
  geography and climate of, 56–60
  methods used in coarse-filter
    approach for, 97–101
  overview of, 6, 8f
  overview of vegetation trends in,
    172–175
  overview of vulnerability assessment
    for, 55, 71–72
  paleoclimate of, 58–60
  precipitation trends (contemporary)
    in, 62t, 65–68
  results of habitat types and landforms
    of, 101–103, 102t
  Scanning the Conservation Horizons
    framework and, 10t
  science-management partnerships
    and, 270
  snowfall and runoff trends
    (contemporary) in, 68–69
  temperature, precipitation, and
    land use trends in projected by
    TOPS-BCG model, 127–129,
    129t
  temperature trends (contemporary)
    in, 61–65, 62t, 63f, 64f
  vegetation trends (projected) in,
    176–185, 177f, 178f, 179f, 180f
  vulnerability and, 101, 103–106
  workshops and, 261
Great Smoky Mountains (GRSM)
    National Park. *See also* Appalachian
    LCC

fish management and conservation
    in, 244–246
  overview of impacts in, 4
  vegetation trends in, 160, 161f
  workshops and, 44t
Great Smoky Mountains National
    Park Fishery Management Plan,
    245
Great Smoky Mountains PACE
  temperature, precipitation, and land
    use trends in projected by TOPS-
    BCG model, 139–140
  vegetation trends in, 154, 160–164,
    161f
Greater Yellowstone Area Land
    Managers Adaptation Workshop,
    45t
Greater Yellowstone Area Whitebark
    Pine Strategy, 270, 271
Greater Yellowstone Coordinating
    Committee (GYCC)
  collaboration and, 266
  communications approach and, 45t
  land managers and, 46
  Landscape Climate Change
    Vulnerability Project and, 27
  management of whitebark pine in
    GYE and, 310
  Northern Rockies Adaptation
    Project and, 270
Greater Yellowstone Ecosystem (GYE)
  biodiversity trends in, 335–342
  case studies of adaptation approaches
    in, 268–272
  changes in ecosystem processes of,
    332–335
  defining impact and, 361
  ecological integrity trends in,
    326–327, 345–347, 345t
  ecology of whitebark pine in,
    303–305, 304f
  environmental predictors of tree
    species distributions in, 193–194

history of land use and climate change in, 328–332

management implications of projected vegetation trends in, 204–206, 205f, 206f

management of whitebark pine in, 310–320, 315–316t

map of, 327f

paleoecological synthesis of changes in, 198–199

predictions for, 199–201

prioritization and implementation in, 264

research to reduce uncertainty in, 201–204

status of private lands in, 345t, 346

status of public lands in, 345–346, 345t

tree species distribution models and, 194

tree species modeled in, 191–193, 191f, 192t

vegetation trends across elevation gradients in, 188–191, 194–198, 195–196t, 197f, 198f, 206–207

vulnerability to land use change in, 98, 106, 111, 112f

Greater Yellowstone PACE, 180–181

Greater Yellowstone Whitebark Pine Monitoring Working Group, 27, 311

Grey wolves, 337, 339–340, 340f

Grizzly bears, 3, 339–340, 340f

Gross, John E., 7

Gross primary production (GPP)
in Appalachian LCC (projected), 139, 141t
in Great Northern LCC (projected), 133–134, 141t
overview of, 121, 121t

GRSM. *See* Great Smoky Mountains National Park

GYCC. *See* Greater Yellowstone Coordinating Committee

GYE. *See* Greater Yellowstone Ecosystem

Habitat fragmentation. *See* Fragmentation

Habitat modification, trout and, 233

Habitat restoration, trout and, 241

Habitat suitability
in Appalachian LCC, 212, 214f
in GYE, 337–342, 338t, 339f
management in GYE and, 204–205, 205f, 206f

Habitat types
in Great Northern LCC using coarse-filter approach, 103–105, 104t
landforms and, 101–103, 102t, 104t

Habitats, 95–96. *See also Specific habitats*

Hansen, Andrew J., 9, 358

Hemlock forests, 226–227

Hemlock wooly adelgid, 149, 165, 226–227

High Divide Collaborative, 47

Hillslopes, 98

Human density, 329, 330f, 331f, 343t

Human modification estimates
Appalachian Landscape Conservation Cooperative and, 215, 216f
Climate-Smart Conservation framework and, 23
coarse-filter approach in Great Northern LCC and, 97
GYE and, 335–336, 337f
intensification and, 353–354

Hybridization of fish species, 3, 233–234, 241, 243

Hydrological cycling
in Appalachian LCC (projected), 137–138

Hydrological cycling (*cont.*)
    effects of climate change on aquatic
        environments and, 235–237
    in Great Northern LCC (projected),
        133
    in GYE, 193, 332–333, 333f
    indicators of, 120–121, 121t
    monitoring and reporting of in GYE,
        343–344t
    in Rocky Mountain National Park,
        290–291

I&M program. *See* Inventory and
        Modeling program
Idaho-Montana High Divide, 98
Impact. *See* Potential impact
Impervious surface area, 124t,
        125–126, 128, 129t, 134
Increasive function, 100
Indicators. *See also Specific indicators*
    Climate-Smart Conservation
        framework and, 23–24, 24f
    coarse-filter, 23–24
    ecosystem, 120–121, 121t
    fine-filter, 24, 118
    of habitat suitability in GYE, 192t
Information System (IRMA), 28–29,
        29t
Infrastructure, 354, 359
Inherent sensitivity, vulnerability and,
        212
Insecticides, 312
Integration of science and management
    challenges of, 34–43
    creating and promoting framework
        for, 47–48
    LCCVP's collaborative approach
        and, 34–43, 41f
    overview of, 33–34, 48
    partnerships and, 357–358
    products for, 43–47
Integrity. *See* Ecosystem integrity
Interagency Whitebark Pine
        Monitoring Protocol, 311

Intergovernmental Panel on Climate
        Change (IPCC), 35
Intermountain West Joint Venture
        Wetland Landscape, 98, 110
Invasive species. *See also* Salmonids,
        nonnative
    in GYE, 341–342
    impacts of, 149
    monitoring and reporting of in GYE,
        343t
    in Rocky Mountain National Park,
        281, 293
    vegetation trends in eastern U.S.
        and, 165, 166
Inventory and Analysis database, 193
Inventory and Monitoring Program
        (NPS I&M), 6, 7, 46, 193,
        288
IPCC. *See* Intergovernmental Panel on
        Climate Change
IRMA. *See* National Park Service
        Information System

Jantz, Patrick, 8, 358

Kelly, Virginia, 358
Knowledge accumulation, pace of,
        283–284

Ladder of Engagement, 42
Lag effects, 190
LAI Climatology, 124t, 126
Land allocation, 124t, 125–126, 184
Land facets (physiography), 96, 98,
        108f
Landforms-based analysis. *See* Coarse-
        filter approach
Landscape Climate Change
        Vulnerability Project (LCCVP)
    approach of, 15, 23–24
    case studies of adaptation approaches
        and, 268–272
    Climate-Smart Conservation
        framework and, 19

collaborative approach to climate adaptation, 39–43, 41f
conservation goals review, management option evaluation and, 26–27
data and information transfer and, 28–29, 29t
ecological models and forecasting in, 24–25, 24f
federal land management and, 34
overview of, 7, 18–19, 19f, 30
partnerships and, 27–28
steps of, 20–22, 20f
study area and approach, 22–23
vulnerability assessment and, 25–26, 26t

Landscape conservation cooperatives (LCCs), 6, 47, 267–268. *See also Specific LCCs*
Landscape facets, 96
Landscape integrity index
    Great Northern LCC and, 95, 97, 99–100
    human development and, 54, 335
Landscape permeability, 97, 99
Landscape vulnerability assessments, 219
LCCs. *See* Landscape conservation cooperatives
LCCVP. *See* Landscape Climate Change Vulnerability Project
Leaf area index, 124t, 126
Legislation, as limiting management options, 4, 362
Life stages, lag effects and, 190
Little Bighorn, Battle of the, 329
Little Ice Age, 59
Loch Vale watershed, 290–291

Mammals, large, 4, 336–338, 338t, 344t
Management options (adaptation approaches). *See also* Adaptive management; Integration of science and management
    approaches for, 257–258
    case studies of, 268–272
    challenges and opportunities of, 263–265, 265–268, 314–320
    evaluation and prioritization of, 262–263
    identification of, 18, 19f, 21, 258–262, 260–261t
    implementation of, 18, 19f, 21, 263–265
    issues of, 358–359
    legal limitations on, 4
    overview of, 255–256
    partnerships and, 357–358
    prioritization of, 263–265
    progress in, 354–357, 356f
    in Rocky Mountain National Park, 282–284
    for trout and nonnative salmonid species, 239–242
    vulnerability assessments for, 257–258
    for whitebark pine in GYE, 310–320, 315–316t
Medieval Warm Period, 59, 89
Melton, Forrest, 8
Microhabitats, trout and, 236–237
Microrefugia, whitebark pine and, 202–203
Mitigation, 18, 35
MOD12Q1 Land cover, 124t
Model uncertainty, 90
Modeling. *See also* Ecosystem models; *Specific models*
    Climate-Smart Conservation framework and, 24–25
    Great Northern LCC and, 60–61
    lack of understanding of, 35–36
    LCCVP's collaborative approach and, 46
    limitations and complexities of, 90, 190

Moderate Resolution Imaging
  Spectroradiometer (MODIS),
  123, 124t, 126
Monahan, William B., 9, 358
Monitoring
  effectiveness of in GYE, 342–345,
    343–344t
  management of whitebark pine in
    GYE and, 310, 311
Mount Rainier, 110
Mountain hemlock, 179f, 180, 180f
Mountain pine beetle. *See also*
    Whitebark pine
  in GYE, 335
  overview of, 2–3
  in Rocky Mountain National Park,
    281, 291–292, 292f
  whitebark pine in GYE and,
    305–306, 308t, 314

NASA (National Aeronautics and
  Space Administration), 6, 7, 23
NASA Ames Research Center, 23, 25
NASA Earth Exchange (NEX), 23, 25,
  60, 81
NASA Earth Science Program, 6, 7
NASA Moderate Resolution Imaging
  Spectroradiometer (MODIS),
  123, 124t, 126
NASA TOPS. *See* Terrestrial
  Observation and Prediction
  System
National Congress on Conservation
  Biology, 44t
National Conservation Training Center
  on Climate-Smart Conservation
  and Vulnerability Assessments, 42
National Elevation Dataset (NED),
  124t, 126
National Environmental Policy Act
  (NEPA), 34, 270, 359
National Fish, Wildlife, and Plant
  Adaptation Strategy (NFWP-
  CAS), 259, 260–261t, 265

National Land Cover Database, 124t,
  125
National Park Service, 270, 355. *See
  also* Inventory and Modeling
  program
National Park Service Climate Change
  Action Plan, 7, 264
National Park Service Information
  System (IRMA), 28–29, 29t
National Park Service Inventory and
  Monitoring Program (NPS I&M),
  6, 7, 46, 193, 288
National Parks Omnibus Management
  Act of 1998, 288
National Wildlife Federation, 257
Native Americans, 328–329
Natural fluctuations, 90
Natural Resource Condition
  Assessment, 295
Naturalness, *Climate-Smart
  Conservation* framework and, 23
The Nature Conservancy (TNC),
  264
NED. *See* National Elevation Dataset
NEPA. *See* National Environmental
  Policy Act
Net primary productivity, 121
NEX. *See* NASA Earth Exchange
NEX-DCP30 data
  Appalachian Landscape
    Conservation Cooperative and,
    81–83, 86t, 88–89
  northern Rocky Mountains and, 60
  Rocky Mountain National Park and,
    286
  TOPS-Biogeochemical Cycle model
    and, 124, 124t
  vegetation trends in eastern U.S.
    and, 157, 167
NFWP-CAS. *See* National Fish,
  Wildlife, and Plant Adaptation
  Strategy
Nitrogen Deposition Reduction Plan,
  290

North American Land Change Monitoring System, 100
North Cascades and Pacific ranges, 98, 110
North Central Climate Science Center, 27, 46, 71, 270, 355, 356f
Northern Institute of Applied Climate Science, 211
Northern Rockies Adaptation Project (NRAP), 261, 268–269, 270
NorWest, 264
NPS I&M. *See* National Park Service Inventory and Monitoring Program
NRAP. *See* Northern Rockies Adaptation Project

Okanagan Valley, 98
Olliff, S. Thomas, 358
Omnibus Management Act of 1998, 288
Oregon Trail, 328
Outdoor recreation, 330, 343t

PACES. *See* Protected area centered ecosystems
Paleoclimate, 58–60, 198–199
Parameter-elevation Regression on Independent Slopes Model (PRISM)
    Appalachian LCC and, 80–83, 82f
    GYE modeling and, 193
    TOPS-Biogeochemical Cycle model and, 124, 124t
    vegetation trends in eastern U.S. and, 157, 167
Parcelization, 211
Partnerships, 357–358. *See also* Collaboration
Pathogens. *See* Pests and pathogens
Pecos National Monument, 261
Peregrine falcons, 341

Pests and pathogens
    Appalachian LCC and, 218–219, 224
    blister rust, 203, 305, 308t, 313–314, 316–317, 319
    chestnut blight, 149
    in GYE, 335
    Hemlock wooly adelgid, 149, 165, 226–227
    native trout and, 232–233, 238
    in Rocky Mountain National Park, 281, 291–292, 292f
    vegetation trends in eastern U.S. and, 149, 165
    whitebark pine in GYE and, 305–310
Pheromones, whitebark pine management and, 312
Physiography, 96, 98, 108f
"Pikas-in-Peril" project, 288
Pines, 2–3. *See also* Whitebark pine
Piscicides, 241, 243, 245
Place, importance of, 360
Plains Landscape Conservation Cooperative, 270
Planning, time required for, 359
Population density. *See* Human density
Potential impact. *See also* Exposure
    in Appalachian LCC, 213, 214f
    defining, 361
    exposure and, 97, 99–100, 117
    in Great Northern LCC (projected), 105f
    sensitivity, exposure and, 117
Prairie Potholes Conservation Cooperative, 270
Precipitation trends
    in Appalachian LCC, 83t, 84f, 85f, 86t, 87–88, 87f, 141t
    in Great Northern LCC, 58–60, 62t, 65–68, 141t
    projected using TOPS-BCG model, 128, 129t

Precipitation trends (*cont.*)
  whitebark pine in GYE and, 307,
    308t
Prioritization, 262–265
Priority Agenda for Enhancing the
  Climate Resilience of America's
  Natural Resources, 33
PRISM. *See* Parameter-elevation
  Regression on Independent Slopes
  Model
PRISM 800m Monthly Weather
  Surfaces, 124, 124t
Propagule pressure, 215–216
Property taxes, Appalachian LCC and,
  210–211
Protected area centered ecosystems
  (PACES). *See also Specific
  PACEs*
  in Appalachian LCC, 79, 79f, 85,
    86t
  Climate-Smart Conservation
    framework and, 22, 27
  defined, 8f
  ecosystem models and, 120–121
  in Great Northern LCC, 60
  moving towards, 362
Protection, whitebark pine in GYE
  and, 310, 311–312, 318–319

RandomForest R package, 158–159,
  194
Range-Wide Strategy for Whitebark
  Pine (Keane et al.), 317
RCPs. *See* Representative concentration
  pathways
Realized niches, defined, 150
Refugia. *See* Climate refugia
Remote sensing, 8
Renfro, Jim, 358
Representative concentration pathways
  (RCPs), 60–61, 81, 124, 157
Resilience, 217–218, 218f, 224
Resistance, 217–218, 218f, 224–225

Restoration treatments, management
  of whitebark pine in GYE and,
  310, 312, 313f
Riparian corridors, 96
Rocky Mountains. *See also* Canadian
  Rocky Mountains; Great
  Northern LCC; Greater
  Yellowstone Ecosystem; *Specific
  parks and regions*
  Climate-Smart Conservation
    framework and, 22–23
  global change in, 353–354
  overview of impacts on, 2–4
Rocky Mountain National Park.
  See also Great Northern LCC
  adaptive management in, 284–289,
    285f
  defining impact and, 361
  exotic species in, 293
  forests in, 291–293, 292f
  future of, 293–297
  issues in, 279–282, 296f
  management challenges and
    opportunities in, 282–284
  overview of management in,
    277–278
  water and air quality in, 290–291
  workshops and, 44t
Rocky Mountain PACE, 137, 296f,
  358
Rodman, Ann, 70–71
Rotenone, 241, 243
Runoff. *See* Snowfall and runoff trends
Rust. *See* Blister rust

Salazar, Ken, 355
Salmonids, nonnative. *See also* Trout
  adaptation vs. mediation measures
    and, 239–242
  change of aquatic environments and,
    235–236
  effects of climate change on,
    237–238

in GYE, 341–342
overview of, 233–234
workshops and, 261
SAP4.4, 258–259, 260–261t
Scale, 56–57, 360, 362–363. *See also*
Coarse-filter approach; Fine-scale
climate projections
Scanning the Conservation Horizon
(Glick et al. 2011), 7, 19, 257,
272
Scanning the Conservation Horizons
framework, 9, 10t
Scenario uncertainty, 90
Science, smart, 359–360
Science and management (integration
of)
challenges of, 34–43
creating and promoting framework
for, 47–48
LCCVP's collaborative approach
and, 34–43, 39–43, 41f
overview of, 33–34, 48
partnerships and, 357–358
products for, 43–47
Science Assessment Product 4.4
(SAP4.4; 2008), 258–259,
260–261t
SDM. *See* Species distribution
models
Seasonality changes, 69, 140–142
Secretarial Order 3289, 33, 47, 355,
356f
Seeds
dispersal potential and, 216–217
propagule pressure and, 215–216
whitebark pine in GYE and,
303–304, 307–308, 312–314,
317, 319–320
Sensitivity
adaptive capacity and, 217–218,
218f
exposure and, 117
vulnerability and, 20–21, 25–26, 26t

SERGoM (Spatially Explicit Regional
Growth Model), 25, 124t, 125
Settlers, 328–329
Seven "Rs," 259
Shenandoah National Park, 44t. *See also*
Appalachian LCC
Shenandoah National Park PACE
temperature, precipitation, and land
use trends in projected by TOPS-
BCG model, 139–140
vegetation trends in, 154, 160–164,
161f
Shoshone people, 328
Simulations, 8
Smart science, 359–360
SNOTEL sites, 68
Snow water equivalent
in Appalachian LCC, 137–138, 141t
in Great Northern LCC, 133, 141t
in GYE, 332
overview of, 121, 121t
Snowfall and runoff trends
in Appalachian LCC, 137–138, 141t
in Great Northern LCC, 58–60,
68–69, 133–134, 136–137, 140,
141t
in GYE, 332–333
overview of, 121, 121t
Social complexity, 38–39
Soil moisture
in Appalachian LCC, 138, 141t
climate refugia and, 57
elevation in GYE and, 197
in Great Northern LCC, 134, 141t
overview of, 121, 121t
vegetation trends in eastern U.S.
and, 158
Soil properties
climate refugia and, 57
TOPS-BGC model and, 124t, 126
vegetation trends in eastern U.S.
and, 158
Solar heat load, 57

Solar radiation, 158
Songbirds, 341
Southeastern warming hole, 80, 84
Special Report on Emissions Scenarios
　(SRES; IPCC), 175
Species distribution models (SDM)
　Appalachian LCC and, 212–219
　elevation in GYE and, 197
　limitations of, 202
　overview of, 150–152, 189–190,
　　212
　summary of studies of eastern forests
　　using, 151t, 152–154
Species vulnerability assessment,
　220–224, 221–222t, 223f
SRES. *See* Special Report on Emissions
　Scenarios
Stakeholders, 38–39, 362
State of the Alpine report, 295
State Soil Geographic Dataset
　(STATSGO), 124t, 126, 158
STATSGO2 database, 124t, 126
Storm events (extreme), 89
Story lines of ecological interactions,
　183
Sugar maple, 220, 221–222t, 223f
Suitable habitat space
　eastern tree species and, 152–153,
　　152f, 153t, 164
　elevation in GYE and, 194–198,
　　195–196t
Supermale fish, 246
Susceptibility, adaptive capacity and,
　217–218, 218f

Target identification, 18, 19f, 20–21
Temperature trends
　in Appalachian LCC, 83–87, 83t,
　　84f, 85f, 86t, 87f, 141t
　effects of climate change on aquatic
　　environments and, 235–236
　effects of on native trout, 236–237,
　　238

　in Great Northern LCC, 58–65, 62t,
　　63f, 64f, 97–99, 141t
　in GYE, 330–332, 332f
　projected using TOPS-BCG model,
　　127–128, 129t
　vegetation trends in eastern U.S.
　　and, 164–165
　whitebark pine in GYE and, 307,
　　308t
Tercek, Michael, 70–71
Terminology, lack of understanding of,
　34–35
Terrestrial Observation and Predic-
　tion System (TOPS). *See also*
　TOPS-Biogeochemical Cycle
　model
　ecosystem indicators modeled using,
　　120–121, 121t
　for linking science and management,
　　25
　overview of modeling approach,
　　122–123, 123f
　Rocky Mountain National Park and,
　　286–287, 287f
Teton Range, 3f
Theobald, David M., 8, 358
Third National Climate Assessment,
　119
Threat maps, 225
3-R Framework, 271
Timpanogos Cave National
　Monument, 264
TNC. *See* The Nature Conservancy
Topographic metrics, 158
TopoWx, 60
TOPS. *See* Terrestrial Observation and
　Prediction System
TOPS-Biogeochemical Cycle (TOPS-
　BGC) model
　data sets and validation of, 123–127,
　　124t
　ecosystem responses in Appalachian
　　LCC and, 137–140

ecosystem responses in Great
Northern LCC and, 129–137,
130f, 131t, 132t, 135–136t
ecosystem vulnerabilities predicted
using, 140–144, 141t
limitations of, 142–143
overview of, 122–123, 123f
projected changes in temperature,
precipitation, and land use,
127–129, 129t
Translocations, native trout and,
241–242
Tree improvement, whitebark pine
management and, 310, 313–314,
317
Trees. *See* Gross primary production;
Vegetation trends; *Specific trees*
Trojan Y Chromosome (TYC)
program, 246
Trout. *See also* Salmonids, nonnative
adaptation vs. mediation measures
and, 239–242
case studies of adaptation approaches
and, 270–271
changing aquatic environments and,
235–236
effects of climate change on,
236–237
in GYE, 341–342
indirect effects of climate change on,
237–238
overview of native, 232–235, 234f
in Rocky Mountain National Park,
282
threats to, 3
uncertainty and, 239
workshops and, 261
TYC program. *See* Trojan Y
Chromosome program

Uncertainty
difficulty in decision-making and, 36
modeling and, 90

research to reduce in GYE, 201–204
in trout responses to climate change,
239
United Nations Convention on
Biodiversity, 356, 356f
Upwelling, native trout and, 237
U.S. Forest Service, 264, 270, 355.
See also Inventory and Analysis
database
U.S. Geological Survey (USGS). *See*
North Central Climate Science
Center
U.S. STATSGO2 database, 124t, 126

Vegetation trends. *See also* Gross
primary production; Whitebark
pine
climate predictors of, 157
climate suitability models and,
175–176
Delaware Water Gap PACE case
study and, 225–228
in eastern tree species, 160–167,
161f, 162t, 163t
ecological consequences of in
northern Rocky Mountains,
181–182
environmental predictors of tree
species distributions in GYE and,
193–194, 194
in GYE, 338t, 342
monitoring and reporting of in GYE,
344t
in northern Rocky Mountains,
172–175
overview of, 149–154, 151t, 152f,
153t, 167
overview of impacts of climate on
across elevation gradients in GYE,
188–191, 194–198, 195–196t,
197f, 198f, 206–207
overview of in Appalachian LCC,
210–212, 228

Vegetation trends (*cont.*)
  paleoecological synthesis of changes
    in GYE and, 198–199
  predictions for in GYE, 199–201
  projected, management implications
    of in GYE, 204–206, 205f,
    206f
  projected in northern Rocky
    Mountains, 176–185, 177f, 178f,
    179f, 180f
  research to reduce uncertainty in, in
    GYE, 201–204
  in Rocky Mountain National Park,
    286, 291–293, 292f
  soil properties and, 158
  study area and approach, 154
  topographic metrics and, 158
  tree species distribution models and,
    158–159
  tree species modeled, 155–157,
    155f, 156t
  for tree species modeled in GYE,
    191–194, 191f, 192t
Velocity. *See* Biome velocity; Climate
    change velocity
Vital Signs Summary report, 295
Vulnerability
  challenges of assessing, 118
  coarse-filter approach in Great
    Northern LCC and, 101,
    103–106
  defining in Appalachian LCC,
    212–213, 214f, 217–219, 218f
  potential impact, adaptive capacity
    and, 117
Vulnerability assessments
  Appalachian LCC and, 78–79, 91,
    219, 220–224, 221–222t,
    223f
  climate suitability and, 176
  Climate-Smart Conservation
    framework and, 18, 19f, 20–21,
    25–26, 26t

  Great Northern LCC and, 55,
    71–72, 95–97, 175–185
  landscape, for Appalachian LCC, 219
  species, for Appalachian LCC,
    220–224, 221–222t, 223f

Washburn Expedition, 328
Washington Connected, 47
Water cycling. *See* Hydrological
    cycling
Water quality, 333–334
Water tower of West, 328
Watershed-based adaptive capacity
    estimates, 100–101, 264
Western Governors' Association, 355,
    356f, 358
Western Native Trout Initiative, 271
Wetness, topographic, vegetation
    trends in eastern U.S. and, 158
Whitebark and Limber Pine
    Information System, 193
Whitebark pine. *See also* Mountain pine
    beetle
  case studies of adaptation approaches
    and, 269, 271
  defining impact and, 361
  ecology of in GYE, 303–305, 304f
  in GYE, 335
  management of in GYE, 204–205,
    205f, 206f, 310–320, 315–316t
  in northern Rocky Mountains,
    178–182, 179f, 180f
  overview of, 302–303
  overview of in GYE, 191
  research questions and
    considerations for in GYE,
    202–204
  in Rocky Mountain National Park,
    286, 291–293, 292f
  smart science and, 360
  threats to in GYE, 305–310, 308t
Whitebark pine blister rust. *See* Blister
    rust

Whitebark Pine Subcommittee of
   Greater Yellowstone Coordinating
   Committee, 27, 44t, 311, 314,
   317, 318t
Wildland ecosystems, overview of, 1
Wildlife, 337–342, 338t, 339f. *See also*
   Specific animals
Wildlife Conservation Society, 264
Wilson (E. O.) Foundation, 282
Wind, 165
Wooly adelgid, 149, 165, 226–227
Workshops. *See also Specific workshops*
   of Continental Divide Research
   Learning Center, 277–278, 283
   integrating science with federal land
   management and, 44–45t, 47–48
   management frameworks and, 261,
   268–272

Wyoming Landscape Conservation
   Initiative, 47, 98, 106, 110, 113

Yale Framework, 259, 260–261t
Yellowstone National Park. *See also*
   Great Northern LCC; Greater
   Yellowstone Ecosystem
   Climate-Smart Conservation
   framework and, 47–48
   designation of, 329
   fish management and conservation
   in, 242–244
   overview of issues of, 4
   visitation trends, 330, 331f
   workshops and, 44t
Yellowstone PACE, 135–136
Yellowstone River flows, 262